Chaos in Astrophysics

NATO ASI Series
Advanced Science Institutes Series

A series presenting the results of activities sponsored by the NATO Science Committee, which aims at the dissemination of advanced scientific and technological knowledge, with a view to strengthening links between scientific communities.

The series is published by an international board of publishers in conjunction with the NATO Scientific Affairs Division

A	Life Sciences	Plenum Publishing Corporation
B	Physics	London and New York
C	Mathematical and Physical Sciences	D. Reidel Publishing Company Dordrecht, Boston and Lancaster
D	Behavioural and Social Sciences	Martinus Nijhoff Publishers
E	Engineering and Materials Sciences	The Hague, Boston and Lancaster
F	Computer and Systems Sciences	Springer-Verlag
G	Ecological Sciences	Berlin, Heidelberg, New York and Tokyo

Series C: Mathematical and Physical Sciences Vol. 161

Chaos in Astrophysics

edited by

J. R. Buchler
Physics Department, University of Florida, Gainesville, U.S.A.

J. M. Perdang
Astrophysical Institute, Liège, Belgium

and

E. A. Spiegel
Astronomy Department, Columbia University, New York, U.S.A.

D. Reidel Publishing Company

Dordrecht / Boston / Lancaster / Tokyo

Published in cooperation with NATO Scientific Affairs Division

Proceedings of the NATO Advanced Research Workshop on
Chaos in Astrophysics
Palm Coast, Florida, U.S.A.
9-11 April 1984

Library of Congress Cataloging in Publication Data

NATO Advanced Research Workshop on Chaos in Astrophysics (1984 : Palm Coast, Fla.)
   Chaos in astrophysics.

   (NATO ASI series. Series C, Mathematical and physical sciences; vol. 161)
   "Proceedings of the NATO Advanced Research Workshop on Chaos in Astrophysics,
Palm Coast, Florida, U.S.A., 9–11 April 1984"–T.p. verso.
   "Published in association with NATO Scientific Affairs Division."
   Includes index.
   1. Chaotic behavior in systems—Congresses.   2.   Astrophysics—Congresses.
I. Buchler, J. R. (J. Robert)   II.   Perdang, J. M.   III.   Spiegel, E. A. (Edward A.)
IV.   North Atlantic Treaty Organization. Scientific Affairs Division.   V.   Title.
VI.   Series: NATO ASI series. Series C, Mathematical and physical sciences; vol. 161.
QB466.C45N38   1984   523.01   85-18362
ISBN 90-277-2125-4

Published by D. Reidel Publishing Company
P.O. Box 17, 3300 AA Dordrecht, Holland

Sold and distributed in the U.S.A. and Canada
by Kluwer Academic Publishers,
190 Old Derby Street, Hingham, MA 02043, U.S.A.

In all other countries, sold and distributed
by Kluwer Academic Publishers Group,
P.O. Box 322, 3300 AH Dordrecht, Holland

D. Reidel Publishing Company is a member of the Kluwer Academic Publishers Group

---

All Rights Reserved
©1985 by D. Reidel Publishing Company, Dordrecht, Holland.
No part of the material protected by this copyright notice may be reproduced or utilized
in any form or by any means, electronic or mechanical, including photocopying, recording
or by any information storage and retrieval system, without written permission from the
copyright owner.

Printed in The Netherlands.

CONTENTS

FOREWORD ix

LIST OF PARTICIPANTS xiii

LIST OF TALKS xv

A MATTER OF STABILITY: PROFESSOR PAUL LEDOUX
J. Perdang 1

IRREGULAR STELLAR VARIABILITY
J. Perdang 11

COSMIC ARRHYTHMIAS
E. A. Spiegel 91

A PERTURBATIVE APPROACH TO STELLAR PULSATIONS
J. R. Buchler 137

CHAOS AND NOISE
T. Geisel 165

CLUES TO STRANGE ATTRACTORS
J. Guckenheimer 185

INFORMATION ASPECTS OF STRANGE ATTRACTORS
P. Grassberger 193

ON THE RAPID GENERATION OF MAGNETIC FIELDS
S. Childress and A. M. Soward 223

ORDERED AND CHAOTIC MOTIONS IN HAMILTONIAN SYSTEMS AND THE
PROBLEMS OF ENERGY PARTITION
L. Galgani 245

THE TRANSITION TO CHAOS IN GALACTIC MODELS OF TWO AND THREE
DEGREES OF FREEDOM
G. Contopoulos 259

NONLINEAR NONRADIAL ADIABATIC STELLAR OSCILLATIONS:
NUMERICAL RESULTS FOR MANY-MODE COUPLINGS
W. Däppen 273

CHAOTIC OSCILLATIONS IN A SIMPLE STELLAR MODEL – A MECHANISM
FOR IRREGULAR VARIABILITY
O. Regev and J. R. Buchler 285

X-RAY BURSTERS – THE HOT ROAD TO CHAOS?
O. Regev and M. Livio 295

COMPRESSIBLE MHD TURBULENCE: AN EFFICIENT MECHANISM TO HEAT
STELLAR CORONAE
M. Pettini, L. Nocera and A. Vulpiani 305

INDEX 317

...weil die Natur auch selbst im Chaos
nichts anderes als regelmässig and ordentlich
verfahren kann.

>   Immanuel Kant   1755
>   Allgemeine Naturgeschichte und
>   Theorie des Himmels

FOREWORD

The period of an oscillator tells us much about its
structure. J. J. Thomson's deduction that a particle with the
e/m of an electron was in the atom is perhaps the most stunning
instance. For us, the deduction of the mean density of a star
from its oscillation period is another important example. What
then can we deduce about an oscillator that is not periodic?
If there are several frequencies or if the behavior is chaotic,
may we not hope to learn even more delicate vital statistics
about its workings? The recent progress in the theory of
dynamical systems, particularly in the elucidation of the
nature of chaos, makes it seem reasonable to ask this now.
This is an account of some of the happenings of a workshop at
which this question was raised and discussed. We were
interested in seeing ways in which the present understanding of
chaos might guide astrophysical modelling and the
interpretation of observations. But we did not try to conceal
that we were also interested in chaos itself, and that made for
a pleasant rapport between the chaoticists and astrophysicists
at the meeting. We have several introductory papers on chaos
in these proceedings, particularly on the analysis of data from
systems that may be suspected of chaotic behavior. The papers
of Geisel, Grassberger and Guckenheimer introduce the ways of
characterizing chaos and Perdang illustrates how some of these
ideas may be put into practice in explicit cases.

As the title indicates, this meeting was primarily devoted
to astrophysical chaos. Here we make a distinction between
astronomy and astrophysics. In celestial mechanics, the
awareness of chaos goes back to Poincaré [1], where it remained
dormant till his ideas were given a new life in the enlarged
context of stellar dynamics by Hénon and Heiles [2]. The
subsequent developments of Hamiltonian chaos (or stochasticity)
have by now fed back on the stellar dynamicist who has accepted
the payment and the interest. Though this is not of immediate
concern to the astrophysicist, such work is too related and too
interesting to be omitted completely. It is introduced here in
the paper of Contopoulos and makes a surprising appearance in

pulsation theory in the papers of Perdang and Däppen, and shows up again in Galgani's paper in a novel form that should intrigue the astrophysicist.

The astrophysical applications of the new discipline are really just gleams in the eyes of the organizers at this stage. One of our aims is to point the way to areas where the ideas of dynamical systems theory may have a significant impact on astrophysical models and to begin to formulate such models. The most natural place to begin looking into such possibilities is in the realm of stellar variability. Indeed, the first appearance of chaotic behavior in the astrophysical context was an attempt to understand the origin of solar oscillations as an acoustic instability [3]. The proposed model, like the Lorenz model [4], was devised using convection theory as a guide; the astrophysical incarnation is based on doubly-diffusive convection.

Such models, including an analogous one for radial stellar pulsation [5] that it led to, are based on rough physical arguments. In recent years, bifurcation theory has developed to the point where a correct representation of the behavior of a system near the onset of instability can be given in terms of ordinary differential equations if one has a complete knowledge of the solutions of linear theory. In the case of stellar pulsation this is the realization of a program that goes back to Eddington [6]. We can now write down amplitude equations that are capable of accurately describing the pulsations, at least for mild enough instability. A general procedure to derive such amplitude equations and a discussion of the general ideas behind them are contained in Spiegel's paper. Buchler's paper gives an alternative way of deriving such amplitude equations and he presents an application to realistic Cepheid models. The solutions that such equations spawn are properly the realm of dynamical theory, but their description of stellar pulsations is the stuff of this volume and Perdang gives us a large account of the raw material that we have to confront.

A prominent place among the apparently chaotic processes in astrophysics is occupied by the magnetic activity cycles of the sun and some other stars. Childress and Soward give us an introduction to some of the richness of the dynamo theory.

Finally, the papers of Regev and Buchler, Regev and Livio, Spiegel, and Pettini, Nocera and Vulpiani contain additional specific applications to astrophysical problems.

It has been a great pleasure to dedicate this workshop to our friend and teacher, Professor Paul Ledoux, on the occasion

FOREWORD

of his seventieth birth year. His active participation in the workshop has been a great inspiration to us.

The workshop would not have been the same without the dedicated help of our charming hostess in the person of Daniele Buchler to whom we owe the creation of a very warm and pleasant atmosphere, conducive to scientific and social interactions among the participants. These proceedings would never have appeared if it were not for Sharon Bullivant and Deborah Welch and we thank them for their efficient help.

Finally, we thank the Director of the Scientific Affairs Division of NATO for granting us the financial support for a very fruitful Advanced Research Workshop.

1. H. Poincaré, 1892, Les Méthodes Nouvelles de la Mécanique Céleste, Gauthier-Villars, Paris.
2. M. Hénon and C. Heiles, 1964, Astrophys. J. **69**, 73.
3. D. W. Moore and E. A. Spiegel, 1966, Astrophys. J. **143**, 871.
4. E. N. Lorenz, 1963, J. Atmosph. Sci. **20**, 130.
5. N. H. Baker, D. W. Moore and E. A. Spiegel, 1966, Astron. J. **71**, 845.
6. P. Ledoux and Th. Walraven, 1956, Handb. d. Physik, **LI**.

Professor Paul Ledoux

LIST OF PARTICIPANTS

M. Barranco
Fisica Atomica y Nuclear
Universidad de Barcelona
Barcelona
SPAIN

R. Bless
Astronomy Department
University of Wisconsin
Madison, Wisconsin
USA

J. R. Buchler
Physics Department
University of Florida
Gainesville, Florida
USA

S. Childress
Courant Institute
New York University
New York, New York
USA

P. Coullet
Mécanique Statistique
Université de Nice
Nice
FRANCE

G. Contopoulos
Astronomy Department
University of Athens
Athens
GREECE

W. Däppen
Institute of Astronomy
Cambridge
GREAT BRITAIN

J. Faulkner
Astronomy Department
University of California
Santa Cruz, California
USA

J. Fry
Physics Department
University of Florida
Gainesville, Florida
USA

L. Galgani
Dipartamento di Matematica
Universita di Milano
Milano
ITALY

T. Geisel
Physics Department
Universität Regensburg
Regensburg
F. R. GERMANY

M. J. Goupil
Observatoire de Nice
Nice
FRANCE

P. Grassberger
Physics Department
Universität Wuppertal
Wuppertal
F. R. GERMANY

J. Guckenheimer
Mathematics Department
University of California
Santa Cruz, California
USA

L. Howard
Mathematics Department
Florida State University
Tallahassee, Florida
USA

P. Ledoux
Institut d'Astrophysique
Université de Liege
Cointe-Ougrée
BELGIUM

F. Makino
Institute of Space
  & Astronomical Sciences
Tokyo
JAPAN

J. Perdang
Institut d'Astrophysique
Université de Liege
Cointe-Ougrée
BELGIUM

M. Pettini
Osservatorio di Arcetri
Firenze
ITALY

O. Regev
Physics Department
Technion
Haifa
ISRAEL

M. Signore
Ecole Normale Superiéure
Paris
FRANCE

N. Simon
Physics Department
University of Nebraska
Lincoln, Nebraska
USA

E. Spiegel
Astronomy Department
Columbia University
New York, New York
USA

J. P. Zahn
Observatoire du Pic
   du Midi
Toulouse
FRANCE

LIST OF TALKS

COSMIC ARRYTHMIAS
E. A. Spiegel

STABLE ROUTES TO CHAOS
P. Coullet

THE NATURE OF HYDRODYNAMIC INSTABILITIES IN ASTROPHYSICS
J. P. Zahn

IRREGULAR STELLAR VARIABILITY
J. Perdang

A PERTURBATIVE APPROACH TO STELLAR PULSATIONS
J. R. Buchler

MATHEMATICAL METHODS FOR ASTROPHYSICAL PROBLEMS
L. Howard

INFORMATION ASPECTS OF STRANGE ATTRACTORS
P. Grassberger

ORDERED AND CHAOTIC MOTIONS IN HAMILTONIAN SYSTEMS AND THE PROBLEMS OF ENERGY PARTITION
L. Galgani

ON THE RAPID GENERATION OF MAGNETIC FIELDS
S. Childress

CHAOS AND NOISE
T. Geisel

CLUES TO STRANGE ATTRACTORS
J. Guckenheimer

THE TRANSITION TO CHAOS IN GALACTIC MODELS OF TWO AND THREE DEGREES OF FREEDOM
G. Contopoulos

CHAOTIC OSCILLATIONS IN A SIMPLE STELLAR MODEL - A MECHANISM FOR IRREGULAR VARIABILITY
O. Regev

DWARF NOVA OUTBURSTS - A HOT DAM INSTABILITY RELATED TO COUPLED OSCILLATORS
J. Faulkner

X-RAY OBSERVATIONS OF THE RAPID BURSTER WITH THE TEMMA
SATELLITE
F. Makino

NONLINEAR NONRADIAL ADIABATIC STELLAR OSCILLATIONS: NUMERICAL
RESULTS FOR MANY-MODE COUPLINGS
W. Däppen

X-RAY BURSTERS - THE HOT ROAD TO CHAOS?
O. Regev

COMPRESSIBLE MHD TURBULENCE: AN EFFICIENT MECHANISM TO HEAT
STELLAR CORONAE
M. Pettini

A MATTER OF STABILITY: PROFESSOR PAUL LEDOUX

This NATO ARW is held in celebration of Professor Paul Ledoux's seventieth birthday. It is fitting that these Proceedings should contain a brief account of his scientific career.

Paul Ledoux's birthyear (1914) is marked by two astronomical events regarding **variable stars** and **stellar stability**. In 1914 the Astrophysical Journal published a contribution by Harold Shapley, entitled 'On The Nature and Cause of Cepheid Variation.' In that paper, the author pinpointed the main observational arguments in support of an intrinsic pulsational mechanism and, indirectly, in favor of an internal instability responsible for the cyclic variability of these stars (Shapley 1914). The same issue also contains an extensive book review on the first part of Hagen's monumental history of stellar variability (Parkhurst 1914). In its definitive form, the latter work (Hagen and Stein 1921-1924) remains a vital source-book on the early observations and theoretical ideas about variable stars. Those two catchwords, **variable stars** and **stellar stability**, will be guiding Paul Ledoux's whole scientific life.

The actual selection of his field of research is perhaps best traced through his linkage with three of his teachers. The man who was the first to recognize Paul Ledoux's predestination to astronomy was Professor Pol Swings. He introduced the young undergraduate at the University of Liege into astrophysics and stimulated his zeal for theoretical research. The turning point in Paul Ledoux's life seemingly was a lunch party at Professor Swing's home, one Saturday afternoon in 1937. On that day, Pol Swings, with the invaluable assistance of Leon Rosenfeld, managed to convince Paul Ledoux, who apparently took some time in making up his mind, to study theoretical astrophysics with Professor Rosseland in Oslo.

Paul Ledoux's second, and most influential teacher then became Svein Rosseland, director of the newly founded Institute of Theoretical Astrophysics at Blinderen, near Oslo. (It may

be of some relevance to remind you that Rosseland turned to
Astronomy after learning quantum mechanics from Niels Bohr.)
Strongly marked by the sway of Sir Arthur Eddington, Rosseland
undoubtedly also felt a linkage with the expert in fluid
mechanics, Vilhelm Bjerknes, a former student of Henri
Poincaré and collaborator of Heinrich Hertz, who in turn,
was a student of Hermann von Helmholtz. He must also have felt
the influence of the expert in celestial mechanics, Carl
Stormer, also a former student of Poincaré, who was also
working at the Institute. These twigs of the genealogical tree
of Paul Ledoux's intellectual inheritance elucidate, at least
in part, his permanent interest in stellar fluid dynamics,
classical mechanics and perhaps his penchant for setting up
Schrödinger Equation-like eigenvalue problems to describe the
various classes of stellar stability and oscillation problems.

Paul Ledoux's stimulating training at Rosseland's
Institute from February 1939 to May 1940, was interrupted while
Belgium was invaded by German troops. Through the
contingencies of Word War II, he was the guest of Professor
Bertil Lindblad at Stockholms Observatorium in Saltsjöbaden
(May 1940 to September 1940). Professor Swings advised Ledoux
to continue his studies in Chicago. Following a three-month
round-the-world tour through Finland, Siberia and Japan, Ledoux
arrived at Yerkes Observatory, working under a third teacher
and friend, Professor Subrahmanian Chandrasekhar (December 1940
to September 1941). A student of the prominent
thermodynamicist Ralph Howard Fowler and Sir Arthur Eddington,
Chandrasekhar was also deeply influenced by the writings of Sir
James Jeans. The close association with Chandrasekhar made
Paul Ledoux receptive to mathematical instruments not in the
standard toolkit of the theoretical astrophysicist. It may
also explain his keen interest in statistical mechanics.

By the end of 1941, Paul Ledoux's scientific
apprenticeship was complete. In retrospect, it seems that he
set himself the clearly defined life task to develop a
theoretical basis for the interpretation of the short timescale
variability in stars.

In the broader astronomical community, Professor Ledoux is
now best known for his encyclopaedic reviews on stellar
variability and stability. His 242 page paper 'Variable
Stars,' contributed to Handbuch der Physik in 1958 in
collaboration with Théodore Walraven, still ranks today, a
quarter century after its publication, among the most cited
works on stellar variability. The companion paper 'Stellar
Stability,' which appears in the same issue, figures as a
classic on general stability theory and is well known outside
the circles of astronomy (Thompson and Hunt 1973). The

distinguishing feature of these and his many other review papers - over a dozen so far, is that they are so much more than mere reviews. Most of them are genuine research papers, developing new ideas, pointing towards new problems and suggesting new solutions.

I have attempted here to survey Paul Ledoux's main original contributions to the field of theoretical astrophysics. Broadly speaking, his research may be separated into two classes, mirroring to some extent the complementary tendencies and influences of his teachers.

On the one hand, we encounter a series of papers of a predominantly mathematical character in which new techniques are being adapted to tame a number of difficult problems arising in stability questions. Indeed, Paul Ledoux played a pioneering part in the development of at least five mathematical approaches which now seem standard in stellar stability theory.

1. While at Yerkes, he contributed a paper to the Astrophysical Journal under the neutral heading 'Radial Pulsations in Stars'. Presenting for the first time a **variational principle** for linear radial stellar pulsation frequencies, this paper spearheaded the attack on a variety of problems (frequencies of non-radial oscillations, the effect of rotation,...; cf. Chandrasekhar 1963, 1964, Lynden-Bell and Ostriker 1967, Schutz 1980). Although co-authored by Chaim Pekeris, this paper was not the result of a collaboration in the etymological sense of the word. It came out as the amalgamation of two independent papers on the same topic, one written by Pekeris and the other by Professor Ledoux, both reaching the editor of the Astrophysical Journal, Professor Chandrasekhar, at the same time.

2. During World War II, following his stay at Yerkes, Paul Ledoux was alternatively serving in the Belgian Armed Forces and in the RAF Meteorological Service in the U.K. and Central Africa. While in England, he had the opportunity of attending several meetings of the Royal Astronomical Society where he conferred with Professors Cowling and Milne. It was during this period that he thought of adapting the **virial theorem** to compute stellar oscillation frequencies. Over twenty years later, this novel approach, exhibited in a modest note to the Astrophysical Journal, grew into an impressive technique leading to new physical insights, in particular the context of the theory of self-gravitating figures (cf. Chandrasekhar 1969, Collins 1978).

3. In 1960, presumably as a logical continuation of his

two major review papers, Paul Ledoux anticipated the relevance
for stellar evolution of the concept of secular stability.
Hinted at by Russell (Russell et al. 1927) and schematically
formalized by Jeans (1928), secular stability was shown to be
captured by a **non-selfadjoint eigenvalue problem**, involving a
spectrum of an infinity of decay or e-folding times. The order
of magnitude of the latter is essentially fixed by the
Helmholtz time. This important result, for Ledoux's secular
eigenvalue problem is the first non-selfadjoint problem we
encounter in the literature of stellar stability, even though
published in a local Belgian journal, directly stimulated the
flurry of numerical secular stability tests of stellar
evolutionary models in the 70's.

4. In the early 60's a numerical calculation of high
order radial displacement eigenfunctions of the adiabatic
oscillations of stars, carried out by one of his first
collaborators, Arsene Boury, and an undergraduate, Monique
Breton, puzzled Paul Ledoux's inquisitive mind. He then
embarked on the application of the type of **asymptotic expansion
technique** as used in studies of the Schrödinger equation to
adiabatic stellar oscillations. This procedure enabled him to
explain the shape of the asymptotic eigenfunctions theoretically, and by the same token, to obtain a representation
formula for the asymptotic radial stellar oscillation
frequencies. These results were published in two short notes
in the bulletin de l' Académie Royale de Belgique. In 1964,
in collaboration with an undergraduate, Paul Iweins, the
asymptotic frequency formula was extended to non-radial
oscillations. Unfortunately, this research was never
published. Rediscovered by Yuri Vandakurov (1967), the non-
radial asymptotic frequency formula is known as 'Vandakurov's
Relation' in recent literature. With the observational
detection of the solar 5 minute oscillations, Paul Ledoux's
seminal work on linear asymptotics has been germinating and has
lately given rise to a new subfield of the theory of stellar
oscillations, in which Ledoux remains active.

5. Drawing on the intellectual heritage of Poincaré,
and of Jeans, Ledoux, in collaboration with Maurice Gabriel,
managed to rejuvenate the old formalism of **Linear Series** to
track the class of thermal instabilities of formal stellar
models (late 60's). This technique triggered a systematic
stability analysis of families of realistic stellar models in
hydrostatic and thermostatic equilibrium, especially in the
hands of Paczynski and coworkers (1972, 1977). In particular,
the latter author demonstrated the complementary character of
the new method with the direct method of computing secular
eigenvalues.

A MATTER OF STABILITY: PROFESSOR PAUL LEDOUX

In recognition of his outstanding contributions to the theory of stellar stability, Professor Ledoux was awarded the Belgian Prize for Applied Mathematics.

Besides his abilities as an applied mathematician, Paul Ledoux has also contributed an impressive number of basic physical ideas to the theory of stellar evolution in general and to stellar stability and oscillations in particular.

1. As early as 1941, The Astrophysical Journal issued a note under the innocuous heading 'On the Vibrational Stability of Gaseous Stars,' concealing a conclusion of far reaching astrophysical relevance. In that work, Paul Ledoux proved that a star is bound to turn vibrationally unstable as a consequence of its thermonuclear energy release, once its mass exceeds a **limit mass** M_L, around 100 $M_\odot$. Using improved stellar physics, Schwarzschild and Härm (1959), almost twenty years later, confirmed Ledoux's qualitative conclusion, lowering the value of the 'Ledoux mass' by some 40%. Following the interpretation of the latter authors, the 'Ledoux mass' is now generally identified with the upper mass limit of actually observable stars. It is a typical feature of Paul Ledoux's cautious character and of his critical mind that he never strongly backed that interpretation (cf. the more recent work by Appenzeller 1970, Talbot 1971 and Ziebarth 1970).

2. During his second stay at Yerkes (March 1946 to March 1947), Paul Ledoux introduced a concept of utmost relevance to the theory of stellar evolution, namely what is now known as **semi-convection**. In an acute analysis published in the Astrophysical Journal in 1947 ('Stellar Models With Convection and With Discontinuity of the Mean Molecular Weight') and, in an expanded form in his Thèse d' Agrégation (1949), Paul Ledoux stresses a mathematical flaw in the construction of composite stellar models. On the one hand, the border of the convective core of an energy-generating star coincides with the point of instability of the radiative gradient. On the other hand, at the interface of the convective core and the radiative envelope, the mean molecular weight suffers a discontinuity. If one attempts to fulfill both conditions, then the requirement of continuity of the luminosity, ie., the condition of the energy conservation, is violated at the junction of the core and the atmosphere. To resolve this paradoxical situation, Ledoux blows up the core-envelope interface into a finite, stationary transition zone in which a partial mixing of the stellar matter is taking place. Within this zone of spatially varying molecular weight, the local radiative gradient is equal to the adiabatic gradient.

3. During the same stay at Yerkes, Paul Ledoux also

initiated a systematic mathematical investigation of the
incidence of uniform **rotation** on the oscillation frequencies of
stars. He discovered a new class of oscillation modes, with
pulsation periods proportional to the rotation period of the
star. The astronomical interest of these modes was only fully
appreciated some thirty years later, when the modes were
rediscovered and christened r-modes by Papaloizou and Pringle
(1978, 1981).

Paul Ledoux also demonstrated that the $(2\ell + 1)$ - fold
degeneracy of the conventional non-radial adiabatic modes is
fully lifted by a uniform rotation, each frequency of degree ℓ
being split up into a $(2\ell + 1)$ - tuplet of equidistant
frequencies. This reasearch was published in his
Thèse d' Agrégation (1949).

Following a stay at Princeton (1951), he contributed an
important paper to the Astrophysical Journal ('The Non-Radial
Oscillations of Gaseous Stars and the Problem of Beta Canis
Majoris'), in which he applied his theoretical result to the
puzzle of Beta Cepheid-variability. He points out that the
beat phenomenon observed among this class of stars can be
interpreted as the effect of a rotational splitting of a non-
radial p-mode of degree ℓ = 1. Today, Ledoux's interpretation
of the Beta Cepheid-phenomenon remains as popular as it was at
the time of publication over thirty years ago.

4. In 1950, in collaboration with his first graduate
student, Madame E. Sauvenier-Goffin, he embarked on a
discussion of the oscillation frequencies and the **vibrational
stability of white dwarfs.** That subject, I suspect, was of no
more than academic interest to most astronomers at that time.
Entitled 'Vibrational Stability of White Dwarfs,' Paul Ledoux's
work prepared the way for a theoretical understanding of the
recently discovered two classes of pulsating white dwarfs (Cox
1982).

5. Paul Ledoux invested a great deal of effort in
understanding **stellar convection.** He made several attempts to
put Schwarzschild's intuitively derived convection criterion on
an unassailable mathematical basis. In his
Thèse d' Agrégation, he points towards a connection between
dynamical instability with respect to non-radial disturbances
and the Schwarzschild convection requirement. Later, in his
great review paper 'Variable Stars,' this requirement is
refined; he demonstrates the occurrence of unstable g-modes if
Schwarzschild's convection condition is obeyed. The latter
result foreshadows the rigorous proofs by Lebovitz (1965) and
Kaniel and Kovetz (1967) of the equivalence of Schwarzschild's
condition with the dynamical instability of the stellar g-

modes.

Together with Martin Schwarzschild and Ed Spiegel, Paul Ledoux published 'On the Spectrum of Turbulent Convection' (Astrophysical Journal 1961). In this analysis, the convective motions are expanded in a series of the unstable g-mode eigenfunctions. This technique is essentially the same as the procedure adopted by Lorenz (1963) in his epoch-making paper on convection which marked the beginning of the era of dissipative chaos in physics.

I have omitted Paul Ledoux's various contributions to dynamical and gravitational stability, to non-radial oscillations, to magnetic stars. Instead, I have stressed Ledoux's visionary gift to "...look into the seeds of time, and say which grain will grow and which will not." Virtually every topic he touches is turned, sooner or later, into fertile new ground.

His scientific achievements brought him widespread recognition, both at the national and international level. By 1964 he had received the prestigous Belgian Prix Francqui. He is an Eddington Medallist of the Royal Astronomical Society, London (1972) and a recipient of the J. Jansen Medal of the Académie des Sciences, Paris (1976). In 1974, he was elected an Associate of the Royal Astronomical Society and in 1980 became a Fellow of the American Association for the Advancement of Science. He was also elected an Associé Etranger de l' Académie des Sciences, Paris.

At the age of 45, Paul Ledoux was appointed Full Professor at the Institut d' Astrophysique of the University of Liege . In addition to his routine teaching duties and a heavy administrative burden, Ledoux took advantage of his academic situation to build up what might be called the 'Liege School of Theoretical Astrophysics.' Over the years, he managed to recruit a staff of young scientists who carried out research under his direct or indirect leadership.

In charge of an impressive schedule of 'compulsory' and 'optional' lecture courses, he has always been one of the most appreciated teachers in the Physics and Mathematics Departments, both as a lecturer and as an examiner. His 'optional' lecture course on theoretical astrophysics has been especially popular among the undergraduates. Attended by a large faction of the physics and mathematics students, this course, which covered such broad subjects as statistical mechanics, fluid dynamics, and radiation theory, seems to have influenced the careers of many students. His reputation as a good and kind man as well as excellent teacher, in addition

to his scientific stature, attracted up to ten undergraduates a
year to supervise and assist in the preparation of
dissertations, under his patient, concerned and efficient
guidance. Frequently, a student's dissertation, always on some
topic of scientific actuality in the field of theoretical
astrophysics, developed into a valuable piece of research, due
in great part, I am sure, to Professor Ledoux's invaluable
assistance.

Professor Ledoux has been a steadfast worker throughout
his career. His official retirement in 1983 did not
essentially alter his life-style. He continues to regularly
show up at the Institut d' Astrophysique; freed from his
administrative constraints, he now devotes much of his time to
discussions with former students and collaborators, encouraging
and critiquing their scientific work, sharing his advice and
broad experience with others.

Paul Ledoux's association with stellar stability and
stellar variability now spans more than four decades. The
contributors of this Workshop, joined by many of his friends
and former students, congratulate him on his many remarkable
achievements in this field and wish him further success and
good health in the years ahead. We all rejoice in the
opportunity to continue to learn from him in the future.

J. Perdang

REFERENCES

Appenzeller I., 1970, Astron. Astrophys. 5 355-371, 9 216-220.
Chandrasekhar S., 1963, Astrophys. J. 138 896-897, 139 664-674.
Chandrasekhar S., 1969, 'Ellipsoidal Figures of Equilibrium'
　　Yale Univ. Press, New Haven and London.
Collins G.W., II, 1978, 'The Virial Theorem in Stellar
　　Astrophysics' Pachart, Tucson.
Cox J.P., 1982, Nature 299 402.
Hagen J.G., Stein J., 1921-1924, 'Die Veränderlichen Sterne'
　　vol.1 & 2, Specola Astronomica Vaticana 5 & 6.
Jeans J.M., 1928, 'Astronomy and Cosmogony' Cambridge
　　University Press.
Kaniel S., Kovetz A., 1968, Physics of Fluids 10 1186-1193.
Lebovitz N.R., 1965, Astrophys. J. 142 229-242.
Lorenz E.N., 1963, J. Atmosph. Sci. 20 130-141.
Lynden-Bell D., Ostriker J.P., 1967, Month. Not. Roy. Astron.
　　Soc. 136 293-310.
Paczynski B. ,1972, Acta Astronomica 22 163-174.

Paczynski B., Kozlowski M., 1972, Acta Astronomica 22 315-325.
Paczynski B., Rozyczka M., 1977, Acta Astronomica 27 213-224.
Papaloizou J.C., Pringle J.E., 1978, Month. Not. Roy. Astron.
    Soc. 182 423-442, 1981, 195 743-753.
Parkhurst J.A., 1914, Astrophys. J. 40 483-485.
Russell H.N., Dugan R.S., Stewart J. Q., 1926, 'Astronomy'
    Ginn, Boston.
Schutz B., 1980, Month. Not. Roy. Astron. Soc. 190 7-20.
Schwarzschild M., Härm R., 1959, Astrophys. J. 129 637-646.
Shapley H., 1914, Astrophys. J. 40 448-465.
Talbot R.J., 1971, Astrophys. J. 163 17-27, 165 121-138.
Thompson J.M.T., Hunt G.W., 1973, 'A General Theory of Elastic
    Stability,' Wiley, NY.
Vandakurov Yu. V., 1967, Astronon. Zh. 44 786-797 (English
    trans.: 1968, Soviet Astron. 11 630-638).
Ziebarth K., 1970, Astrophys. J. 162 947-962.

IRREGULAR STELLAR VARIABILITY

J. Perdang

ABSTRACT

The main varieties of observed irregular stellar variability among 'simple' stars are reviewed and some arguments in favor of the endogenous nature of this variability are supplied. Moreover, an attempt at cataloguing the physically distinct species of theoretical chaotic motions compatible with the structure of the stellar hydrodynamics and thermodynamics is presented. Striking statistical similarities between specific types of observed disordered stellar rhythms and certain categories of theoretical chaotic oscillations are stressed. Manifestly, at this early state of the art, these coincidences between observation and 'simple' theory should be regarded with caution.

I. INTRODUCTORY COMMENTS

Variability is a relative stellar attribute. A star is said to be stable or variable according to the meshsize of the observer's fishing net. At a low level of observational accuracy and technical sophistication, its variability may be concealed. With increasingly more refined instrumentation, and very often also for more extended time series, the same star may reveal a time-dependence, which in turn, according to the degree of precision, may appear as periodic, multi-periodic or irregular.

The sun is a typical instance. Regarded as a model of stability some twenty-five years ago, the sun is classified nowadays by all observers, I believe, as a variable star. In the early sixties a five-minute periodicity of the surface velocity field was discovered, which over the last decade was resolved into a huge number of distinct periods. Likewise,

regular variables, once thought of as perfect clocks, are now
reported by some observers as featuring 'noisy' period
fluctuations.

In the first part of this paper I shall review the
observational evidence for cyclic, non-periodic intrinsic
variability among non-eruptive stars. I shall concentrate on
variables undisturbed by complicated physical mechanisms such
as interaction with a companion, magnetic fields, rotation
etc. Cataclysmic variables, such as U Gem, Z Cam, SU UMa
stars, Ap and Am stars, such as α CVn stars, etc. lie outside
the scope of this review.

Theoretical inferences on the variability of a stellar
model are equally relative. They largely reflect the sharpness
- or bluntness - of our analytical tools. In a perturbation
scheme, for instance, the non-linear oscillation of a model may
be found to be strictly periodic or multi-periodic, depending
on the ingredients of the formalism we are using. A refined
analysis of the same model, however, relying on the theory of
dynamical systems, may enable us to establish that the
oscillation is a 'stochastic' or 'chaotic' motion.

The second part of this paper will be devoted to the
theoretical evidence more recently gathered in favor of this
latter type of time-behavior in non-linear oscillations of
simple stellar models. Unsuspected even ten years ago, such a
behavior of complex aperiodic intrinsic fluctuations may
provide new constraints on the structure of models of non
strictly periodic variables.

II. AN INVENTORY OF CYCLIC, APERIODIC VARIABLES

When plotted on the Hertzsprung-Russel (HR) diagram, the
observed intrinsic cyclic stellar pulsators are found to be
concentrated in several well defined areas. The positions of
the latter enable us to partition these variables into four
groups.

(I) The strip approximately normal to the main sequence
and ranging from an absolute visual magnitude +1 to - 6 carries
the most regular pulsating stars. I shall refer to this strip
as the Regular Sequence.

(II) The band to the right of the Regular Sequence and
approximately parallel to the main sequence, roughly 9 or 10
magnitudes above the latter, will be referred to as the Red
Sequence. The stars covering this band all display different
types and degrees of irregular variability.

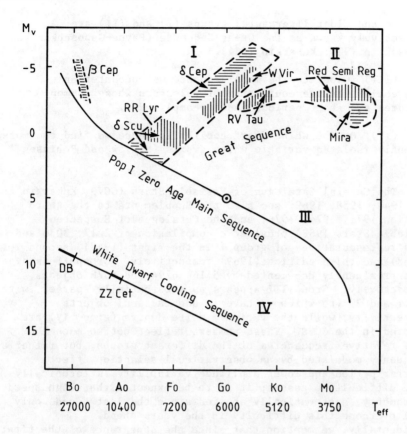

Fig. 1.- Location of the main types of variable stars in the HR diagram. Horizontal hatching indicates Pop I stars; vertical hatching corresponds to Pop II stars.

In the older literature, groups (I) and (II) are collectively known as the Great Sequence (Payne-Gaposchkin and Gaposchkin 1938; Kukarkin 1954).

(III) Close to the main sequence we encounter an apparently heterogeneous array of pulsators whose common features are short periods and low amplitudes.

(IV) On the white dwarf cooling sequence we find the more recently isolated variable white dwarfs (McGraw and Robinson 1976).

The General Catalogue of Variable Stars (GCVS, Kukarkin et al. 1948, 1958, 1969; see also the Supplements to the third edition 1971, 1974, 1976, and the Catalogue of Suspected Variable Stars 1982) provides a compilation of 2217, 3036 and 5139 representatives of group I in the first (1948), second (1958) and third edition (1969) respectively. Group II is even more prominently documented by 5116, 6794 and 8578 objects respectively. Group III appears as substantially sparser, with 6, 11 and 23 (to which we have to add the sun) objects respectively, while the variable white dwarfs (group IV) are ignored in the GCVS. These numbers reflect not so much the true relative frequencies of the different groups, but rather a frequency modulated by an observational selection effect. Shorter period and lower amplitude variability is technically more difficult to assess; it is to be expected that high speed photometry, systematically applied since the last decade only, will overcome this difficulty in the years ahead.
Incidentally, we mention that since the appearance of the first edition of the GCVS the known number of variables has been increasing by a factor of 1.3 to 1.7 per 10 years for groups I and II. For the observationally less easily accessible variability of group III pulsators, this factor is about 2. By comparison, the total number of known variables was increasing on average by a factor of 1.8 per 10 years over the period 1844 through 1954 (Ledoux and Walraven 1958, Fig. 1).

A plot of the known numbers of variables of the Great Sequence, corrected for selection effects, versus period P reveals a complex structure. Around P = 30 days this curve shows a well marked minimum separating the Regular (shorter periods) from the Red (longer periods) Sequence. The plot further displays four maxima near P = .5, 5, 16 and 275 days, together with three shoulders close to .3, 1.2 and 75 days. As suggested already by Payne-Gaposchkin (1954), the observed frequency curve of variable stars can be interpreted as a superposition of 7 Gaussian distributions. Each such Gaussian shape defines a specific class of pulsators within groups I and II.

1. The Regular Variables

In order of decreasing periods, the variables of group I fall into the following three types.

a. The Classical Cepheids

These pulsators have periods in the range 1 - 80 d in our Galaxy; the shortest period of this class is found in HR7308 (Breger 1981), 1.49 d; the longest periods are encountered in BP Her (Makarenko 1972), 83.1 d, and SV Vul, 45.2 d. The Magellanic Clouds, on the other hand, contain Cepheids of periods up to 200 d (cf. the GCVS).

Classical Cepheids are Population I stars concentrated in the galactic plane, of spectral type F5 to G5, effective temperatures 6600 to 5500K, and visual magnitudes - 1 to - 6. Light amplitudes typically range from .1 to 2 magnitudes; velocity amplitudes range from 30 to 60 km/s, short period pulsators showing smaller, and long period pulsators larger amplitudes. Relative radius semi-amplitudes ($\delta R/R$) are almost independent of the period and close to .05. Fig. 2 shows the light curve of the prototype of these pulsators, δ Cephei, whose variability was discovered in 1784 by Goodricke (cf. Gilman 1978). Goodricke assigned this star a period of 5d 8h 37.5 min, an estimate which compares favorably with the modern value (5d 8h 47 min, cf GCVS).

The morphology of the light curves changes regularly with the period, as was first noticed by Ludendorff (1919) and Hertzsprung (1926). The light curve is highly asymmetric, with asymmetry factors ε - defined by the time interval between minimum and maximum divided by the period - around .2 - .4 for longest and shortest periods; for intermediate periods, close to 10 d, the light curve is symmetric ($\varepsilon = .5$). Moreover, a secondary hump appears on the light curve for periods in the range 7 to 13 d. This hump lies on the descending branch for $P \lesssim 9.5$ d. It progresses regularly towards the main maximum with increasing P, coincides with the latter for $P \sim 9.5$ d and moves away from the maximum on the descending branch for $P \lesssim 9.5$ d. The prototype of these 'bump Cepheids' is ζ Gem. Discovered by Schmidt, in 1847 (cf. Argelander 1848), a first estimate of this variable's period was given in Argelander (1849), 10 d 3.6 h; the most recent period determination of this star is 10.15 d (Szabados 1983); the asymmetry factor ε is .5, i.e., the light curve is symmetric.

We observe that from an evolutionary point of view classical Cepheids are 3 to 10 $M_\odot$ stars of typical composition

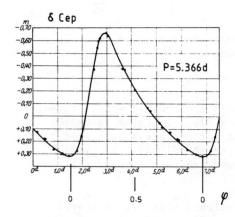

Fig. 2. Photoelectric lightcurve of the classical Cepheid δ Cephei; φ is the astronomical phase.

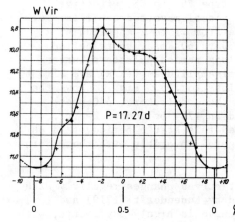

Fig. 3. Photographic lightcurve of W Virginis.

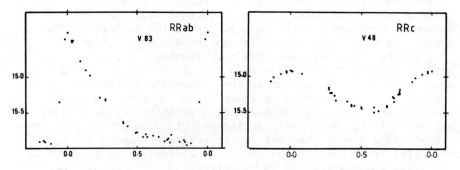

Fig. 4.- Lightcurves of different RR Lyrae variables of the globular cluster NGC 32o1 (after Cacciari 1984); variable V83 is an RRab object, while V48 is an RRc object.

$X = .602$, $Y = .354$, in a phase of central helium burning and hydrogen shell-burning. These stars perform horizontal loops in the HR diagram. When crossing the 'instability strip' – which occurs provided that the chemical composition is that of extreme Population I stars (cf. Hoffmeister 1967) – these stars undergo oscillations. The position of the instability strip in the HR diagram is determined, in principle, by standard vibrational stability analyses (cf. Cox 1980).

b. The W Virginis Stars

With periods in the range of the periods of the classical Cepheids, these variables are Population II stars predominantly encountered in globular clusters. Of spectral types F3 to G0 and effective temperatures of 6800 to 6000, these pulsators are distributed along a line roughly parallel to the classical Cepheid line in the HR diagram. Their absolute visual magnitudes range from -1 to -4. Light amplitudes are of the order of 1 magnitude. The radial velocity curves have a typical range of 50 km/s; they display a discontinuous expansion phase indicative of the formation of a shock in the surface regions of these stars. The relative surface semi-amplitudes, $\delta R/R$, of the order of .1 to .3, are significantly larger than the surface amplitudes of classical Cepheids. Fig. 3 illustrates the variability of the prototype of this class of stars, W Vir, of period of 17.27 d (cf. GCVS).

From an evolutionary viewpoint, this class of Population II pulsators is divided into two subtypes. The longer period stars, $P \gtrsim 8$ d, have been identified as asymptotic red giant branch stars in a stage of helium shell-burning (Schwarzschild and Härm 1970, Mengel 1973). With typical masses around .6 $M_\odot$ and chemical composition $X = .732$, $Z = .001$, these stars suffer a helium shell instability that triggers a leftward loop in the HR diagram, crossing the 'instability strip'.

The shorter period stars of this class, $P \lesssim 8$ d, have been interpreted as post horizontal branch stars, i.e., stars in a phase of helium core-burning (Kraft 1972). These pulsators are also known in the literature as BL Herculis stars.

We observe that in contrast to classical Cepheids, which show a one-to-one correspondence between period and shape of the light curve, W Virginis variables seemingly exhibit a multi-valued period-light curve relation. In fact Kukarkin and Rastorgouev (1973) report the instance of two variables in the globular cluster ω Cen, V60 and V92, that have identical periods, $P = 1.35$ d, but different light curves; the light amplitude of V60 is 1.2 mag, while V92 shows a much smaller variability of .5 mag.

c. The RR Lyrae Variables

The maximum around .5 d in the frequency curve of the observed periods identifies a class of Population II pulsators known in the literature as cluster variables, short period Cepheids or RR Lyrae stars. Among the variables of group I, this class is not only the most populated one; the number of known RR Lyrae stars is also growing most rapidly. The different editions of the GCVS list in fact 1720, 2426 and 4433 such pulsators respectively, against 497, 610 and 706 classical Cepheids and W Virginis stars.

The locus of the RR Lyrae in the HR diagram is a horizontal band over the spectral range A2 to F0, or an effective temperature interval from 9800 to 7200 K, extending from 1 to 0 in visual magnitude. According to the morphology of the light curves, the GCVS subdivides these variables into RRab and RRc stars. RRab pulsators, combining Bailey's (1899) types (a) and (b), are characterized by asymmetric light curves (ε = .1 - .2) of amplitudes in the range .6 to 1.5 mag and periods up to 1 d. RRc variables have symmetric light curves, with smaller amplitudes around .5 mag, and with periods down to 90 min. clustering around 8 h. Fig. 4 illustrates the light curves of two RR Lyrae variables of the globular cluster NGC 3201, an RRab star, V83, of period of 13.1 h and an RRc star, V48, of period of 8.19 h (Cacciari 1984).

The radial velocity curves of RR Lyrae variables are reminiscent of the velocity curves of the W Virginis stars; they exhibit a discontinuous expansion of amplitude around 100 km/s.

RR Lyrae variables are horizontal branch stars. In evolutionary terms this means that they are in a phase of helium core-burning and hydrogen shell-burning, the core being convective. They have typical masses of .6 to .8 $M_\odot$ with probable core masses around .475 $M_\odot$, and chemical composition X = .732, Z = .001 (cf. Sweigart and Demarque 1973).

The variables of group I all show a high degree of regularity. However, already the earlier observers reported changes in the cycles of these stars. The first strong statement I encountered among the earliest writers on variable stars dates back to Harding, in 1831. This author attempted to filter out 'general laws of stellar variability' from the available observations of his time. Harding states as **5th law of variability** that 'the period of the light curves is not rigorously constant, but it is subject to observable irregularities...' (quoted in Hagen 1921). Shapley (1914), in his epoch-making paper on the physical nature of Cepheid stars, uses the observational fact of deviation from strict periodicity as one of his major arguments against the double star interpretation of Cepheid variability. In particular, he calls attention to the irregularities in the radial velocity curve of ζ Gem, reported by various observers (Campbell 1901, Russell 1902, Plummer 1913). He also mentions that the shape of the light curve of the Cepheid W Sgr (P = 7.59 d) is undergoing irregular changes (Curtiss 1905), and that similar results hold for the RR Lyrae stars SU Dra (P = .66042 d) and SW Dra (P = .56967 d) (Sperra 1910), as well as for SW And (P = .44227 d); for the latter, the asymmetry is reported to vary in a random fashion over a time scale of a few days.

To present a more systematic discussion of the character of the deviations of the light or velocity curves from strict periodicity, we represent the latter functions, denoted y(t), in the following form

$$y(t) \equiv S(\phi,t) \ . \qquad\qquad \text{II.1.1}$$

In this expression S is a shape function, periodic in the phase ϕ ($\in [0,1)$, with the astronomical definition $\phi = t/P \mod(1)$), the second argument t being held fixed; the period P, entering the definition of the phase, depends on the epoch t; moreover, the structure of the shape function depends on time through the second argument of S.

Schematically we shall distinguish the following types of behavior of the light or velocity curve:

(1) The period and shape are <u>smooth, slowly varying</u> functions of the epoch t. Then we are entitled to set

$$P(t) = P(t_o) + \Delta t \ \partial/\partial t_o P(t_o) + O(\Delta t^2), \qquad \text{II.1.2a.}$$

$$y(t) = S(\phi,t_o) + \Delta t \ \partial/\partial t_o S(\phi,t_o) + O(\Delta t^2), \qquad \text{II.1.2b.}$$

with $\Delta t = t-t_o$, t_o being a reference epoch. The period representation (a) describes what is known in the literature as <u>secular period changes</u>. Mere dimensional reasons indicate that several possible causes for slow variations are conceivable:

(a) The evolution of the star, proceeding on a nuclear timescale typically much longer than the dynamical pulsation time.

(b) The interferences between the dynamics and the thermal behavior of the star, if the thermal time-scale is long in comparison with the pulsation period.

(c) Presumably also the non-linear interaction between different oscillation modes, the energy flowing from one mode onto a neighbor mode on a non-linear coupling time-scale.

(2) The period and shape are __fluctuating__ from one cycle to the next one, these fluctuations occurring about an average period and an average shape. Formally, we set

$$P(t) = \bar{P} [1 + \eta\, p(t)] \, , \qquad\qquad \text{II.1.3a.}$$

$$y(t) = \bar{S}(\phi) + \eta \bar{S}_M\, s(\phi,t) \, , \qquad\qquad \text{II.1.3b.}$$

where $\bar{P}$ and $\bar{S}(\phi)$ denote the averages over the epoch

$$\{\bar{P}, \bar{S}(\phi)\} = \lim_{T\to\infty} (1/2T) \int_{-T}^{+T} dt\, \{P(t), S(\phi,t)\}. \qquad \text{II.1.4}$$

The small dimensionless parameter η measures the relative order of magnitude of the fluctuations; $p(t)$ and $s(\phi,t)$ are dimensionless rapidly fluctuating functions with t, of zero average (cf Eq. II.1.4), which are bounded in modulus; $\bar{S}_M$ denotes the maximum of the average shape function. Mathematically the fluctuating components $p(t)$ and $s(\phi,t)$ may be **periodic** or **multi-periodic** in time, in which cases they can be expanded in an ordinary or a multiple Fourier series. Such alternatives arise if the star is not just a mono-mode pulsator, but if several dynamical oscillation modes are simultaneously excited.

In principle, the fluctuations $p(t)$ and $s(\phi,t)$ may also be entirely **erratic.** A statistical characterization of these functions then appears as more significant than a detailed temporal description. The latter, when derived from an observational run, has at best an interpolation value; it remains useless for predictive purposes.

The statistical approach to the problem of astronomical period fluctuations seems to have originated with Newcomb (1901) in the specific framework of sunspot periodicity. Eddington and Plakidis (1929) resorted to this tool to analyze the variability of group II pulsators, while Sterne (1934) extended the statistical method to the periods of all classes of variable stars.

Define the local period of cycle i, P_i, of an observational time-series as the time interval between the i + 1, the maximum of the light or velocity curve and i the maximum. Instead of the maximum, we can likewise adopt the minimum or some other well defined phase of the curve. Rewrite then Eq. II.1.3a. in discrete form

$$r_i = \bar{P}[1 + \eta p_i], \quad i = 1,2,\ldots,c \qquad \text{II.1.3'}$$

c being the total number of cycles of the run. An exhaustive statistical description of the local periods is then recorded in the joint probability $\mathcal{P}_{1,2,\ldots,c}(p_1, p_2, \ldots p_c)\, dp_1\, dp_2 \ldots dp_c$ that gives the likelihood of having the fluctuation p_1 in the range $(p_1, p_1 + dp_1)$, the fluctuation p_2 in the range $(p_2, p_2 + dp_2)$ etc. Two tentative assumptions are then made: (a) The fluctuations of the periods of any pair of different cycles are uncorrelated; the probability density $\mathcal{P}_{1,2,\ldots,c}(p_1, p_2, \ldots, p_c)$ then reduces to a product $\mathcal{P}_1(p_1)\, \mathcal{P}_2(p_2) \ldots \mathcal{P}_c(p_c)$ of probability densities of the individual cycles. (b) The individual cycles are statistically equivalent; the individual probability densities $\mathcal{P}_i(p)$ are therefore independent of the cycle number i, or $\mathcal{P}_i(p) = \mathcal{P}_j(p) \equiv \mathcal{P}(p)$.

With these assumptions we can formulate two observationally testable statistical properties of the local set of periods:

(a) The standard deviation s_1 of the local period P_i from the average value $\bar{P}$, for any cycle $i = 1,2,\ldots,c$ is given by

$$s_1^2 = (\bar{P}\eta)^2 \int dp_1\, dp_2 \ldots dp_i \ldots dp_c\, p_i^2$$
$$\mathcal{P}_{1,2,\ldots,i,\ldots,c}(p_1, p_2, \ldots, p_i, \ldots, p_c)$$
$$= (\bar{P}\eta)^2 \int dx\, x^2 \mathcal{P}(x), \qquad \text{II.1.5}$$

and therefore independent of the cycle number.

(b) The standard deviation s_n of the sum of n successive periods $P_i, P_{i+1}, \ldots, P_{i+n-1}$, from the statistical average $n\bar{P}$ obeys

$$s_n^2 = (\bar{P}\eta)^2 \int dp_1\, dp_2 \ldots dp_i\, dp_{i+1} \ldots dp_{i+n-1} \ldots dp_c \times$$
$$(p_i + p_{i+1} + p_{i+2} + \ldots + p_{i+n-1})^2 \mathcal{P}_{1,2,\ldots,c}(p_1, p_2, \ldots, p_c)$$

or

$$s_n^2 = n\, s_1^2, \qquad \text{II.1.6}$$

irrespective of the initial cycle number i.

If observation establishes that this latter property (Eq. II.1.6) holds for some class of variables, then this result tells us that the irregular period fluctuations of the stellar clock arise 'because of imperfections in the star's regulating mechanism' (Sterne 1934). Of course, the statistical description does not yet resolve the problem of tracking the physical origin of the irregularities in the periods. While it does not even answer Belserene's (1973) key question 'Can the theoreticians envisage this much disorder in the pulsation...?', it does help restrict the a priori conceivable range of interpretations of the erratic period behavior. For instance, if observation supports relation (II.1.), then it is manifest that the period fluctuations cannot be swept aside as observational noise. Indeed, random observational errors lack the cumulative property borne out by Eq. (II.1.6). If the standard error of measurement of the local period P_i of the i th cycle (in the sense of the standard deviation from the true period $\bar{P}$) is e*, then the standard error of estimate of the time interval $P_i+P_{i+1}+\ldots+P_{i+n-1}$ remains e*, under the assumption that the phenomenon under investigation is rigorously periodic with period $\bar{P}$.

We should perhaps stress that the mere existence of intrinsic erratic fluctuations does not automatically entail a behavior of type (II.1.6); a more general dependence of the standard deviation s_n on the number of cycles n, or equivalently on the time interval $t = n\bar{P}$, may hold, such as

$$s_n = A\, t^a \,, \qquad\qquad\qquad \text{II.1.6'}$$

with A and a observationally given parameters, $a \neq 1/2$ (and of course $a \neq 0$).[1]

With a fluctuation time-scale by assumption of the order of the dynamical time-scale of the star, we have two mechanisms at our disposal to account for these fluctuations:

(a) Pure dynamical (adiabatic) effects. Energy is rapidly exchanged between different oscillation modes through a non-linear mode coupling mechanism.

(b) An interaction between dynamical and thermal effects, on condition that the thermal time-scale becomes of the order of the pulsation period. The oscillation energy is partially transformed into thermal energy and vice versa, as a result of a non-linear coupling between the dynamics (the oscillation equations) and the thermodynamics (the heat transport equation) of the star.

Alternatives (1) and (2) are concerned with gentle alterations of the stellar rhythms; the stellar clocks get out of step, but only mildly so. A more radical form of disorder in the light or velocity curve (Eq. II.1.1) is witnessed in the following alternative:

(3) The shape function is undergoing violent changes over sucessive cycles; the cycles themselves may disappear and reappear.

I presume that the most plausible physical mechanism capable of generating such a wild irregularity is the dynamical-thermal interaction mentioned under (2.b). A large part of mechanical energy is being absorbed by the thermal degrees of freedom, which causes a partial disappearance of the pulsation; thermal energy is then again fed back into the mechanical degrees of freedom; a necessary condition is that the thermal time is of the order of the period. Formally however, it is not impossible that purely dynamical non-linear effects might also bring about a high degree of disorderly behavior, with the exception of producing complete disappearance of the oscillations.

Remarkably, all three variants of irregularity in the variability of the pulsators of group I have been reported already by the earlier observers.

(1) Relying on the observational data from 1785 onward, Hertzsprung (1919) found a secular period variation for δ Cephei, satisfying Eq. (II.1.2) with the following parameter values

$$P = 5.3663770 \text{ d in } 1883 \quad,$$

$$\partial/\partial t \, P = -9.16 \times 10^{-7} \text{ d/y} = -7.9 \times 10^{-2} \text{ s/y} \quad.$$

Using observations extending over the time-span 1881 to 1920, Bemporad (1921) estimated a secular period change in the Cepheid variable T Monocerotis, conforming to the parameters

$$P = 27.00313 \text{ d on Julian day } 2410011 \quad,$$

$$\partial/\partial t \, P = +4.17 \times 10^{-5} \text{ d/y} = +21.9 \text{ s/y} \quad.$$

A recent study of the secular change of SV Vul (cf. above), together with an attempt at a theoretical interpretation of the observed trend has been published by Fernie (1979). Observation yields $\partial/\partial t \, P = -(254 \pm 10)$ seconds per year. The evolutionary period change of a 10 $M_\odot$, X = .71, Z = .015 star crossing the Cepheid instability strip appears to be remarkably close to the empirical rate of change. Fernie stresses, however, that this agreement may largely be due to a coincidence: the evolutionary rate of change of the period of a stellar model is highly sensitive to the precise chemical composition as well as to other model parameters. Szabados (1981 cf. also 1983) has carried out a systematic survey of the

secular period changes of classical Cepheids, using virtually
all of the observations published so far on the variability of
the stars under investigation. His main conclusion is that the
rate of change rises steeply with period. On average, a
shorter period pulsator, of period ~ 3.5 d, undergoes a period
change $|\partial/\partial t\ P| = 2.5 \times 10^{-9}$ - which, incidentally, compares
favorably with Hertzsprung's old estimate for δ Cephei; for a
long period pulsator, of period ~ 45 d, one has on average
$|\partial/\partial t\ P| = 6.0 \times 10^{-4}$ - of the order of Fernie's estimate of
the period change of SV Vul. We might mention that for T Mon
Szabados derives a rate of period increase of $+ 1.1 \times 10^{-5}$,
which is ~ 1/3 of Bemporad's estimate. From a theoretical
point of view, Szabados's analysis demonstrates that canonical
stellar evolution is not in conflict with the observed period
changes of classical Cepheids.

Secular changes in W Virginis stars have been discussed by
Kwee (1969), while in RR Lyrae variables period changes have
received considerable attention by a number of authors. Martin
(1938) appears to have been the first to carry out a complete
survey of the secular trend in the RR Lyrae stars of a globular
cluster, namely ω Cen. He finds as an average value of $\partial/\partial t\ P$
over all RRab stars of this cluster $+ 5.1 \times 10^{-10}$ s/y, with an
observational dispersion of 8.8×10^{-10} s/y. A more recent
discussion of ω Cen by Belserene (1973) essentially confirms
these results. The orders of magnitude of the rate of period
change of RRab variables in other clusters (M5, Oosterhoff
1941; M3, Szeidl 1965; cf. also Rosino 1973) are all comparable
to the ω Cen values. For instance, for M5 the average $\partial/\partial t\ P$
is $- 1.5 \times 10^{-10}$ s/y with an observational spread of
7.6×10^{-10} s/y. As transpires from the large dispersions, RRab
variables may suffer both secular period lengthening and
shortening. Sweigart and Renzini (1979) have compared the
period change of theoretical horizontal branch stars with the
observational values. The latter are found to be an order of
magnitude larger than the period changes derived from canonical
evolution. Renzini and Sweigart conclude, therefore, that
canonical evolution theory of these stars requires revision.
We wish to re-emphasize here that secular period changes are
not the exclusive attribute of the nuclear evolution of a star;
mechanisms (2.b) and possibly also (2.c) may likewise be
responsible of a slow period drift.

(2) <u>Multi-periodicity</u> among cluster variables as well as among
classical Cepheids is a fairly common phenomenon. In the
context of formulation (II.1.3) it reveals itself through the
occurrence of periodic or possibly multi-periodic functions
$p(t)$ and $s(\phi,t)$. Thus Blazhko (1926) discovered that the non-
secular residual of the period change of the RR Lyrae variable
XZ Cygni can be represented in a Fourier series

$$p(t) = \sum_{n=1}^{\infty} p_n \cos(n\omega't + \phi_n) \quad . \qquad \text{II.1.7}$$

On limiting this series to its first two terms, Blazhko determined the associated period $P' = 2\pi/\omega' \simeq 57.39$ d; the main period of this star is .46659 d. Approximately 1/3 of all RR Lyrae stars are now found to be multi-mode oscillators (Preston 1964). Typically, the effect of a secondary period manifests itself in an oscillatory variation of the light curve, with a periodicity of about 100 times the main period; for instance, in RR Lyr (P = .567 d) the second period is P' = 40.8 d $\simeq$ 72 P (Detre and Szeidl 1973). This effect, reminiscent of a beat phenomenon, is known as the Blazhko effect. Tsesevich (1969) points out, however, that the actually observed alterations in the light curves of these stars are substantially more complex than a mere interference phenomenon of two periodic waves. We should perhaps mention also that there is a growing belief among variable star observers that multi-periodic variability is to be expected in <u>all</u> RR Lyrae stars, provided that they are observed over a <u>long</u> enough time-span at a sufficient level of precision (Balazs-Detre and Detre 1962).

Regarding classical Cepheids, in the older literature we encounter a number of attempts at describing long-term period changes by periodic interpolation formulae of type (II.1.7). For ζ Gem Becker (1924) finds $P' \simeq 143$ years; AY Sgr is assigned the periods P = 6.74 d and P' = 130 d (Hoffmeister 1923). However, these long periodicities have not been confirmed in later studies. On the other hand, a particular class of Cepheids have been isolated by Oosterhoff (1957), now referred to as Beat Cepheids, or Double Mode Cepheids, which have the following pulsational properties. Of the 11 objects of this class (listed in Cox 1982) all show main periods in the range $2.1 \leq P_o \leq 6.3$ d. The period ratio of the two periodicities, P_o/P_1, is surprisingly close to .7

$$.6967 \leq P_o/P_1 \leq .7105$$

(Stobie 1977, Schaltenbrand and Tammann 1971). Moreover, in the period range of 2 to 3 days, 40% of all Cepheids are Beat Cepheids. The analysis of the light curves in terms of a multiple Fourier series demonstrates that a mere linear superposition of two component waves of periods P_o and P_1 respectively does not adequately represent the data (cf. in particular the example U TrA cited in Stobie 1977). Therefore, one is led to infer that a non-linear mode interaction mechanism must be operative in these stars (cf. also Faulkner and Shobbrook 1979). With the exception of AX Vel, the longer period is always associated with the higher amplitude; the amplitudes themselves are, as a rule, reported to be constant

over the period of observation of these stars (cf. for instance
the discussion of V367 Scuti in the open cluster NGC 6649, by
Madore et al. 1978), with the exception of two Beat Cepheids:
TU Cas displays a decaying amplitude of the shorter periodicity
(Niva 1979, Hodson et al. 1979); while U TrA exhibits an
opposite behavior, the amplitude of the short periodicity being
growing (Faulkner and Shobbrook 1979). The latter observations
are not inconsistent with a mode-switching phenomenon,
especially since in U TrA the total oscillation energy seems to
remain constant.

Irregular period fluctuations in classical Cepheids,
although not very common, have been reported by Szabados (1980,
1983) for ζ Gem. Plotting the deviations of the observed
epochs of maxima of the light curve (photoelectric and
photographic data), OH, from the calculated epochs of the
maxima using the secular interpolation formula (II.1.2a), C,
Szabados finds excursions of several percent which cannot be
attributed to observational noise (cf. Fig. 5). A statistical
analysis of these deviations, of the type outlined above,
remains to be carried out. It is perhaps worth stressing that
among the 25 northern long-period Cepheids studied by Szabados,
only one specimen seems to exhibit intrinsic noisy period
fluctuations.

Among the RR Lyrae variables, a specific statistical study
of the period fluctuations has been performed by Sterne (1934)
on the prototype RR Lyr itself. Sterne finds, instead of
relation (II.1.6), the following behavior for the standard
deviation s_n

$$s_n^2 = n\, s_1^2 + 2e^2 \qquad \text{II.1.6''}$$

with the numerical values s_1 = (.000725 ± .000048) d and e =
(.00895 ± .00073) d, with $\overline{P}$ = (.56684186 ± .0000056) d. The
correction to formula (II.1.6) is interpreted as an
observational error effect (cf. the discussion presented above
[2]). Sterne's result, if confirmed, appears to be a strong
indication of intrinsic noisy period fluctuations among the RR
Lyrae stars.

Irregular fluctuations among the RRc stars have been
reported by various authors, on the basis of O-C plots (cf. for
instance Szeidl's exhaustive discussion of the RR Lyrae stars
in M3, 1965). However, a quantitative statistical study of
these fluctuations is lacking. Finally, in W Virginis stars,
noisy fluctuations have also been detected (Kwee 1967, Osborn
1969, Coutts 1973), though not thoroughly studied.

(3) Various observers have witnessed <u>drastic changes</u> in the

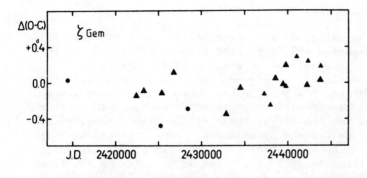

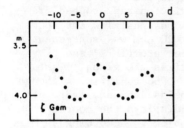

Fig. 5.- Period fluctuations of ζ Gem shown on an O-C plot (Szabados 1980), and short stretch of a recent light-curve (Gragg 1983).

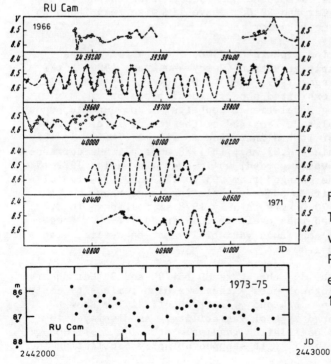

Fig. 6.
The irregular variability of RU Cam (Zaitseva et al 1973, Gragg 1983).

light curves of several group I pulsators.

RU Camelopardalis, usually classified as a W Vir variable (cf. however Szabados 1983 and Lloyd Evans 1983), was discovered to be variable by Ceraski in 1907; Blazhko determined its period (22.27 d) and reported a light variation from 9.1 to 8.0 mag. Carrying out a total of 24 observations, Sandford (1927) reported that the velocity curve, with a range of 30 km/s, was not the mirror image of the light curve, as it should be for bona fide RR Lyrae stars. From searches of the older literature, he also concluded that this object was exceptionally faint in 1888, 1894, and 1896; this was the first evidence for disordered variability in RU Cam. In 1966 Demers and Fernie reported that the light variation had dropped from 1.2 mag to less than .2 or .1 mag, the residual low amplitude fluctuation being entirely irregular. RU Cam was subsequently studied by Huth (1967), Broglia and Guerrero (1972) and Zaitseva et al. (1973). Fig. 6 summarizes the visual light variation of this star over the period 1966 to 1971. Although the information on the light curve remains lacunary over this time stretch, the figure convincingly displays the complex character of the variability of this object. According to Lloyd Evans (1983) extreme irregularity is correlated with an overabundance of carbon. RU Cam has, in fact, been known to be a carbon star since 1928 (Sandford 1928); other carbon W Vir stars, such as DI Car and RV Nor, have likewise been found to show an erratic time behavior.

The short period Cepheid HR 7308, P = 1.49107 d, has received considerable attention (Percy and Evans 1980, Burki and Mayor 1980, Breger 1969, 1980). Discovered by Breger in 1969, this object exhibited a pulsational behavior unprecedented among Population I Cepheids. From June to August 1966 the light amplitude remained at the steady level of .06 mag (in V); in 1967 it had increased to .17 mag, while in 1969 it had dropped again to .03 mag; in 1978 an amplitude decrease from .3 to .1 mag was recorded, while over the period 1979-1980 the light amplitude changed from .05 to .15 mag. The radial velocity semi-amplitude underwent similar large range variations. From 20.0 km/s in July 1978 it fell down to 3.7 km/s in May 1979 and rose again to 10.9 in July 1980. Breger concludes that the amplitude variability may be cyclic, with a periodicity of 970 ± 40 d.

Two other Cepheids, the stars HD 161796 and 89 Her deserve a few comments. Both objects – they are high luminosity early F supergiants (M_v = – 6.4 and – 7.1 respectively) – show a definitely disordered type of variability. As displayed in Fig. 7, 89 Her was oscillating regularly at a low amplitude level (~ .2 mag) in 1977; it stopped abruptly oscillating in

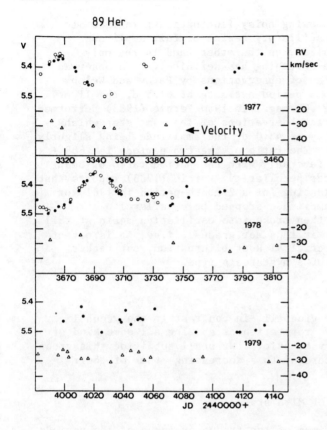

Fig. 7. The irregular variability of 89 Her (Fernie 1981).

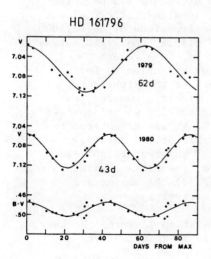

Fig. 8. Mode switching in HD 161796 (Fernie 1983).

1978, but continued showing noisy fluctuations; in 1980 it resumed its regular oscillatory regime at full amplitude (Fernie 1981). HD 161796, on the other hand is the only Cepheid in which mode-switching has definitely been observed (cf. Fig. 8). A series of observations by Percy and Welch (1981) in 1979 led to a period estimate of 62.5 d, the light amplitude level being .08 mag. In 1980 Fernie (1983) performed a series of photometric observations of the same star which exhibited a periodicity of 43.0 d, the amplitude being slightly smaller than in 1979. The ratio of the two periods is .688 ± .026, which compares favorably with the period ratio of Beat Cepheids. As a further peculiarity, Fernie (1983) reports that HD 161796 stopped pulsating for a 2 month-period in 1981, or more precisely, its amplitude dropped below a level of $\lesssim$ .005. The star then broke into oscillation again at full amplitude. The onset is thus not gradual, i.e., it is not of the nature of a supercritical Hopf bifurcation, but rather discontinuous, i.e., of a hysteresis type.

2. The Red Variables

The variables of group II, in contrast to the group I pulsators, do have the reputation of showing all some kind of disordered oscillatory behavior. We shall subdivide this group into three types, according to a decreasing order of their average cycle lengths.

a. The Long-period, or Mira stars

The dominant maximum of the frequency curve of the periods of variable stars, near 275 d, isolates this type of stars. The different editions of the GCVS list 3025, 3657 and 4566 of these objects. With light amplitudes exceeding 2.5 mag, the cyclic light changes of these giants, located at the cool end of the Red Sequence (Fig. 1), with local periods of 80 to 1000 d, were recorded since 1596. Figs. 9 display the light curves of two well studied Mira stars. Fig. 9a. is a recent compilation of the visual brightness of R Aquilae by Davis (1982), while Fig. 9b. is an old record of the light curve of χ Cygni extending over a period of almost a hundred years (Merrill 1938). The latter star's variability was discovered by Kirch in 1687; the variability of R Aql was discovered by Argelander in 1856 (cf Hagen 1921 for historical details). Notice the enormous range of the visual apparent brightnesses of both stars; for χ Cyg ($\bar{P}$ = 406.6 d) the extreme values are 3.5 and 14.5 mag; for R Aql ($\bar{P}$ = 329.3 d) the maxima and minima lie around 6 and 11.5 mag respectively.

Mira stars are predominantly Population I stars. However, Kukarkin (1954) has shown, on the basis of an extensive kinematic analysis, that approximately 20% of them are globular cluster stars (Population II). For a large number of these variables direct radius measurements have been carried out, by means of speckle interferometric and occultation techniques; individual radii lie in the range of 200 to 600 $R_\odot$ (cf. Willson 1982). Indirect mass estimates have been made which are based on a theoretically obtained period-mass-radius relation; Willson (1982) gives as lower and fuzzy upper limits .7 and $\gtrsim 2.5$ $M_\odot$. From an evolutionary point of view, these stars are asymptotic giant branch stars (cf. above).

Irregularities in the variability of the Mira stars are twofold, as is already obvious from a mere visual inspection of a sufficiently extended time series of the luminosities of these stars (cf. Fig. 9b): (1) The shape of the light curves does not repeat from cycle to cycle. Deviations occur especially at maximum brightness, while the minima are, in general, more stable. Moreover, it is found that sharper maxima are undergoing larger fluctuations than broader maxima. The prototype, O Ceti (Mira, $\bar{P}$ = 331.6 d) shows maxima fluctuating over a range of some 4 mag, and so does V Delphini ($\bar{P}$ = 533.3 d). (2) The cycle length is undergoing seemingly irregular fluctuations, characteristically of some 10%; for instance X Cam, of $\bar{P}$ = 142.3 d, exhibits local periods from 130 to 160 d. In addition to these period irregularities, secular changes in the periods have also been found. Hagen (1921) credits Olbers as having been the first to search for a trend in the period change of χ Cyg in the form of Eq. (II.1.2a). For a compilation of secular period variations of various Mira stars we refer to Ludendorff (1928, sec. 24); besides the linear trend, Ludendorff also includes periodic components for a large number of stars. For R Aql the following secular parameters are listed: P = 329.63 d in 1890, $\partial/\partial t$ P = - .485 d/y = - 1.33 x 10^{-3} s/y if the maxima of the light curve are used; a slightly different result is obtained from the minima. Notice that this trend is about 7 orders of magnitude larger than the secular period change in RR Lyrae.

Detailed statistical studies of the period fluctuations of the long-period variables were initiated by Eddington and Plakidis (1929), Plakidis (1932) and Sterne (1934). The general conclusion of these analyses is that the standard deviation s_n conforms to formula (II.1.6''), except for T Gem and S UMa. That the coefficient e is linked to observational errors, is borne out by the result that e is correlated with the widths of the maxima: broad maxima, which are difficult to date, invariably lead to larger e values than sharp maxima. For instance, for RR Scorpii Sterne (1934) has computed the following statistical parameters: $\bar{P}$ = 279.44 ± .44 d, s_1 = 4.66 ± .35 d and e = 6.43 ± .25 d. Notice that the relative

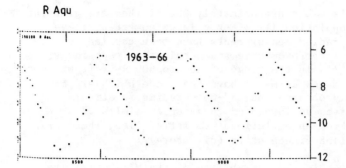

Fig. 9.- Lightcurves of R Aql (Davis 1982) and χ Cyg (Merrill 1938).

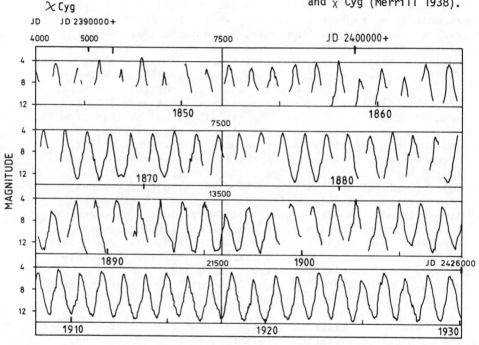

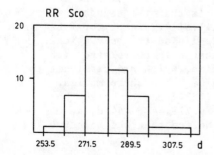

Fig. 1o.- Histogram of local periods of RR Scorpii (Sterne 1934).

standard deviation s_1/P is here 1.7%, which is over an order of magnitude larger than for RR Lyrae stars. Fig. 10 shows the histogram of the local periods of this star; the Gaussian shape reasonably approximates the empirical data. These results are fairly typical for all Mira variables. The relative standard deviations of all of these stars that have been investigated lie in the range of .7 to 2.3%. More recent statistical studies have been performed by Sandig (1948), Schneller (1950) and Fischer (1969). Schneller points out in particular that the parameter e in Eq. (II.1.6'') cannot be interpreted as a measure of observational errors alone; comparing the observational data of the Mira star S UMa, collected by some 20 observers, Schneller estimates an extrinsic standard deviation of $\sim$ 4.5 d, while the parameter e is found to be $\sim$ 8 d. It seems therefore that the latter statistical coefficient contains also information on endogenous period fluctuations.

A number of Long-Period variables are reported to be doubly-periodic. Examples are SV And, with a primary mean period $\bar{P}_o$ = 316 d and a secondary period $\bar{P}_1$ = 930 d, with $\bar{P}_1/\bar{P}_o \sim 3$ (Wood 1975); the variable V4 in the cluster 47 Tuc has periods $\bar{P}_o$ = 82 d and $\bar{P}_1$ = 165 d, or $\bar{P}_1/\bar{P}_2 \sim 2$ (Willson 1982).

Irregular shape fluctuations, in particular amplitude changes, of a time-scale of 20-50 years have been recorded in R Tri, R Aur, R Cam, T Cas, U Per, S UMa, R Vir (Wood 1975).

b. Red Semiregular Variables

Located in an area of the HR diagram of visual magnitude extending from 0 to - 3 mag and spectral range K5 to M5, these stars are identified on the frequency diagram of the periods by the shoulder at 75 d; their period spread extends from 30 to 1000 days. The photometric criterion adopted to distinguish Red Semiregulars from Long-Period variables is based on the light amplitude of the variability: by definition, the amplitude is less than 2.5 mag for Semiregulars. The successive editions of the GCVS have listed 1046, 1675 and 2221 stars of this category. A typical representative is µ Cephei[3]. According to the observations of Argelander this pulsator had a cycle of 400 to 460 days in the middle of the last century, while by the end of the century the cycle had increased to 1000 days; the related light amplitude is about .5 mag; an illustration of the complexity of the light curve of this star can be found in Payne-Gaposchkin and Gaposchkin (1938). Another example of a semiregular variable is Betelgeuse (α Orionis); this supergiant exhibits 140 to 300 day-cycles, with a light amplitude varying from .2 to 1.2 mag. Although the radius of this star has been measured directly since the earliest applications of the interference technique to astronomical objects (Michelson and Pease 1921), and although surface details of this object have been resolved by speckle methods (cf. Bates 1982, Murdin and Allen 1979), convincing direct evidence of a variable radius is, to the best of my knowledge, still lacking. A statistical study of the period fluctuations of the semiregulars has been conducted by Lacy (1973) who reports that the period distribution of the individual stars obeys a Gaussian law with a very large standard deviation; period fluctuations may exceed 100%.

c. RV Tauri stars

These objects of spectral class G0 to K5 (6000 to 4500 K) and visual magnitude in the range − 3.5 to − 5 mag form a fairly homogeneous, though sparse group of variables. The three editions of the GCVS report 72, 92 and 104 stars of this category, respectively. With cycle lengths in the range of 50 to 150 days, light variations of .8 to 2 mag and velocity amplitudes of 30 to 40 km/s, these stars − predominantly Population II objects − exhibit a characteristic shape of the light curve (Preston et al. 1963). Deep minima are alternating with shallow minima, the alternation being randomly interrupted. Fig. 11 illustrates this behavior for AC Herculis, an F8 star of mean cycle length $P \simeq 76$ d and light amplitude of ~ 1 mag. (Mantegazza 1983). Local periods are defined for these objects as the intervals between two successive deep minima. As already observed in Ludendorff (1928), the radial velocity curve of certain RV Tauri stars shows a periodicity equal to half the period of the light curve, or equivalently, equal to the interval between two successive minima (cf. R Scuti). Moreover, Rosino (1951) calls attention to the fact that the semiregulars, the RV Tauri and possibly also the RR Lyrae stars form a homogeneous family as far as their period-luminosity relation is concerned − on condition, however, that half the photometric period is taken as the relevant period for the RV Tauri variables. This property might suggest that the time interval between successive minima is close to a linear oscillation mode of these stars, while the light curve seems to display a period doubling phenomenon. The local periods are undergoing fluctuations of the same order as the period fluctuations in Mira stars. The characteristic new type of irregularity of the RV Tauri objects, namely the unpredictable switch-over of the alternations of deep and shallow minima, has not received, so far, a statistical treatment.

The two broad groups of pulsators discussed so far, the Regular Sequence and the Red Sequence, show relatively large amplitude ranges. The standard theoretical framework for

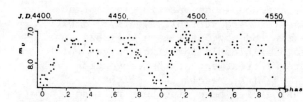

Fig. 11.- Lightcurve of the RV Tauri star AC Her (Mantegazza 1983).

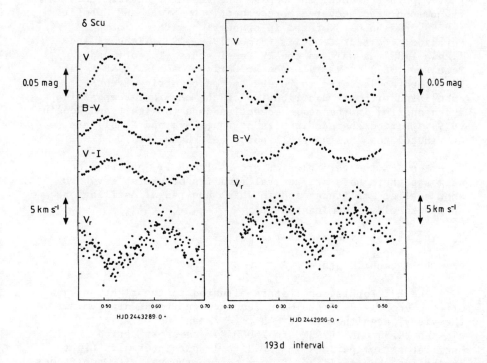

Fig. 12.- Variability of δ Scuti (Balona et al 1981).

modelling the variability of these stars is provided by the
theory of radial oscillations (cf. the reviews by Cox 1976,
1980). In a few cases where a coupling between convection and
pulsations is to be expected (RR Lyrae, RV Tauri, Mira stars
etc.), attempts at including non-radial convective motions have
been made by Deupree (1976, 1977). Such interactions are found
to destroy the precise periodicity of the cycles. Non-periodic
light curves have also been found in asymptotic giant branch
star models representative for Mira variables; performed with a
standard radial hydrodynamic code by Ostli et al. (1982), the
calculations exhibit irregularities in the light curves which
are attributed to an improved treatment of the convection
zone. On the other hand, we might recall that slight
excursions from strict periodicity have also been noticed
already in the earliest hydrodynamic calculations by Christy
(1964, 1967, 1968) modelling regular variables; here the noisy
contributions to the velocity curve are attributed to complex
physical mechanisms occurring in the surface zones of the
models (shock waves). Besides these numerical studies it has
been recognized a long time ago by Woltjer (1937) that
resonances among the dynamical linear oscillation modes of a
star are instrumental in shaping the velocity curve. This
phenomenon, rediscovered more recently by Simon and Schmidt
(1976), has been advocated in connection with the Bump
Cepheids, the Beat Cepheids (Simon 1979, 1980, Takeuti and
Aikawa 1980) as well as the RV Tauri stars (Takeuti and
Petersen 1983). 'Rapid' secular period changes in RR Lyrae
have been conjectured to be due to random mixing events
associated with a redistribution of the chemical composition
and occurring in the semi-convective zone (Sweigart and Renzini
1979). Period changes in Mira stars (R Hya and R Aql) are
attributed to a helium shell pulse (cf. Cahn 1980).

Besides the variables of groups I and II, observation
suggests that there are several families of variables which
show smaller amplitudes and in which non-radial oscillations
appear to play a dominant part.

3. The variables close to the Main Sequence

a. The β Cepheid Stars

The hot Population I stars situated to the right and just
off the upper end of the Main Sequence in the HR diagram, in
the visual magnitude range − 3 to − 7 mag and of spectral
classes B0.5 to B5 (20000 to 25000 K) are short period
oscillators, of periods in the 3 to 7-hour interval. Light
variations are small ($\lesssim$ .1 mag), while the velocity amplitudes
go from 5 to 50 km/s; a few exceptions of higher velocity
amplitudes are known. The GCVS reports 6, 11 and 23 β Cephei

stars in the three successive editions. As discussed in the
reviews by Aizenman (1980) and Lesh and Aizenman (1975, 1978),
about 50% of these stars show a beat phenomenon, of beat period
of several days. Moreover, observation indicates that β Cephei
are slow rotators; with equatorial velocities of the order of
15 to 60 km/s, these pulsators are exceptional, the typical
equatorial velocities of other stars of the spectral range
of β Cepheids being 165 km/s. The existence of the beat
phenomenon led Ledoux (1951) to conjecture that non-radial
oscillations must be involved, radial frequencies being too
widely spaced to account for the observed beat period.
Circumstantial evidence for non-radial variability was also
given by Walker (1954) who, on the basis of Baade's test, was
led to rule out purely radial pulsations. Percy (1980) has
compiled the recently gathered evidence for non-radial effects
based on 7 independent methods. From a collection of
16 β Cepheids investigated by at least one of these methods, 9
stars are found to be non-radially oscillating; for two
stars, β CMa and 12 Lac, non-radiality has been assessed on the
basis of several methods; for four stars, β CMa, β Cep, 12 Lac
and BW Vul quadrupole modes have been identified.

Shobbrook (1979) reports that the frequency spectra of a
number of β Cephei stars are changing over periods of decades
and sometimes of years, indicating a secularly varying shape
function of their light or velocity curves (cf. Eq. II.1.2b).
Young and Furenlid (1980) and Cherewick and Young (1975) have
detected random amplitude fluctuations both in the light and
the velocity curves of BW Vul. Over the period 1968 to 1972
the oscillation amplitude of α Vir decayed and the star has now
virtually stopped oscillating (Lomb 1978). In the non-radial
triple mode pulsator 16 Lac all three amplitudes dropped to
about half their value of 1965 in a period of 12 days
(Jarzebowski et al. 1979). Secular changes have been reported
in the pulsational behavior of δ Ceti (Lloyd and Pike 1984),
which are not likely to be due to evolutionary effects.

The canonical model of a β Cepheid star, conforming to the
observational information of α Vir[4], is a massive star of 10
to 20 $M_\odot$ and of standard chemical composition of a Population I
object, just moving off the Main Sequence. The precise
instability mechanism, however, responsible for the driving of
the observed oscillations, remains to be isolated (cf. Percy
1980 for a list of 5 currently proposed excitation processes).

b. The δ Scuti Stars[5]

These stars of spectral type A8 to F2 and visual magnitude
3 to 1 mag are located at the lower end of the Cepheid
instability strip, just above the Main Sequence. The periods

are less than 8 hours; stars closest to the Main Sequence have short periods, typically of 1 hour, and small V amplitudes, ~ .02 mag; periods and amplitudes increase with distance of these stars from the Main Sequence; upper amplitudes lie around .8 mag (Breger 1979). The number of known δ Scuti stars has been increasing dramatically over the past years. The different editions of the GCVS record 0, 5 and 17 such variables, respectively. Breger (1979), adopting a broader definition of this class of stars, including the RRs, objects of the GCVS (also known as short period RR Lyrae or Dwarf Cepheids in the literature), lists some 130 δ Scuti pulsators. It is now thought that this category of variables is the second most numerous family of pulsators of our galaxy, after the variable white dwarfs.

From a theoretical point of view, δ Scuti stars are mostly Population I objects, of masses in the range 1.5 to 3 $M_\odot$ ($Z \simeq .02$), moving towards the giant region. The distinguishing pulsational feature of these variables is their multi-periodicity; there are just a few pulsators of this class in which so far only one period has been detected, namely ρ Pup, AD CMi and BD + 43° 1894 (Yamaska et al. 1983). The prototype, δ Sct, has been shown to be bi-periodic. Fig. 12 illustrates the variability of this star at two epochs separated by a 193 day interval; the change in the amplitudes has been shown to be due to the presence of frequencies of 5.160765 and 5.354018 cycles per day by Balona et al. (1981). These authors have also been able to demonstrate that the two components are to be identified with one radial and one non-radial quadrupole oscillation. The variable V474 Mon has been found to show 3 evenly spaced oscillation frequencies, of 7.217, 7.346 and 7.475 cycles per day respectively. Stobie and Shobbrook (1976) have interpreted this triplet as a rotationally split quadrupole mode. Some authors have denied the reality of well-defined periods of certain δ Scuti variables. Thus Morguleff et al. (1976) suggest that the star 14 Aur possesses a randomly fluctuating periodicity, similar to the behavior of the cycle lengths of Mira stars; the local periods apparently obey a Gaussian distribution law (cf. the histogram in Fig. 13a), centered at .085 d, with deviations from .06 to .12 d. An analogous behavior has been reported for 44 Tau (Morguleff et al. 1975); as shown in Fig. 13b, the periods of this δ Scuti star are randomly distributed over the range .09 to .17 d, with a most frequent cycle of .125 d. Wizinovich and Percy (1979) have claimed, however, that the observations of this latter object are reasonably well represented by a double-periodic pulsation; the analysis of these authors is based on a least squares fit of the data, which yields the periods .1449 and .1120 d ($P_o/P_1 = .773$, different from the value for Beat Cepheids); of course, one may object that a least squares period determination necessarily produces periods; these computed periods cannot be regarded as intrinsic periods of the star unless an independent proof of existence of multi-periodicity has been given. Irregularities in the behavior of θ Tuc have been detected by Stobie and Shobbrook (1976); by means of a Fourier analysis of the light curve of this variable, these authors conclude that coherent periods are lacking.

Besides short-timescale fluctuations, also secular variations have been recorded in some δ Scuti stars. Stobie et al. (1977), discussing 1247 observations of 21 Mon, conclude that over the time spans 1970-1971 and 1972-1973 this pulsator had no common periods. This might be an instance of mode switching.

In his review paper, Breger (1979) has pinpointed the major problem that besets the characterization of δ Scuti variability in particular, and, in fact, of stellar variability in general: The paradigm the observer is adopting is 'cyclic' variability, i.e., in mathematical terms 'quasi-periodic' oscillations (cf. Section III). 'What does the poor observer do? He plots his data, finds regularities (cycles), and determines the average length of a cycle. It is quite easy to see regular cycles despite the irregularities present in some δ Scuti stars.' But 'are δ Scuti stars really periodic?'

c. The Sun

Several types of oscillation are currently being observed on the sun. According to their period ranges, these oscillations are conveniently divided into 3 groups.

(1) In the long period range, $P \gtrsim 2$ h, a velocity oscillation of period of 160.01 minutes was first detected by Severny et al. (1976); established via a Doppler shift measurement of the 5123.7 A line, this oscillation has been confirmed by other groups (Scherrer et al. 1980, Grec et al. 1980). With a velocity amplitude of 20-50 cm/s, this oscillation is 5 orders of magnitude weaker than a velocity amplitude of an ordinary variable star.

(2) In the intermediate period range, of 100 to 10 minutes, 'apparent radius fluctuations' have been detected by the SCLERA group since 1973 (Hill et al. 1976); while initially Hill's observations suggested relative radius amplitudes of 10^{-4}, the latest measurements point towards a value of 4×10^{-7} for this parameter (Bos and Hill 1983).

(3) In the short period range, around 5 minutes, we can

roughly distinguish two types of well studied oscillations (a) velocity fluctuations of typical horizontal wavelength of the order of 1 Mm, first discovered by Leighton (1963) and carefully analyzed by Deubner (1975) and Deubner et al. (1979); and (b) whole solar disk velocity and light variations, recorded in a 6 day uninterrupted run of observations carried out at the terrestrial South Pole (Grec et al. 1980), and by the Solar Maximum Mission (SMM) satellite respectively (Woodard and Hudson 1983). Fig. 14a shows the power spectrum of the South Pole data; 75 peaks have been isolated and identified, by comparison with solar model frequencies, as radial, or non-radial modes of low harmonic degree ($l = 0,1,2,3$) and high order ($k = 15$ to 35); these peaks are approximately evenly spaced, with an average distance of 68 μ Hz; the amplitude of these velocity fluctuations is 10 cm/s, i.e., within an order of magnitude of the 160 min oscillation amplitude. The power spectrum of the relative luminosity variations as detected in the SMM experiment is shown in Fig. 14b, in which some 20 peaks have tentatively been identified.

To increase the frequency resolution, Claverie et al. (1981) have combined a series of consecutive data strings into a single set. This procedure has led them to identify frequency splittings, of average spacing of .75 μ Hz. The physical origin of the splitting is thought to be solar rotation; Claverie et al. conclude that the core rotation is 2 to 8 times as rapid as the visually observable surface rotation. Gough (1984) arrives at a similar conclusion on the basis of an analysis of recently obtained data by Harvey (cf. the proceedings of the Boulder conference 1983).

The solar five-minute band as observed in whole disk measurements is currently interpreted in terms of linear non-radial acoustic modes (p-modes) of low harmonic degree and high radial order (Christensen-Dalsgaard and Gough 1982). The Deubner-type observations are linked to p-modes of low radial order ($\lesssim 10$) and high harmonic degree (l several hundred); the corresponding eigenfunctions being trapped in the surface region of the sun - at a depth greater than a few times 10^3 km the amplitudes of these oscillations are vanishingly small - such measurements provide information on the outer layers of the sun. The SCLERA data are thought to isolate p-modes of low radial order and low harmonic degree, exploring therefore the deep interior of the sun. No generally accepted interpretation of the 160 min oscillation has been offered so far. For some time regarded as an artifact - 160 min is 1/9 day - the precise period, 160.0095 ± .001 min, and especially the phase coherence of this oscillation over the whole period of observation of eight years now, militate in favor of a genuine oscillation mode of solar origin. Among the <u>linear</u> solar modes gravity

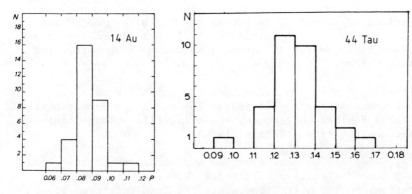

Fig. 13.- Histograms of local periods of two δ Scuti stars, 14 Aurigae (Morguleff et al 1976) and 44 Tauri (Morguleff et al 1975).

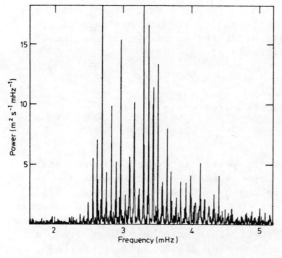

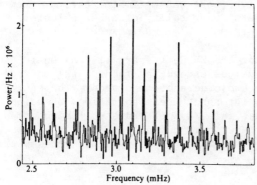

Fig. 14.- Solar power spectra in the 5-minute range; (a) South Pole spectrum obtained from a 6-day continuous run (Grec et al 1980) and (b) SMM power spectrum from a 137-day run (Woodard and Hudson 1983).

oscillations (g-modes) alone can be matched with the 160 min periodicity; Christensen-Dalsgaard (1982) has argued that a g-mode of harmonic degree less than 5 should be responsible for this observation.

Finally, I should mention that the detection of a further class of g-modes of harmonic degree $\ell=1$ and 2 has been reported at the 1983 Boulder conference, whose identifications remain somewhat controversial (Gabriel 1984).

Following the detection of low-amplitude and short-period coherent solar velocity oscillations, a search for analogous oscillations in the solar type stars has been initiated by Traub et al. (1978). These authors, investigating the stars of the solar neighborhood, found no statistically significant peak at the 3σ level in the power spectra of their velocity measurements; nevertheless, at the 2σ level, the star α C Mi (Procyon), of spectral type F5 IV-V, exhibits a peak at a period of 57 ± 1 s, with velocity amplitude of 15 m/s. Incidentially, Procyon is the only dwarf star, i.e., star of a structure similar to the sun's, studied by Traub et al. Observational work in this direction is now being continued by Fossat and coworkers (cf. Gelly et al. 1984).

4. The White Dwarf Variables

Oscillating white dwarfs come in two flavors.

a. The ZZ Ceti stars

Discovered nearly twenty years ago by Landolt, pulsating white dwarfs of spectral type DA were shown to define a homogeneous class of variables by McGraw and Robinson (1976). With luminosities less than 10^{-2} $L_\odot$ and effective temperatures in the 12,000 to 10,000 K range, these objects are difficult to trace. So far only 17 ZZ Ceti stars have been discovered, all in the solar neighborhood; however, taking account of the observational selection effect, one can estimate that this new class of variables outnumbers the totality of other pulsators of our Galaxy by a factor of 100.

The observed periods of the ZZ Ceti stars lie in the 100 to 1200 s interval. Such values exceed the fundamental periods of typical white dwarfs by two orders of magnitude. Therefore, theorists have been led to interpret these pulsations as being (non-radial) linear g-modes. All ZZ Ceti stars so far observed have been found to be multi-periodic pulsators; light amplitudes are typically in the range .003 to .3 mag (cf Kepler et al. 1983).

The prototype ZZ Ceti has been reported to have 4 periods occurring in doublets of 212.77 and 213.13 s, and 274.25 and 274.77 s respectively; the corresponding semi-amplitudes are .0077 and .0044, and .0044 and .0034 mag respectively (Stover et al. 1980). One notices that the period ratio of the average values of the two doublets is close to 5/4. The rate of period change has been found to obey $|\partial/\partial t\ P| < 10^{-13}$, an upper limit consistent with the change of a g-mode of a model evolving on the white dwarf cooling sequence.

For the variable GD 385 observed since 1978, Kepler (1984) found a period doublet of 256.127 and 256.332 s and a singlet of 128.115 s, at half the value of the doublet; the relative semi-amplitudes are .0114, .0109 and .0037 respectively. Additional periods were reported previously by Fontaine et al. (1980) and Vauclair and Bonazzola (1981), which were not detected by Kepler. On the other hand, the latter author found evidence for an amplitude variability on a time-scale of several years; the semi-amplitude of the doublet, at a level of .035 in 1978-1981, dropped to .022 in 1982. The closeness of the two longer periods manifests itself as a beat phenomenon, of a beat period of 89 h. At the moment, no mechanism has been suggested for the rapid amplitude variability. The doublet is interpreted as being the result of a rotational splitting.

The variable G117-1315A exhibits 6 periods ranging from 107.6 to 304.4 s. The rate of change of the periods obeys $|\partial/\partial t\ P| < 2 \times 10^{-14}$ s/y (Kepler et al. 1982).

Using 65 hours of high speed photometry data gathered in 1980-1982, Kepler et al. (1983) detected a period triplet of 109.08684, 109.27929 and 109.47242 s, evenly spaced in frequency, and of relative half-amplitudes .0031, .0011 and .0031 respectively. Fig. 15 exhibits the pulsation of this variable. The photometric record obtained in July 1980 shows an obvious periodicity of ~109 s (Fig. 15.a). The power spectrum confirms this periodicity; Fig. 15.b illustrates the structure of a spectrum averaged over three continuous runs; notice that it does not resolve the peak at 9.4 mHz into a triplet; this spectrum shows a conspicuous amount of power at the low frequency end, with a broader peak around .5 mHz (~30 min).

The frequency splitting, reported for 6 ZZ Ceti objects, is interpreted as a rotational effect. The required rotation periods lie in the 1/2 to 30 h range, which is compatible with current ideas on white dwarf rotations (Winget and Fontaine 1982). From a theoretical point of view, ZZ Ceti stars have masses in the .4 to .8 $M_\odot$ span; their chemical composition is stratified, with a carbon-oxygen core, a helium layer on top of

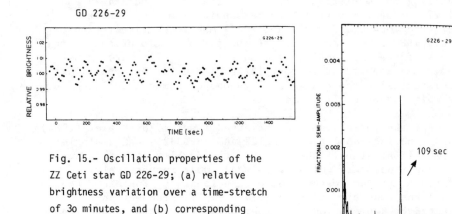

Fig. 15.- Oscillation properties of the ZZ Ceti star GD 226-29; (a) relative brightness variation over a time-stretch of 30 minutes, and (b) corresponding power spectrum (Kepler et al 1983).

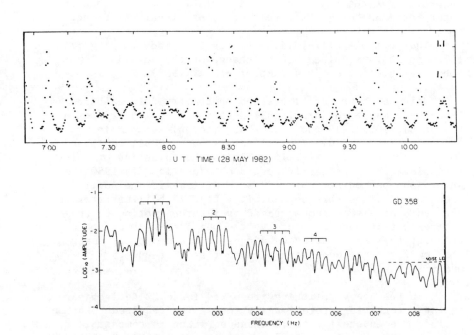

Fig. 16.- The pulsation of the DB white dwarf GD 358; (a) brightness variation, and (b) power spectrum (Winget et al 1982).

the core with a mass less than 1% of the total mass, and a pure hydrogen surface layer of mass estimated to lie in the range 10^{-14} to 10^{-4} times the total mass (Winget and Fontaine 1982). Such models exhibit vibrational instabilities of g-modes, driven by the ionisation mechanism of hydrogen and in part, at least, also of helium.

b. Variable DB White Dwarfs

Aside from the pulsating DA white dwarfs, discovered observationally, numerical stability analyses of models of white dwarfs of spectral type DB have revealed that the latter may oscillate as well (Winget 1981). The characteristic feature of DB white dwarfs is the absence of a hydrogen surface layer; in these stars, the atmosphere on top of the carbon-oxygen core is made up of almost pure helium; the driving mechanism of the pulsations is here the second ionisation of helium which becomes operative in the effective temperature interval of 30,000 to 20,000 K.

Being less numerous than DA white dwarfs – Cox (1982) estimates that only 10% of the white dwarfs are of spectral type DB – oscillating DB white dwarfs have been traced observationally only recently, and so far only one member of this class has been detected. This object, known as GD 358, exhibits a complex light curve (Fig. 16.a) showing an amplitude of .3 mag; the corresponding power spectrum (Fig. 16.b) has an equally involved structure; Winget et al. (1982) claim to have isolated 28 periods ranging from 2 minutes up to a quarter of an hour; these periods seemingly appear in groups of 4 or 5-tuplets reported to be evenly spaced in frequency.

In the context of the DA white dwarf GD 385 Fontaine et al. (1980) and Vauclair and Bonazzola (1981) conjectured that non-linear oscillation effects are responsible for the complicated variability of that star. Likewise, a mere inspection of the power spectrum of the DB star GD 358 suggests that a linear oscillation theory is insufficient to explain the true nature of the observed pulsations.

III. THE THEORY OF NON-LINEAR STELLAR OSCILLATIONS

Even ten years ago every experimenter finding ... a complex aperiodic oscillation ... rejected it from considerations referring to impurities of the experiment, chance external effects and such like. Now it is clear ... that these complex oscillations can be linked up with the very essence of the situation. They can be determined by the fundamental equations of the problem and are not random effects. They can, and must, be studied along with

>the classical stationary states and periodic processes of
>the problem.
>
>   Arnold 1984

The brief and admittedly incomplete review of endogenous aperiodic oscillations of stars given in the previous chapter is intended to stress that such a variability appears to be a pervading phenomenon among perhaps all categories of stellar pulsators. Until recently, theorists did not seriously pay attention to irregular variability. Although the currently available hydrocodes do produce nonstrictly periodic velocity curves in many cases, the consensus seems to have been that deviations from periodicity are indicative that the model oscillation has not yet relaxed towards its limit cycle. On the other hand, when rapid period variations are badly needed - as in the case of the RR Lyrae stars (cf. above) - theorists have attempted to 'save the phenomena' by introducing randomly operating new physical processes (cf. Sweigart and Renzini 1979).

It is the objective of this chapter to emphasize that complicated <u>ad hoc</u> mechanisms are not required to generate an irregular aperiodic time behavior of a star. The latter may arise spontaneously as a result of the non-linear structure of the stellar hydrodynamics. While the role of non-linearities in shaping the velocity curve has been fully appreciated, following Eddington's (1918, 1919) seminal papers on intrinsic stellar variability and Woltjer's (1935 and later notes) clarifying canonical transformation approach, the possibility of disordered pulsations induced by non-linear effects, hinted at already by Wesselink (1939), has not been followed up until recently.

I shall review the recent development achieved at three levels of approximation. Following the path traced out by Woltjer, I shall first discuss adiabatic radial non-linear oscillations. I shall next comment on the extension of that formalism to non-radial non-linear oscillations. In a final brief section I shall examine a few aspects of non-adiabatic non-linear pulsations. In the context of the adiabatic approximation we have to be well aware that we miss a fundamental aspect of the oscillation problem. Our equations fail to tell us which modes are actually excited. Therefore, even if we can demonstrate the existence of irregular adiabatic fluctuations, the latter may be 'Garden of Eden patterns' of the full non-conservative stellar oscillation equations: these disordered fluctuations may be unattainable in realistic non-adiabatic evolution. On the other hand, if we can prove that no irregular adiabatic fluctuations can occur for a given range of stellar oscillation parameters, then we are automatically sure

that no such short time-scale fluctuations can arise in the
full oscillation problem, as long as the non-conservative
effects are small, or equivalently, the thermal time is long as
compared to the dynamical time-scale.

1. Adiabatic Radial Non-linear Oscillations.

Woltjer (1935) was the first to point out that the stellar
adiabatic non-linear motions of radial symmetry around a state
of hydrostatic equilibrium can be described in terms of a
Hamiltonian particle system of a finite number F of degrees of
freedom. The theoretical interest of this observation resides
in its opening a gate to a mathematical framework that enables
us to characterize the nature of the possible conservative
oscillations in an exhaustive way.

a. The Theoretical Framework

Under local conservation of entropy the second principle
of thermodynamics entails as a corollary, that the equilibrium
state of an isolated physical system is given by the
stationarity condition of its total potential energy (cf.
Perdang 1977). By analogy with conventional mechanics we are
then led to conjecture that the dynamical state of this system
can be obtained from the stationarity condition of a
Lagrangian, defined as the difference between the kinetic
energy and the potential energy. In the problem of stellar
oscillations one can easily verify that this procedure leads in
fact to the conventional non-linear adiabatic equations of
motion.

Since we are not interested here in the underlying
equilibrium state, but only in the bounded motions around that
state, we set

$$L_{osc} = K - V_{osc}, \qquad \text{III.1.1}$$

L_{osc}, K and V_{osc} representing the total Lagrangian, the kinetic
energy and the total potential energy of the conservative
motions inside the star. With the mass variable, m, taken as
the independent 'space' variable [6], these functionals are
given by

$$K = \int_o^M dm |\dot{r}(m,t)|^2/2 , \qquad \text{III.1.2}$$

$$V_{osc} = - [\; G\int_o^M dm\; m/r(m,t) - G\int_o^M dm\; m/r_E(m)]$$

$$+ [\int_o^M dm\; U(S(m), v(r(m,t), d/dm\; r(m,t)))$$

$$- \int_o^M dm\; U(S(m), v(r_E(m), d/dm\; r_E(m)))] . \qquad \text{III.1.3}$$

In this expression S denotes the specific entropy, U and v stand for the specific thermal (internal) energy and volume respectively, G is the gravitational constant; the subscript E indicates an equilibrium distribution and M is the mass of the star. The dot stands as usual for the time derivative. The equations of motion in field form are then found through the variational principle

$$\delta \int_{t_o}^{t_1} dt\, L_{osc} = 0 , \qquad \text{III.1.4}$$

the Lagrangian being a functional of $r(m,t)$, and $d/dm\, r(m,t)$; the endpoints t_o, t_1 of the integration interval are to be held fixed. The associated Euler equations are discussed in Ledoux (1958). Notice that the choice of the independent variable m takes care of conservation of mass.

The variational principle in field form (III.1.4) is transformed into a variational principle of particle mechanics by expanding the finite displacement field

$$\delta\, r(m,t) = r(m,t) - r_E(m) \qquad \text{III.1.5}$$

in the linear adiabatic eigenfunctions of the equilibrium model (Woltjer 1935)

$$\delta\, r(m,t)/r_E(m) = \sum_{i=1}^{\infty} q_i(t)\, \xi_i(m) . \qquad \text{III.1.6}$$

Since the system of eigenfunctions $\xi_i(m)$ of the linear adiabatic oscillations around a stable equilibrium state is known to be complete, expansion (III.1.6) is rigorous. For the purposes of explicit computations we now approximate this expansion by a finite sum of F linear modes. Substitution of the latter into Eqs. (III.1.2,3) yields

$$K = 1/2 \sum_{i=1}^{F} \dot{q}_i^2 \equiv K(\dot{q}_i) , \qquad \text{III.1.2'}$$

$$V_{osc} = 1/2 \sum_{i=1}^{F} \omega_i^2 q_i^2 + 1/3! \sum_{i,j,k=1}^{F} V^{(3)}_{(ijk)} q_i q_j q_k +$$

$$1/4! \sum_{i,j,k,\ell=1}^{F} V^{(4)}_{(ijkl)} q_i q_j q_k q_l + \cdots$$

$$= V^{(2)} + V^{(3)} + V^{(4)} + \cdots \equiv V_{osc}(q_i) , \qquad \text{III.1.3'}$$

so that the Langrangian (III.1.1) now transforms into an ordinary function in the 2F unknowns q_i, $\dot{q}_i$, $i = 1,2,\ldots,F$. The simple form of the kinetic energy (III.1.2') arises if we normalize the eigenfunctions as follows

$$\int_0^M dm \, r_E(m)^2 \xi_i \xi_j = \delta_{ij} \, , \quad i,j = 1,2,\ldots,F \, . \qquad \text{III.1.7}$$

The coefficients ω_i represent the linear oscillation frequencies of the star. The potential energy (III.1.3') has been expanded in a power series in the time-dependent functions $q_i(t)$.

The evolution of the functions $q_i(t)$ is given by the Euler equations of (II.1.4),

$$\ddot{q}_i = - \partial/\partial q_i \, V_{osc}(q_j) \, ,$$

or

$$\ddot{q}_i + \omega_i^2 q_i = \sum_{j,k=1}^{F} D_{ijk} q_j q_k + \sum_{j,k,\ell=1}^{F} D_{ijk\ell} q_j q_k q_\ell + \ldots, \qquad \text{III.1.8}$$

$i = 1,2,\ldots,F.$

Equivalently we can write these equations

$$\dot{q}_i = \partial/\partial p_i \, H(q_j,p_j) \, , \quad \dot{p}_i = - \partial/\partial q_i \, H(q_j,p_j) \, , \qquad \text{III.1.9}$$

with

$$H(q_j,p_j) = 1/2 \sum_{j=1}^{F} p_j^2 + V_{osc}(q_j) \, . \qquad \text{III.1.10}$$

Eqs. (II.1.8) are seen to correspond to the Lagrangian equations of a discrete mechanical system of F degrees of freedom, and more specifically to a system of F non-linearly coupled harmonic oscillators. Eqs. (III.1.9), on the other hand, are the associated Hamiltonian equations. The expansion coefficients $q_i(t)$ in expression (III.1.6) are seen to play the parts of generalized coordinates and their time derivatives appear as generalized momenta $p_i(t)$.

The procedure adopted has two advantages. (a) The original partial differential equations of motion are reduced to a system of coupled ordinary differential equations; the oscillation properties of the star are encoded in the set of global parameters ω_i, $V^{(3)}_{(ij)}$, $V^{(4)}_{(ijk)}$,..., where $V^{(f)}_{(ij\ldots n)}$ is symmetric in all of its indices; for low amplitude oscillations, the lowest order non-linear contribution $V_{(ij)}^{(3)}$ is expected to describe the motion with sufficient accuracy. The numerical integration of the resulting coupled oscillator equations (III.1.8) is then trivial. (b) The very existence of a Hamiltonian form (III.1.10) of the adiabatic bounded stellar motions enables us to resort to the Poincare-Birkhoff-Siegel-Kolmogorov-Arnold-Moser results on Hamiltonian oscillations to anticipate the different types of solution and to interpret the numerical integrations (Perdang 1979, 1983).

We summarize here the most relevant properties of Hamiltonian theory.

(1) A Hamiltonian of F degrees of freedom $H(q_i,p_i)$ is said to be <u>integrable</u> if it possesses F smooth enough independent isolating integrals $G_j(q_i,p_i)$ which are in involution.

The condition of being isolating means that each integral $G_j(q_i,p_i)$ decreases by one unit the dimension of the accessible phase space; the presence of F isolating integrals then implies that the motion in the 2F-dimensional phase space is restricted to an F-dimensional region R^F.

Two integrals G_j and G_k are in involution if their Poisson bracket vanishes

$$(G_j,G_k) \equiv \sum_{i=1}^{F} (\partial/\partial q_i G_j \; \partial/\partial p_i G_k - \partial/\partial p_i G_j \; \partial/\partial q_i G_k) = 0. \quad \text{III.1.11}$$

A Hamiltonian of F degrees of freedom is <u>non-integrable</u> if it fails to obey the requirements of integrability. In stellar oscillation Hamiltonians of form (III.1.10,3') this failure is due to an insufficient number of isolating integrals[7]. A Hamiltonian of F degrees of freedom is <u>quasi-integrable</u> if it is the sum of an integrable Hamiltonian $H_o(q_i,p_i)$, and a small smooth 'generic' perturbation (i.e., a function of 'non-zero probability' among the class of smooth functions) $\varepsilon\, H_1(q_i,p_i)$:

$$H(q_i,p_i,\varepsilon) = H_o(q_i,p_i) + \varepsilon\, H_1(q_i,p_i) \quad , \quad \text{III.1.12}$$

ε being a sufficiently small parameter.

(2) The nature of the motion of an <u>integrable</u> Hamiltonian system is fully captured by the <u>Liouville-Arnold theorem</u> (cf. Arnold 1976 for a precise formulation).

If the accessible region R^F in phase space is connected and bounded, then this region is topologically equivalent to an F-torus; the motion is quasi-periodic.

The first part of the theorem amounts to showing that a canonical transformation to action-angle variables, $q_i, p_i \to \phi_i, J_i$, can be found, such that the new Hamiltonian becomes a function of the actions alone. Such a form of the Hamiltonian is called a <u>normal form</u>. The equations of motion then reduce to

$$\dot{\phi}_i = \Omega_i(J_j) \quad , \quad (a)$$
$$\dot{J}_i = 0 \quad , \quad (b)$$

III.1.13

with $\Omega_i(J_j) = \partial/\partial J_i \, H(J_j)$; the angles ϕ_i are defined modulo 2π. These equations are equivalent to those of an F-dimensional billiard whose boundary is the surface of an F-dimensional cube with opposite faces being identified; the latter configuration is precisely an F-torus.

The second part of the theorem flows from the integration of Eqs. (III.1.13), showing that the angles are linear functions of time, the actions being constants. Therefore the original generalized coordinates can be expanded in a multiple Fourier series of these angles

$$\{q_i(t), p_i(t)\} = \sum_{k_1, k_2, \ldots, k_F} \{Q^{(i)}_{k_1 k_2 \ldots k_F}, P^{(i)}_{k_1 k_2 \ldots k_F}\}$$
$$\exp i[(k_1\Omega_1 + k_2\Omega_2 + \ldots + k_F\Omega_F)t + \Phi_{k_1 k_2 \ldots k_F}] \quad ,$$

with

$$|Q^{(i)}_{k_1 k_2 \ldots k_F}| \text{ and } |P^{(i)}_{k_1 k_2 \ldots k_F}| \leq A \exp - k B , \quad \text{III.1.14}$$
$$k = \sum_{i=1}^{F} |k_i| .$$

In these expressions the Fourier expansion coefficients $Q^{(i)}_{k_1 \ldots k_F}$ and $P^{(i)}_{k_1 \ldots k_F}$ are functions of the actions J_j; the phases $\Phi_{k_1 \ldots k_F}$ are linear combinations of the integration constants of Eqs. (III.1.13a); the integers $k_1, \ldots, k_F$ take on the values $0, \pm 1, \pm 2, \ldots$; the constants A and B are independent of the k's. A time-dependent function f(t) satisfying a multiple Fourier expansion, with F finite, of form (III.1.14), is termed <u>quasi-periodic</u>; notice that the constraint on the expansion coefficients implies that the derivatives of all orders are then also **quasi-periodic**.

If the Hamiltonian is **pseudo-integrable** [7], then the orbit, instead of being carried by a torus, lies on a surface of genus $g \neq 1$, i.e., a multiply-handled sphere. Such a geometrical constraint prevents the validity of an F-tuple Fourier expansion (III.1.14), F being the number of degrees of freedom of the system.

(3) <u>Quasi-integrable</u> Hamiltonian systems are partially taken care of by the celebrated KAM theorem (Kolmogorov 1957, Arnold 1963, Moser 1962). We merely indicate here the intuitive content of this theorem; for a precise formulation see Arnold (1976). In the absence of the perturbing Hamiltonian, $\varepsilon = 0$, the orbits all lie on tori. If a perturbation is turned on, $\varepsilon \neq 0$, some of these tori survive, although they become distorted. Others explode and are destroyed. The measure in phase space of the latter is finite and increases with ε; for very small ε, therefore, 'most' tori of the non-perturbed Hamiltonian system are just deformed by the perturbation; the orbits carried by these tori remain of course quasi-periodic. The mechanism responsible for the destruction, or the explosion, of tori is a resonance, or an approximate resonance, among the frequencies of the non-perturbed Hamiltonian, i.e., a relation

$$k_1 \Omega_1 + k_2 \Omega_2 + \ldots + k_F \Omega_F \simeq 0 \quad , \qquad \text{III.1.15}$$

k_i being integers (positive, negative and zero). Such a resonance is said to be of order $k = \sum_1^F |k_i|$. Among the infinity of orbits on a resonant torus, a generic perturbation leaves the topology of a finite even number of trajectories unaffected; half of those become stable, the other half unstable. Trajectories of the neighborhood of the unstable orbits explore a region of phase space R^D of dimension $D > F$. Such motions are referred to in the literature as <u>chaotic</u>, <u>irregular</u> or <u>Kolmogorov unstable</u>.

(4) The action of a resonance in a Hamiltonian system of $F = 2$ degrees of freedom has been clarified by Walker and Ford (1969) and especially by Zaslavski and Chirikov (cf. the review by Chirikov 1979).

Suppose the perturbation Hamiltonian H_1 is expanded in a Fourier series of the angles ϕ_1, ϕ_2. Suppose further that we consider a resonant torus of the non-perturbed Hamiltonian

$$k_1 \Omega_1 + k_2 \Omega_2 = 0 \text{ at } J_1^o, J_2^o \quad . \qquad \text{III.1.16}$$

Assuming H_1 even in the angles, we write

$$H(\phi_1,\phi_2,J_1,J_2) = H_o(J_1,J_2) + \varepsilon H_{k_1k_2}(J_1,J_2) \cos(k_1\phi_1 + k_2\phi_2)$$
$$+ \text{ non-resonant Fourier contributions} \qquad \text{III.1.17}$$

Introduce new canonical variables

$$\Phi_1 = k_1\phi_1 + k_2\phi_2; \quad \Phi_2 = K_1\phi_1 + K_2\phi_2;$$
$$r\,I_1 = K_2(J_1-J_1^o) - K_1(J_2-J_2^o); \quad -r\,I_2 = k_2(J_1-J_1^o) - k_1(J_2-J_2^o);$$

with $\qquad r = k_1 K_2 - k_2 K_1,$ \hfill III.1.18

where K_1, K_2 are arbitrary constants chosen such as to avoid vanishing r values. On truncation of the Hamiltonian after the resonant terms, we obtain, in the new generalized coordinates,

$$H(\Phi_1,\Phi_2,I_1,I_2) = H_o(I_1,I_2) + \varepsilon H_{k_1k_2}(I_1,I_2) \cos \Phi_1 + \ldots$$
$$\text{III.1.19}$$

Since Φ_2 has dropped from the (truncated) Hamiltonian, I_2 is an integral of motion; therefore this Hamiltonian is integrable. To investigate the motion near the value of the new action $I_1 \simeq 0$, with $I_2 \equiv 0$, expand H_o in a power series of I_1

$$H_o(I_1,0) = H_o(0,0) + (\partial/\partial J_1 H_o \partial/\partial I_1 J_1 + \partial/\partial J_2 H_o \partial/\partial I_1 J_2)\, I_1$$
$$+ 1/2\, (\partial^2/\partial I_1^2 H_o)\, I_1^2 + \ldots, \qquad \text{III.1.20}$$

where the terms between brackets are evaluated at the resonance. From relation (III.1.16) we see that the linear contribution in I_1 vanishes. The truncated Hamiltonian then becomes, with obvious notations

$$H = 1/2\, H'' \, I_1^2 + \varepsilon h \cos \Phi_1 + \ldots, \qquad \text{III.1.19'}$$

where the coefficients are constants. The motion thus reduces to the motion of a simple pendulum, whose phase space behavior is illustrated in Fig. 17. For $\varepsilon = 0$, the orbits are all straight lines in the Φ_1-I_1 plane, with the exception of the resonant torus $I_1 = 0$; the latter is represented here by a continuous infinity of equilibrium points, $I_1 = 0$, $\Phi_1 = \Phi_1^o \in [0,2\pi]$; alternatively, we may say that at the resonance (H=0) we have $1/2\, H'' \, I_1^2 = 0$, or the line $I_1 = 0$ is in fact a double line; but such a configuration is structurally unstable: under any slight generic disturbance ($\varepsilon \neq 0$) the double line is due to unfold into a stable geometric shape. For the specific perturbation of the Hamiltonian (III.1.19') the unfolding of the double line is given by

$$I_1 = \pm\, [2(H - \varepsilon h \cos \Phi_1)/H'']^{1/2} \qquad \text{III.1.21}$$

(cf Fig. 17b). The maximum elongation of the action I_1 at the resonance (H=0) is then

$$I_1 = (2 \, |\varepsilon \, h/H''|\,)^{1/2} \,, \qquad\qquad\qquad \text{III.1.21'}$$

which measures the width of the resonance zone in action space. Notice that within the resonance zone the character of the oscillation is qualitatively different from the oscillation of the non-disturbed oscillator; the angular variable Φ_1 now exhibits an oscillatory motion of low frequency $\Omega = (\,|H'' \, h \, \varepsilon|)^{1/2}$. On the other hand, outside the resonance area the motion is just quantitatively altered. Of the infinity of equilibrium points $I_1 = 0$, $\Phi_1 = \Phi_1^o$, only two survive in the perturbed problem, namely an elliptic (stable) point surrounded by oval orbits, and a hyperbolic (unstable) point lying on the separatrix (cf. Fig. 17a).

So far we have kept just a single resonance in the Fourier expansion (III.1.17). It is manifest, however, that a resonance (III.1.16) for two given integers k_1, k_2 implies also approximate resonances for an infinity of other pairs of integers k_1', k_2', such that $k_1'/k_2' \simeq k_1/k_2$. Each such resonance defines a double torus of the non-perturbed Hamiltonian system, which, under the action of the disturbance, is broken up as described above. As long as the disrupted tori stay separated in phase space, the single-resonance description remains meaningful. With increasing strength ε of the perturbation two neighboring disrupted tori eventually overlap. To analyze that situation, both interacting resonances must be kept in the truncated Hamiltonian (III.1.17) which becomes

$$H(\phi_1,\phi_2,J_1,J_2) = H_o(J_1,J_2) + \varepsilon \, H_{k_1 k_2}(J_1,J_2) \cos(k_1\phi_1 + k_2\phi_2)$$
$$+ \varepsilon \, H_{k_1' k_2'}(J_1,J_2) \cos(k_1'\phi_1 + k_2'\phi_2)$$
$$+ \text{non-resonant contributions.} \qquad \text{III.1.17'}$$

The latter Hamiltonian is no more integrable. The motion is no longer confined to 2-dimensional tori; nothing prevents it from exploring a 3-dimensional region on the energy manifold; such a motion, which is irregular, is said to be carried by a cantorus, a structure of the nature of a Cantor set.
(5) The stellar oscillation Hamiltonian (III.1.10,3') presents itself as a quasi-integrable Hamiltonian. Regarding the non-linear potential energy contributions $V^{(3)} + V^{(4)} + \ldots$ as corrections to an unperturbed Hamiltonian H_o, namely the Hamiltonian of a harmonic oscillator of F degrees of freedom, we notice that the latter is trivially integrable. However, as discussed at greater length elsewhere (Perdang 1979, 1983), the KAM theorem is not directly applicable to this separation of

the full stellar Hamiltonian[8]. The difficulty is easily
avoided by applying Birkhoff's (1927) procedure consisting in
transforming the stellar Hamiltonian in a formal power series
in the actions

$$H = \sum_{i=1}^{F} \omega_i J_i + 1/2 \sum_{i,j=1}^{F} \omega_{ij} J_i J_j + \text{higher orders,} \qquad \text{III.1.22}$$

by means of successive canonical transformations. Although
Siegel (1954) has proved that this expansion generically
diverges, we can carry out the first step of Birkhoff's
technique, namely up to second order in the actions, keeping
angle variables in the higher orders. This yields

$$H = \underbrace{\sum_{i=1}^{F} \omega_i J_i + 1/2 \sum_{i,j=1}^{F} \omega_{ij} J_i J_j}_{H_o(J_1, J_2)} + H_1(\phi_1, \phi_2, J_1, J_2) \qquad \text{III.1.22'}$$

One finds that this transformation is allowed, provided that
the stellar harmonic oscillators contain no resonances of order
less or equal to 4 (cf Perdang 1983). Intuitively, Siegel's
result means that if we construct Hamiltonians in the form of
power series (III.1.10,3') in the generalized coordinates, and
if we make use of a random number generator to produce the
expansion coefficients $V_{ijk}^{(3)},\ldots$, (defined in some range), then
the likelihood of hitting an integrable Hamiltonian is zero.
For the stellar oscillation Hamiltonian this implies in turn
that for any realistic (i.e., mathematically generic) model,
the oscillation problem is non-integrable.

The combination of the conclusion that the 'typical'
stellar oscillation Hamiltonian (III.1.10,3') is non-
integrable, and the application of the KAM theorem to the
quasi-integrable form (III.1.22') enables us to make the
following statements about the adiabatic pulsations of stars:

(a) In the absence of low order resonances ($k \leq 4$) among
the linear effectively excited modes, we are sure that for
sufficiently small oscillation amplitudes (or small enough
oscillation energies) the non-linearly interacting modes
essentially behave as in the linear approximation; in
other words, practically all motions remain quasi-
periodic.

(b) As we increase the oscillation energy, the 'volume'
occupied by exploded tori in the energy manifold ceases to
be negligibly small, so that we find quasi-periodic
solutions which are qualitatively different from the
harmonic oscillator solutions.

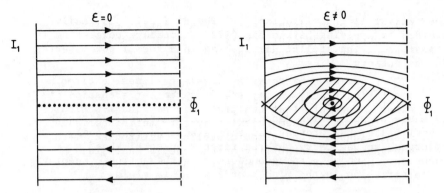

Fig. 17.- Phase space behaviour of the Hamiltonian system (III.I.19)
(a) $\varepsilon = 0$: the line $I_1 = 0$ is a degenerate equilibrium,
(b) $\varepsilon \neq 0$: the degenerate line $I_1 = 0$ explodes to form a cavity.

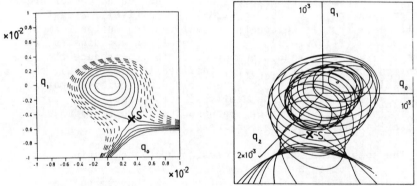

Fig. 18.- Shape of the oscillation energy equipotentials for a polytrope $n = 3$; (a) two-mode and (b) three-mode interaction.

Fig. 19.- Schematic shapes of the Henon-Heiles (left) and Contopoulos-Barbanis (right) potentials.

(c) At an even higher level of oscillation energy, adjacent exploded tori overlap, so that chaotic fluctuations are generated.

(d) If low order resonances among the linear excited modes are taking place, we may expect a similar scenario, except, however, that qualitatively new quasi-periodic solutions may now emerge already for 'infinitesimal' distortions of the harmonic stellar Hamiltonian, with a non-vanishing probability on the energy manifold; likewise, we may expect chaotic solutions to arise at lower energy levels.

b. Exploratory Numerical Experiments

While the quasi-periodic solutions of the stellar oscillation equations (III.1.8) can be investigated by semi-analytical techniques - we can expand these solutions in an F-tuple Fourier series and determine the frequencies and expansion coefficients, either numerically or by means of asymptotic techniques - no such semi-analytical procedures seem to exist to describe the chaotic solutions (cf. however Perdang 1984). The reason is that we lack a general analytical representation formula for chaotic motions, i.e., a counterpart of a Fourier representation (III.1.14) for quasi-periodic motions. Therefore, we are led to study the chaotic stellar fluctuations by direct numerical integration of the equations of motion. Two questions are of particular observational interest: (1) At what oscillation energy level E_T, or at what surface amplitude, do chaotic fluctuations manifest themselves? (2) What is the time behavior of the surface displacement, or of the surface velocity, when the star is oscillating chaotically?

(1) The <u>threshold energy</u> E_T above which a sizeable fraction of orbits of the Hamiltonian system (III.1.8) are chaotic manifestly depends on the geometry of the potential energy surface (III.1.3'). The latter has been studied for a variety of polytropic models, in the lowest order non-linear approximation, (all $V^{(n)}$ are set equal to zero for $n > 3$, Blacher 1981). The typical shapes of the equipotentials we encounter are illustrated in Fig. 18 for two-mode (a), and three-mode (b) interactions respectively. We notice that besides the stable equilibrium state at the origin ($q_i = 0$), we invariably find a single critical point of the nature of a saddle point, S, relatively close to the origin, say at coordinates $q_i = q^*_i$, while all other critical points lie at a much greater distance from the origin (see Perdang and Blacher 1982, 1984 for a variety of examples). Numerical experiments have further shown that the same geometry of the equipotentials

continues to hold true for F-mode interactions, $F \leq 10$, so that we have good reasons to believe that this result remains valid for the coupling of arbitrary numbers of linear modes[9]. For an oscillation energy less than the potential energy at the saddle point S, $E_S \equiv V_{osc}(q_i^*)$, we can convince ourselves that the motion must remain bounded, both in the space of generalized coordinates q_i, and in the space of generalized momenta p_i. Therefore, <u>all</u> trajectories are <u>recurrent</u> in phase space: if an orbit goes through a point P in phase space at time t_o, and if $B_\varepsilon(P)$ is a sphere of radius ε centered at P, then after some finite time T_ε, depending on the size of the sphere, the orbit cuts through the sphere again. This property is in essence Poincaré's (1890) celebrated recurrence theorem. Roughly speaking, the recurrence time T_ε may be pictured as an approximate local period, which for an accuracy factor ε chosen not too small is reminiscent of the concept of a local period as defined for variable stars.

For an oscillation energy higher than the saddle point energy E_S, the orbit may escape to infinity in the lowest order non-linear approximation $V^{(3)} \neq 0$, $V^{(n)} = 0$, $n > 3$. Of course this means that the latter approximation ceases to be acceptable. If higher order non-linear contributions are included in the oscillation potential, it is conceivable that the equipotentials close again at some greater distance from the origin. At the moment detailed numerical information on this question is lacking. We should mention that Whitney (1984), analyzing the non-linear adiabatic pulsations of a 2-zone stellar model without truncating the potential energy, does not encounter any saddle point near the minimum of the potential.

Non-integrable coupled harmonic oscillators have been studied by a number of authors during the past 20 years. The most intensively investigated instances are the Hénon-Heiles and the Contopoulos-Barbanis oscillators of two degrees of freedom. In the Hénon-Heiles (1964) oscillator the cubic potential energy contribution is of the form

$$V_{HH}^{(3)} = a\ r^3\ \sin 3(\theta - \theta_o)\ , \qquad \text{III.1.23}$$

where a and θ_o are constants, and r and θ are polar coordinates. The resulting potential energy, for identical linear frequencies, $\omega_1 = \omega_2$, is schematically shown in Fig. 19a. Besides the minimum at the centre of the coordinate system, this potential exhibits <u>three</u> saddle points S_1, S_2, S_3 lying on a degenerate equipotential (three straight lines forming an equilateral triangle whose centre coincides with the potential minimum). In the Contopoulos (1960, 1963) and Barbanis (1966) oscillator the cubic potential energy part is of the form

$$V_{CB}^{(3)} = a\, q_1 q_2^2 \quad . \qquad \text{III.1.24}$$

The equipotentials, again for equal linear frequencies, are indicated in Fig. 19b. This potential displays two saddle points S_1, S_2, located on a degenerate equipotential (a straight line intersecting a parabola).

Numerical studies have disclosed that for all non-integrable oscillators chaotic motions are observed beyond a threshold energy E_T that is a sizeable fraction of the escape energy E_S. For instance in the Hénon-Heiles case with $\omega_1 = \omega_2 = 1$, $r = 1/3$, we have $E_S = 1/6$, while E_T lies in the range 1/12 to 1/8. We expect therefore that a similar situation continues to hold in the stellar oscillation problem. This, in turn, then implies that the surface amplitude at which chaotic stellar fluctuations set in satisfies in order of magnitude (cf. Eq. III.1.6)

$$\delta R/R_E \lesssim \left| \sum_{i=1}^{F} \xi_i\, q^*_i \right| \quad , \qquad \text{III.1.25}$$

where ξ_i is the surface value of the ith eigenfunction and R_E the equilibrium radius. Numerical estimates of the right-hand side of Eq. (III 1.25) have been made in the framework of the standard polytrope, for two-mode interactions (Perdang and Blacher 1982) and for multi-mode interactions (Perdang and Blacher 1984). For the two lowest modes, the relative surface amplitude (III.1.25) is about 50%; this value drops regularly as we couple higher order modes; for instance, for the 9th and 10th modes (8th and 9th harmonics) this amplitude lies below 9%. On the other hand if more than two modes are interacting, the critical amplitude is lowered further; for the three lowest modes it lies around 30%; if the eight lowest modes are interacting, $\delta R/R_E$ is reduced to 7%. We should keep in mind that all of these estimates are upper limits.

(2) A qualitative idea of the time behavior of the surface displacement

$$\delta R(t)/R_E = \sum_{i=1}^{F} \xi_i\, q_i(t) \qquad \text{III.1.26}$$

in the presence of chaos is already gained if we substitute a comparison oscillator such as the Hénon-Heiles oscillator, to the actual stellar problem, and set the ξ_i values in Eq. (III.1.26) equal to 1. An experiment of this nature is described in Blacher and Perdang (1981). The interesting result is that the chaotic function (III.1.26) is not entirely disordered: The amplitudes remain approximately constant over a few cycles of fluctuation, and there exists a local period,

which in turn is a fluctuating function of time; in the
experiment conducted in Blacher and Perdang (1981) a standard
deviation of nearly 10% in the cycle length fluctuations is
observed. Moreover, a secular trend in the period variation
calculated in the conventional fashion (cf. II.1.2a), shows
that the mean rate of the relative period change is critically
dependent on the length of the available time-stretch of the
'signal'; the trend detected in the period change has nothing
to do with an evolutionary effect, the numerical parameters of
the oscillator being constant. These exploratory experiments
suffice already to exhibit the potential interest of the
concept of chaotic stellar oscillations. In fact, such
oscillations bear an obvious analogy with the observed
pulsations of the variables of group II, and possibly, they may
also be relevant in connection with some more regular
variables.

I shall now summarize the results of numerical
integrations of actual stellar models.

c. Chaotic Stellar Model Oscillations

In Perdang and Blacher (1982, 1984) the Hamiltonian
equations (III.1.8) have been integrated for semi-realistic
expansion coefficients $V_{ijk}^{(3)}$ corresponding to a polytropic model
of index 3. To distinguish quasi-periodic solutions from
chaotic orbits, two tests have been applied, whenever
feasible. On the one hand a power spectrum of the surface
motion of the star (III.1.26) has been computed (cf. Blacher
and Perdang 1981a); if this spectrum exhibits just a few lines,
then we are sure that the motion is quasi-periodic (cf. Eq.
III.1.14); if it exhibits a complex structure which, at higher
spectral resolution becomes increasingly more involved, then
the quasi-periodicity requirements (III.1.14) are violated;
therefore the oscillation must be chaotic. On the other hand,
for two and three-mode interactions the Poincaré surface of
section method is applicable. For $F = 2$ this method consists
in cutting the orbits on the 3-dimensional energy manifold by
the plane $q_2 = 0$ and by plotting the crossing points of the
orbit with the plane q_1, p_1, for $p_2 < 0$. If the actual orbit
is carried by a torus, these intersection points distribute on
the closed curve of intersection of the torus with the q_1, p_1
plane. In other words, the occurrence of closed regular curves
in the so-called surface of section q_1, p_1 is the fingerprint
of a quasi-periodic oscillation. If instead of defining a
closed curve, these intersection points are distributed over
some area of the surface of section q_1, p_1, then the orbit
cannot lie on a torus; therefore this orbit cannot correspond
to a quasi-periodic motion; it must be a 'volume'-exploring
irregular trajectory. The method extends to 3-mode

interactions, in which case one considers the intersection points of the orbit for $q_3 = 0$, $p_3 > 0$ projected onto the 3-dimensional space q_1, q_2, p_2; a quasi-periodic orbit manifests itself through intersection points lying on a surface, while a chaotic orbit occupies a certain volume of the q_1, q_2, p_2 space (cf. Martinet and Magnenat 1981, Contopoulos et al. 1982).

For illustrative purposes, we will consider the 2-mode interaction of the 8th and 9th harmonics. Fig. 20 displays the corresponding surface of section at the escape energy E_S. We see that roughly 1/3 of the area accessible to the motion is populated by randomly scattered points; the latter belong to chaotic orbits. The remainder of the area is filled up by a family of nested closed curves; notice that a few closed curves appear as islands interspersed among the chaotic sea. Therefore, about 66% of the available phase space remains populated by quasi-periodic trajectories, while irregular trajectories occupy only 34%. In other words, if we randomly select an initial condition (at the energy E_S), the probability of hitting a regular solution is nearly twice the probability of catching a chaotic orbit. If we compare with the Hénon-Heiles oscillator (cf. Hénon 1983), at the escape energy essentially all orbits are chaotic. It seems that the presence of a greater number of saddle points near the potential minimum favors a greater amount of chaos.

Fig. 21a illustrates the surface displacement of a chaotic oscillation at the escape energy. The striking property of this time-behavior is its fair degree of regularity. The amplitude of the oscillation remains roughly constant over a number of cycles; the cycle length, however, is fluctuating. As shown in Fig. 22a, the local periods plotted versus the epoch t are scattered around an average value. This diagram is reminiscent of the period fluctuations exhibited by ξ Gem (Fig. 5); it also duplicates the qualitative behavior of the observed cycle length fluctuation of the Mira variables (cf. Campbell and Jacchia 1946). Fig. 22b shows a histogram of the local frequencies (2π/local cycle lengths). Notice the correspondence between this theoretical diagram and the observational local period histograms of typical Mira variables (Fig. 10) or of the δ Scuti stars 14 Au and 44 Tau (Fig. 13). At the moment a closer statistical study of the theoretical local period fluctuations in chaotic stellar oscillations is lacking. Fig. 22a suggests that successive cycle lengths are correlated; I conjecture therefore that the formalism of Brownian functions as devised by Mandelbrot (1977) may lend itself to a probabilistic characterization of the local period - epoch relation. The power spectrum of the surface displacement shown in Fig. 21b is seen to exhibit a large band of frequencies; essentially this figure can be viewed as

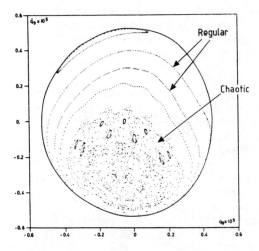

Fig. 2o.- Poincare surface of section for a double-mode interaction (8th and 9th harmonics); regular orbits occupy roughly 2/3 of the area accessible to the motion (Perdang and Blacher 1982).

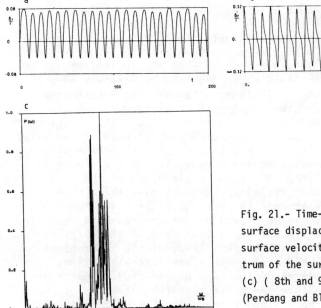

Fig. 21.- Time-behaviour of the surface displacement (a), and the surface velocity (b); power spectrum of the surface displacement (c) (8th and 9th harmonics) (Perdang and Blacher 1982).

representing a high resolution histogram of local frequencies. Note that several prominent peaks stand out of the broad band; these peaks break up into doublets or multiplets as we increase the resolution of the power spectrum.

An observer not aware of the theoretical possibility of chaotic pulsations of stars, when confronted with the displacement or velocity curve of Fig. 21 (given at discrete epochs only), would have little difficulty in assigning such a variability a well-defined period - especially so if the number of data points per cycle are sparse. The slight degree of irregularity could easily be dismissed as observational noise. As a longer time series becomes available, the observer would be tempted to look for additional periods; he would detect period doublets and multiplets.

The finite set of periods isolated in this fashion cannot fully characterize the time behavior of the pulsation - otherwise the pulsation would be quasi-periodic. These periods merely have a formal interpolational character, permitting a quasi-periodic representation of the true motion over a given time-interval of finite length T in the same way as we can make use of standard Fourier series to interpolate the chaotic behavior over the given finite time span. The isolated periods have not much more intrinsic meaning than the Fourier periods T/n, $n=1,2,3,\ldots$, in the Fourier interpolation; they are just 'ghosts of departed quantities' (Berkeley).

While in two-mode coupling experiments the surface maximal relative semi-amplitude $\delta R/R_E$ of chaotic oscillations remains relatively high (cf. above), this value decreases rapidly as we couple more modes. We have found that our numerical results fit the following approximate relationship

$$\delta R/R_E \lesssim F^{-3/2} \quad , \qquad\qquad \text{III.1.25'}$$

if the F lowest radial modes are coupled (Perdang and Blacher 1984). This trend is indicative that variable stars which show multi-periodic behavior may be undergoing chaotic pulsations even if they have small oscillation amplitudes. As more and more modes are coupled, an interesting new phenomenon is observed in the chaotic power spectra. For randomly chosen initial conditions, the power spectrum appears to assume a statistically regular aspect, reminiscent of the observed solar 5-minute spectrum (Fig. 14). It shows a series of approximately equidistant lines with heights conforming to a roughly Gaussian distribution (Fig. 23); although well-defined, these lines have been found to split up into doublets and multiplets with increasing resolution. Besides this bell-shaped structure centered at a high frequency, a second

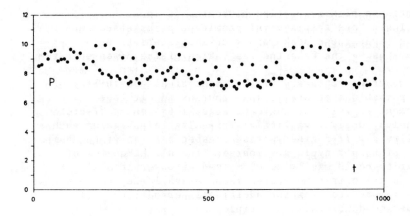

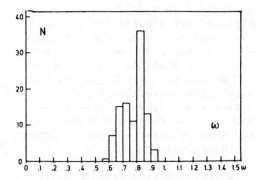

Fig. 22. Local period P as a function of epoch (upper diagram) and histogram of local periods (lower diagram) for the displacement curve of Fig. 21.

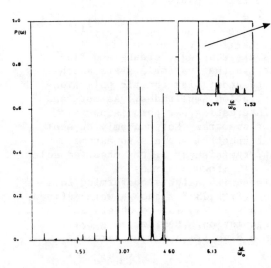

Fig. 23. Power spectrum of a chaotic motion corresponding to F = 1o coupled modes (Perdang and Blacher 1984).

sequence of quasi-equidistant peaks of much lower heights is found close to the origin. The lowest peak is located near .4 times the fundamental frequency in the case of the polytropic model of index 3. A rough conversion into solar values, taking ~ 60 min for the fundamental period, yields ~ 150 min for the corresponding periodicity. This value is remarkably close to the 2 h 40 min solar oscillation, so that one is seriously led to wonder whether in spite of the exceedingly low solar amplitudes non-linear coupling effects among the modes are not playing an essential part in featuring the oscillations of the sun (cf. Perdang and Blacher 1984a).

In this section we have reviewed what we feel are the most relevant properties of chaotic radial conservative stellar oscillations that have been established so far. In a very rough approximation, these oscillations keep a periodic or multi-periodic behavior; the amplitudes and periods show however a noisy character which manifests itself most obviously in the power spectra; though dominated by one or several peaks, the latter exhibit complex structured bands.

A major unresolved problem in this context is to separate this intrinsic noise, which encodes information on the physics of the star, from observational errors. A second unresolved problem is to extract the physical information from the endogenous noise pattern. A statistical attitude towards this question, with the target of isolating joint probability densities such as $\mathcal{P}_{1,2,\ldots,c}(p_1, p_2, \ldots, p_c)$ (cf. section II.1) appears as a promising avenue.

I should emphasize finally that the systematic numerical investigations carried out so far rely on the lowest non-linear order coupling approximation ($V^{(n)} = 0$ for $n > 3$, cf Eq. III.1.3'). This implies that we encounter only a mild species of chaotic fluctuations. In higher order non-linear approximations the potential oscillation energy can have <u>two</u> minima separated by a saddle point. Under these conditions, at an energy level higher than the saddle point energy, the orbit may randomly flip from one potential well to the other one. The surface displacement of the star then alternates between two chaotic patterns. It is not inconceivable that RV Tauri stars and some Semiregulars conform to such a scheme.

2. Adiabatic Non-radial Non-linear Oscillations

The natural extension of the previous formalism consists in generalizing it to cover non-radial oscillations. This step is motivated not only by the direct observational evidence of non-radial motions in the sun, but also by the circumstantial evidence of non-radial effects in most variables of groups III

and IV. The generalization of the variational principle
(III.1.4,3,2,1) to the non-radial pulsation problem has to cope
with a difficulty related to the following circumstance.
Suppose we disturb a star that originally is in a state of
spherically symmetric hydrostatic equilibrium in an absolute
reference frame, by superposing a velocity field on its
structure. In general, such a field can be split up into two
components, one producing a stationary velocity pattern - a
uniform rotation is the simplest instance of such a component -
and a second component generating genuine time-dependent
pulsations - a purely radial expansion or contraction field is
an elementary example of that class of motion. In the
approximation of an infinitesimal disturbance, the first type
of component is represented by a purely toroidal field

$$T(\mathbf{r}) = \mathbf{r} \times \text{grad } \alpha(\mathbf{r}) \quad , \quad \text{III.2.1}$$

(any infinitesimal toroidal velocity field satisfies the
linearized non-radial adiabatic hydrodynamic equations with a
zero frequency); the second type is a poloidal field

$$P(\mathbf{r}) = \text{grad } \beta(\mathbf{r}) + \text{curl curl } (\gamma(\mathbf{r}) \mathbf{r}) \quad , \quad \text{III.2.2}$$

(the frequency eigenvalues associated with poloidal
disturbances are all real oscillation frequencies provided that
we assume that the star is dynamically stable). The non-radial
counterpart of the radial Woltjer type expansion (III.1.6) then
becomes

$$\delta\mathbf{r}(\mathbf{r},t) = \sum_\nu q_\nu(t) \, \delta\mathbf{r}_\nu(\mathbf{r}) \quad , \quad \text{III.2.3}$$

where $\delta\mathbf{r}(\mathbf{r},t)$ is the finite displacement field and $\delta\mathbf{r}_\nu(\mathbf{r})$
represents the full set of conveniently normalized linear
displacement eigenfunctions; $\nu = (k,l,m)$ is a multi-index,
involving besides the radial order k also the spherical
degree ℓ and the azimuthal quantum number m (= 0, ±1, ±2,...,
±ℓ). The eigenfunctions involve both poloidal and toroidal
modes of the spherical equilibrium state. Unfortunately the
latter class of eigenfunctions are mathematically ill-defined
due to the infinite degeneracy of the neutral toroidal
eigenvalues; therefore this approach encounters serious
conceptual difficulties. If we ignore the latter difficulties,
it suffices to introduce expansion (III.2.3) in a variational
principle similar to the radial principle (III.1.4), which in
turn then generates Hamiltonian particle equations.

The procedure just sketched seems to be too general to be
attacked efficiently at the moment. It covers in fact a double
problem: It produces, in principle, the stationary stellar
states involving steady conservative flow patterns, and it

generates the bona fide <u>oscillations</u> around those stationary stellar states.

a. Stationary Stellar States

The goal of the following comments is to stress that the search for stellar configurations with stationary velocity fields is a formidable, mainly unexplored problem. Due to the infinite degeneracy of the toroidal modes, one suspects the existence of an uncountable class of configurations with flow fields of distinct topologies. Systematic investigations have been carried out only for liquid 'stars' - and even in that restricted context only partial answers are available.

The classical work on uniformly rotating incompressible stars of constant density by Maclaurin, Jacobi and Poincaré has established that for an angular momentum parameter $J^2 < J_J^2 = .384$ (in units $a_1 a_2 a_3$ GM, M total mass, a_i, $i = 1,2,3$ principal axes of the ellipsoid) the uniformly rotating body remains spheroidal (Maclaurin spheroid, of dihedral symmetry $D_{\infty h}$) [10]; at $J^2 = J_J^2$ a bifurcation towards a triaxial ellipsoid is observed (Jacobi ellipsoid, of symmetry D_{2h}), which survives up to an angular momentum parameter $J^2 = J_P^2 = .632$; at the latter critical value, a pear-shaped figure branches off the Jacobi ellipsoid (Poincaré figure, of symmetry C_{2v}) (cf. Chandrasekhar 1969). Are we witnessing here the first steps of an infinite cascade of bifurcations, eventually leading to some kind of a stationary 'chaotic' figure for some finite J_∞^2 value? Such a sequence of instabilities would then appear as the gravitational counterpart of the cascade of 'labyrinthine' instabilities observed in magnetic liquids (Rosenzweig 1982).

Moreover, using group-theoretic techniques, Constantinescu et al. (1979) have established that on the Maclaurin branch of spheroidal figures there exists an infinity of bifurcation points at J^2 values $J_3^2 < J_4^2 < \ldots < J_n^2 < \ldots$ towards configurations of dihedral symmetry D_{3h}, D_{4h}, ..., D_{nh}, ... When viewed in an inertial reference frame, these figures show a stationary velocity field, namely a uniform rotation, together with a surface wave travelling with the surface rotation speed. It is likely that many, if not all of these configurations are unstable.

Besides these simplest stationary states, Dirichlet, Dedekind and Riemann proved the existence of a more general

velocity field linear in the space coordinates and compatible
with an ellipsoidal equilibrium figure. This velocity field is
a superposition of a uniform rotation and a vorticity which is
uniform in the rotating reference frame. The rotation and
vorticity axes both lie in a symmetry plane of the ellipsoid.
The transition from the Maclaurin towards the Riemann ellipsoid
takes place at precisely the same critical value J_J^2 at which
the Jacobi ellipsoid branches off (Chandrasekhar 1969). By
analogy, one has to expect that the bifurcation points J_n^2, n =
3,4,... towards the Constantinescu-Michel-Radicati figures are
also bifurcation points towards stationary states with velocity
fields more involved than the Riemann velocity field. This
conjecture relies on the property that the infinity of neutral
toroidal modes ($\omega_{tor,\nu}^2 = 0$ for all ν) of a spherically
symmetric star transform, in the presence of a rotation (with
angular momentum parameter J^2) into a set of non-zero
frequencies, $\omega_\nu^2 (J^2) \neq 0$, for J^2 small. With increasing
angular momentum, each such frequency seemingly goes through a
maximum, decreases and goes through zero at some critical
angular momentum value (cf. Chandrasekhar 1969). At those
critical points the displacement and the velocity
eigenfunctions, which remain characterized by a well-defined
azimuthal quantum number m, produce an infinitesimal neutral
deformation and an infinitesimal stationary flow respectively,
of at least rotational symmetry C_m. In other words, we
encounter here a branching point towards figures of (at least)
C_m symmetry which are equilibrium figures in the rotating
reference frame for zero velocity eigenfunctions, and
configurations with a stationary flow pattern if the velocity
eigenfunctions are non-zero. The first alternative corresponds
to the Constantinescu-Michel-Radicati figures.

These results, though sparse, exhibit already a rich
variety of possible stationary (adiabatic) flow fields in one
simple case, namely near a uniform rotation. If the conjecture
of an uncountable infinity of stationary flow fields turns out
to hold true, then it becomes exceedingly likely that spatially
chaotic stationary flows must exist, which presumably
materialize through an infinite sequence of spatial pitchfork
bifurcations. At the moment these questions await a serious
investigation [11].

b. Genuine Oscillations

In the absence of a satisfactory knowledge of the allowed
stationary velocity patterns, we feel that the most urgently
needed information on non-linear non-radial oscillations
regards pulsations around the radial equilibrium state of a
star with initial conditions chosen such as to avoid the
excitation of any stationary velocity field component. Once

this class of oscillations are thoroughly understood, we may proceed to more complex oscillations, superimposed on other given stationary flows. To achieve this goal, it is not justified to make use of expansion (III.2.3) with the toroidal modes just thrown away. In fact, if we select an initially purely poloidal, finite velocity field, then a toroidal component will develop as well, as a result of the non-linear mode interactions. Therefore, even if we discard stationary velocity field components, a correct approach to the non-radial non-linear pulsations still requires the full expansion (III.2.3).

Instead of the direct expansion of the velocity disturbance, or the displacement field, in terms of their linear eigenfunctions, we found the following approach more economical. If the velocity field is written in Clebsch type form

$$\mathbf{v}(\mathbf{r},t) = \mathrm{grad}\ a(\mathbf{r},t) + b(\mathbf{r},t)\ \mathrm{grad}\ c(\mathbf{r},t)\ ,\qquad \mathrm{III.2.4}$$

with $a(\mathbf{r},t)$, $b(\mathbf{r},t)$ and $c(\mathbf{r},t)$ arbitrary scalar fields, we can convince ourselves that provided $c(\mathbf{r},t)$ is a physical quantity conserved in the motion, the Kelvin-Helmholtz circulation $C(\mathscr{C})$ over any loop lying on a constant c-surface vanishes

$$C(\mathscr{C}) = \int_{\mathscr{C}} \mathbf{v} \cdot d\mathbf{r} \equiv 0,\ \mathscr{C}\ \text{on the surface } c(\mathbf{r},t) = c^{st}.\quad \mathrm{III.2.5}$$

The remaining scalar fields $a(\mathbf{r},t)$ and $b(\mathbf{r},t)$ are free, except that $a(\mathbf{r},t)$ is required to be single-valued over the star. A velocity field obeying (III.2.5), for circulations around loops in the constant c-surfaces, automatically satisfies the conservation of the Kelvin-Helmholtz circulation in adiabatic motion. It is then natural to identify $c(\mathbf{r},t)$ with the specific entropy $S(\mathbf{r},t)$ (cf Däppen and Perdang 1984, 1984a; in the latter paper, we list the main mathematical constraints on the velocity field resulting from the requirement of zero Kelvin-Helmholtz circulation III.2.5). Representation (III.2.4) has the following appealing properties. (a) For an infinitesimal velocity field ($a(\mathbf{r},t)$ and $b(\mathbf{r},t)$ infinitesimal, so that $S(\mathbf{r},t)$ reduces to the radially symmetric equilibrium entropy of the underlying stellar model), the velocity field is seen to be purely poloidal (cf. Eq. III.2.2); therefore it involves no stationary motion. (b) Any finite velocity field of form (III.2.4) ($c(\mathbf{r},t) \equiv S(\mathbf{r},t)$) always suffers a restoring force, as we can observe on substitution into the equations of motion; accordingly, no finite stationary velocity pattern can be associated with such finite disturbances either. Consequently, for our purposes, representation of the non-radial disturbances of the star based on relation (III.2.4) (with c = S), is superior to a general representation of form

(III.2.3) in two respects. (a) In principle, the general representation (III.2.3) implicitly involves 3 arbitrary scalar fields, namely the functions α (**r**,t), β(**r**,t), γ(**r**,t) entering relations (III.2.1,2), while representation (III.2.4) contains only 2 free scalar fields. (b) Representation (III.2.4) filters out genuine oscillations around a stellar equilibrium devoid of stationary motions.

Instead of the free scalar fields a(**r**,t), b(**r**,t) directly entering Eq. (III.2.4) we can select any pair of functionally independent physical fields as the 'dependent variables' of the non-radial oscillation problem, provided that a(**r**,t) and b(**r**,t) are in a one-to-one correspondence with these physical fields. A computationally convenient choice of the 'dependent variables' is supplied by the radial displacement field

$$\delta r \equiv \delta r(S,\theta,\phi;t) = r(S,\theta,\phi;t) - r_E(S) \qquad \text{III.2.6}$$

(projection of the displacement field onto the radial unit vector), and the perturbation of the stratification field, μ (mass per unit entropy and unit solid angle)[12]

$$\delta\mu \equiv \delta\mu(S,\theta,\phi;t) = \mu(S,\theta,\phi;t) - \mu_E(S) \qquad . \qquad \text{III.2.8}$$

As 'independent variables' specific entropy S and the spherical angles θ, φ of a point **r** in the star, together with time t, appear as convenient space-time coordinates. Such a selection of the dependent fields and the independent variables enables one to build a maximum of constraints into the variational formulation. The choice of S as an independent variable takes care of conservation of entropy; mass conservation is trivially included in the dependent field variable δμ; the velocity representation (III.2.4), with a and b expressed as functions of δr and δμ, secures a vanishing Kelvin-Helmholtz circulation over loops on the constant entropy surfaces for all times.

The next step is to expand the dependent fields in a Woltjer type series

$$\begin{bmatrix} \delta r(S,\theta,\phi;t) \\ \delta\mu(S,\theta,\phi;t) \end{bmatrix} = \sum_\nu q_\nu(t) \begin{bmatrix} \delta r_\nu(S)\, Y_\nu(\theta,\phi) \\ \delta\mu_\nu(S)\, Y_\nu(\theta,\phi) \end{bmatrix} \qquad \text{III.2.9}$$

where the components in the right-hand side sum are the standard eigenfunctions of the linear non-radial adiabatic stellar pulsation problem (involving the non-trivial poloidal modes only); ν is the multi-index defined above and $Y_\nu(\theta,\phi) = Y_\ell^m(\theta,\phi)$ is the spherical harmonic function of degree ℓ and rank m conveniently redefined to be real (Däppen and Perdang 1984a). It is worth stressing here that although no toroidal linear modes are included in expansion (III.2.9), substitution of this expansion into the Clebsch form (III.2.4) realistically generates a toroidal component (III.2.1) in the latter field.

If we substitute the Woltjer type expansion (III.2.9) into the variational principle (III.1.4,1), with the kinetic and potential energies now expressed in the appropriate dependent fields δr and $\delta \mu$, and with the choice of the independent variables S, θ, ϕ and t, we obtain a particle Hamiltonian of the form

$$H = 1/2 \sum_\nu p_\nu^2 + 1/2 \sum_\nu \omega_\nu^2 q_\nu^2 + \sum_{\nu\lambda\eta} V^{(3)}_{(\nu\lambda\eta)} q_\nu q_\lambda q_\eta$$

$$+ \sum_{\nu\lambda\eta} K^{(3)}_{\nu(\lambda\eta)} q_\nu p_\lambda p_\eta + \ldots$$

$$= 1/2 \sum_\nu p_\nu^2 + 1/2 \sum_\nu \omega_\nu^2 q_\nu^2 + H^{(3)} + \ldots \quad ; \qquad \text{III.2.10}$$

$H^{(3)} = H^{(3)}(q_\nu, p_\nu)$ is the lowest order non-linear correction to a harmonic oscillator Hamiltonian. We observe that besides the contribution to the potential energy, $V^{(3)} = \sum_{\nu\lambda\eta} V^{(3)}_{(\nu\lambda\eta)} q_\nu q_\lambda q_\eta$, already encountered in the radial oscillation problem, $H^{(3)}$ also involves a correction to the kinetic energy, $K^{(3)} = \sum_{\nu\lambda\eta} K^{(3)}_{\nu(\lambda\eta)} q_\nu p_\lambda p_\eta$ (with coupling matrix symmetric in the last two indices). The equations of motion are of the form

$$\dot{q}_\nu = p_\nu + \partial/\partial p_\nu (H^{(3)} + \ldots) = p_\nu + \sum_{\lambda\eta} c_{\nu\lambda\eta} q_\lambda p_\eta + \ldots$$

$$\dot{p}_\nu = -w_\nu^2 q_\nu - \partial/\partial q_\nu (H^{(3)} + \ldots) = -\omega_\nu^2 q_\nu$$

$$+ \sum_{\lambda\eta} v_{\nu(\lambda\eta)} q_\lambda q_\eta + \sum_{\lambda\eta} k_{\nu(\lambda\eta)} p_\lambda p_\eta + \ldots \qquad \text{III.2.11}$$

the expansion coefficients being symmetric in the multi-indices appearing in brackets.

The general theoretical information on Hamiltonian oscillators summarized in section III.1a remains applicable to Eqs. (III.2.11), so that we may be sure that for realistic stellar models non-radial chaotic oscillations will occur, provided that the pulsation amplitude is large enough.

A systematic numerical discussion of the solutions of the system of coupled mode equations (III.2.11) is still lacking. A preliminary integration in the context of a detailed solar model has been made by Däppen (1984), in which a few radial and

non-radial p-modes with linear periods in the 5-minute range
are coupled. Chaos is found to set in at relative radial
amplitudes of the order of 10^{-3}, i.e., earlier than in purely
radial oscillations. Moreover, the long period, recorded in
the radial experiments, is also recovered; the non-linear
effect in the coupling of the non-radial modes seemingly shifts
the radial peak in the direction of the observed 2h 40 min
oscillation. By the same token, this peak becomes conspicuous
in the power spectrum at a surface amplitude lower than in the
radial experiments (Däppen 1984).

Besides the quantitatively different effects arising in
non-radial chaotic oscillations as compared to the radial
pulsations, there is also an interesting new qualitative
phenomenon that requires investigation. The structure of the
surface pattern is now changing chaotically in time. An
analysis of the evolution of space correlations, with a
comparison with the evolution of correlations in quasi-periodic
non-radial oscillations, seems to be a promising method of
quantifying this question (cf. the similar problem of the
quantum-mechanical eigenfunction in the presence of quantum
chaos, Berry 1983).

Once the nature of the 'pure' non-radial chaotic
oscillations is clearly understood, the method discussed in
this section is to be extended to include velocity disturbances
superimposed on simple stationary velocity fields (rotation
etc.). The formalism devised by Lynden-Bell and Katz (1981)
lends itself to such a treatment, at least for underlying flow
fields with sufficiently simple topologies.

3. Non-adiabatic Oscillations

Historically, the relevance of resolutely non-linear non-
adiabatic effects in stellar variability seems to have been
fully appreciated by Wesselink (1939) - although earlier
authors (Eddington 1919, Woltjer 1937) did attempt to take
account of non-adiabatic corrections to the non-linear
dynamical oscillations. Wesselink proposed the concept of
relaxation oscillations as the archetype of variability for a
broad class of stellar pulsations, non-adiabatic processes
being responsible for the limitation of the amplitudes[13].
The more embracing concept of a limit cycle - the relaxation
oscillation is an asymptotic limit of a limit cycle - as the
appropriate mathematical paradigm of stellar variability seems
to go back to Krogdahl (1955). Modelling the stellar non-
adiabatic mechanisms by a parametrised formal non-linear
damping, namely a viscosity coefficient depending on the
generalized coordinates (Eqs. III.1.8 including a viscous
damping term), this author recognized that the resulting one-

mode stellar pulsation equation is reminiscent of the Van der Pol oscillator and therefore exhibits a limit cycle (cf. also Ledoux and Walraven 1958).

To gain some insight into the possible character of the motions of a star in the presence of a coupling between the hydrodynamics and the thermal behavior, we classify the nature of the defining equations from two points of view: the individual dynamical or thermal equations may be linear (L) or non-linear (N), leading to 4 different alternatives; moreover the coupling may be weak (W), i.e., the dynamical time-scale is very short as compared to the thermal time-scale, or it may be strong (S), meaning that the thermal time becomes comparable to the dynamical time. This implies that we have 8 distinct cases to consider. Among the latter, we can disregard the 2 linear-linear alternatives, essentially dealt with in the standard framework of linear stability theory; being described by purely linear equations, the resulting time behavior can never exhibit chaos. Among the remaining 6 alternatives, the following cases have received particular attention during the past few years.

A. Linear (or weakly non-linear) dynamics and non-linear thermal equations in the presence of weak coupling (LNW)

The method of dealing with this problem is essentially an extension of Landau's treatment of the onset of fluid instabilities (cf. Landau and Lifshitz 1959, section 27). Close to an instability due to the linear mode j of the full system of linearized stellar equations, we just keep that latter mode in the Woltjer type expansion (III.1.6) and set for the corresponding expansion coefficient $q_j(t)$

$$q_j(t) = J_j \exp i \phi_j \, , \qquad \text{III.3.1}$$

where J_j is a <u>slowly</u> varying amplitude, with typical time-scale of the order of the thermal time T_t, and ϕ_j is a <u>rapidly</u> increasing phase, with time-scale of the order of the dynamical time T_d; the amplitude J_j is in general a complex function. Substitute ansatz (III.3.1) into the modal equations, i.e., the system of ordinary differential equations, of non-Hamiltonian structure in the non-adiabatic context, governing the evolution of the generalized coordinates $q_j(t)$ (the non-adiabatic counterpart of Eqs. III.1.8). After multiplying by the complex conjugate amplitude, J^*_j, and averaging over the short-time variability, the phase factor drops out and we are left with an equation of the form

$$d/dt \, J_j^2 = a_2 \, J_j^2 + a_4 \, J_j^4 + \ldots \, . \qquad \text{III.3.2}$$

The latter system, termed the <u>Landau amplitude equation</u>, involves the slow time only. Depending on the signature of the expansion coefficient a_4 (a_2 is positive by the assumption of instability of mode j), the amplitude is eventually limited ($a_4 < 0$), in which case we obtain a limit cycle, or the amplitude keeps growing ($a_4 > 0$); the latter alternative means that higher order expansion terms in Eq. (III.3.2) are needed to treat this problem in a physically consistent fashion.

The construction of Landau equations for two and three-mode interactions (i.e., two or three modes are kept in the Woltjer type expansion and represented by an ansatz of form III.3.1) is discussed by various authors (Vandakurov 1979, 1981, Dziembowski 1982, Buchler and Regev 1983, Dziembowski and Kovacz 1984; Klapp et al. 1984; Buchler and Goupil 1984; Takeuti and Aikawa 1981), adopting different mathematical procedures. An elegant general technique applicable to construct Landau equations for a virtually arbitrary physical system described by partial differential equations, and near a critical (or polycritical) point of onset of an instability (or several simultaneous instabilities) is developed in Coullet and Spiegel (1983). In the first place this paper supplies a clear recipe for determining the modes required in the Woltjer type expansion; the infinity of modes is partitioned into three classes, (a) mildly unstable, (b) weakly stable, and (c) strongly stable modes; the latter, if excited, die away on a time-scale short in comparison with the slow thermal time, and therefore are to be left out of the expansion. A finite number of modes then remain in the truncated expansion, provided that the power of sets (a) and (b) is finite[14]. In the second place the method is, in a sense, the 'dynamical' counterpart of conventional 'static' catastrophe theory; if classes (a) and (b) contain ℓ real roots and m pairs of complex eigenvalues, then for given ℓ + m ('codimension') a full classification of the 'normal' forms of the amplitude equations can be worked out a priori (cf. also Arneodo et al. 1982).

The specific stellar applications of the amplitude equations are reviewed by Buchler (1984) in these Proceedings. I merely wish to comment on the mathematically allowed time-behavior of the stellar surface oscillations compatible with the LNW approximation.

(a) The assumption of linearity (or weak non-linearity) in the dynamics (L), justifying the ansatz (III.3.1), together with the weak coupling hypothesis, implying that the phases vary linearly in time

$$\phi_j = \Omega_j t + \phi_j^o \quad , \qquad\qquad \text{III.3.3}$$

(with the frequencies Ω_j being close to the linear adiabatic frequencies, and in principle slowly changing in time, and with ϕ_i^0 representing initial phases), imply that no short-time chaotic oscillations of the nature discussed in sections III.1 and 2 can arise in the framework of these approximations. On the other hand, secular period changes occurring on the T_t time, i.e., not just due to conventional stellar evolutionary effects, are possible. One should perhaps keep in mind that such a secular mechanism might be operative in RR Lyrae stars, before turning towards more exotic alternatives for explaining the observed period drift.

(b) Already in the simple two-mode coupling problem, the Landau equations which now involve, besides the genuine two amplitudes, say J_1 and J_2, also an equation for the slow time evolution of the phase difference

$$\Delta_{12} = \phi_1 - \phi_2 \qquad \qquad \text{III.3.4}$$

(cf. Buchler and Goupil 1984), make up a system of 3 non-linear ordinary differential equations in normal form. The 3-dimensional phase-space of the latter can accommodate the following attractors: fixed points, limit cycles and strange attractors. In terms of actual stellar variability, fixed points in the Landau equations correspond to double-periodic oscillations which are asymptotically attained, provided that the position of the fixed point in phase-space is arbitrary; the oscillations are mono-periodic if the fixed point lies in a plane $J_1 = 0$ or $J_2 = 0$. Limit cycles correspond to generally doubly-periodic stellar oscillations, with periods of order T_d, and amplitudes changing periodically, with a period of a long time-scale. Such solutions have been found in the context of actual stellar models (cf. Buchler 1984). Strange attractors of the stellar Landau equations remain to be isolated in genuine models. They imply long time-scale irregular amplitude and period fluctuations, which, over short time-spans manifest themselves as secular changes in these oscillation parameters. Over time intervals of the order of T_t, one may expect the amplitude and period fluctuations to be highly irregular, since the standard prototypes of dissipative chaos as exhibited by the Lorenz (1963), the Baker-Moore-Spiegel (1966) and the Rössler (1976) oscillators, show an exceedingly complicated time-behavior. Such dissipation-triggered irregular fluctuations remind us of the strange variability of the Beat Cepheids 89 Her and HD 161796, or of RU Cam - although in these stars the oscillation periods are not very short as compared to the time-scale of the irregular amplitude fluctuations.

B. Linear (or weakly non-linear) dynamics and non-linear thermal equations in the presence of strong coupling (LNS).

This alternative has not yet been investigated in the literature.

C. and D. Non-linear dynamics and non-linear thermal equations in the presence of weak and strong coupling (NNW and NNS).

This situation has been analyzed by Buchler and Regev (1982) in the framework of a rudimentary one-zone model in which He II is undergoing ionization while H and He I are fully ionized. For a well-defined entropy range such a zone admits of 3 hydrostatic equilibria, each equilibrium state corresponding to a different outer radius of the zone. The two models with largest and smallest radius are dynamically stable, while the model with intermediate radius is unstable. The zone parameters are adjusted to generate a vibrational instability in the dynamically stable models. On integrating numerically the coupled hydrodynamic and thermal equations of this formal system, Buchler and Regev (1982) encounter different types of variability: (a) if the coupling is weak ($T_d/T_t \lesssim .1$) then limit cycles around the smaller dynamically stable configuration, or the larger configuration are found, or else a limit cycle encircling both stable configurations is shown to exist; (b) as the coupling gets stronger ($.1 \lesssim T_d/T_t \lesssim 1$), highly irregular fluctuations are found which signal the existence of a strange attractor. These authors have also shown that the equations describing their model system reduce to the Baker-Moore-Spiegel oscillator in a first approximation, so that the presence of a strange attractor is not surprising. Buchler and Regev have noted a strong parameter dependence of the character of the variability: the precise features of the oscillations may drastically change as the precise physical parameters of the configuration are slightly altered.

While Buchler and Regev's one-zone configuration is not granted the status of a realistic stellar model, it does include the main physics of late type giants, i.e., of the variables of group II. The qualitative character of the calculation may therefore be representative for real stars.

A similar even more rudimentary one-zone model has been analyzed by Auvergne and Baglin (1984). In the configuration considered by these authors the non-linearity in the equation of motion is due to the variable degree of ionization alone. Auvergne and Baglin report to find chaotic oscillations for a rather small parameter range. As an interesting novel contribution, they present power spectra of their chaotic oscillations. These spectra show well-defined peaks whose positions, and even more so, whose amplitudes are strongly dependent on the actual run; the corresponding 'periods' are reported to be of the order of the thermal time-scale of the zone.

I should stress that the chaotic behavior traced in the framework of this model is of a physical origin essentially distinct from the dynamical chaos discussed in sections III.1 and 2, or the amplitude chaos arising in the LNW approximation. In the adiabatic approximation, the one-zone model just reduces to a single oscillator and is therefore always integrable. The Landau equation, on the other hand, is just a single equation of type (III.3.2), whose sole attractors are fixed points. Chaos materializes here as a consequence of the strong interaction - reminiscent of a KAM resonance - between the thermal and the detailed mechanical behavior of the star; this contrasts with the amplitude chaos, due to an interaction between the thermal and an <u>averaged</u> mechanical behavior of the star. I wish to emphasize that in a realistic n-zone extension of the Buchler and Regev (1982) configuration in which mode-interactions are also taken into account, we may expect that besides the dissipative chaos due to the thermal-dynamical coupling we also have the alternative of generating adiabatic chaos, possibly superimposed on the dissipative species. A much more interesting variety of adiabatic chaos is bound to show up here, since the potential energy exhibits now two minima separated by a saddle point (cf. the remark at the end of section III.1). At the moment numerical experiments supporting these conjectures are still lacking.

Under the NNS alternative, an attempt at formulating a two-time approach has been made by Perdang (1984). It amounts to setting up Landau equations for chaotic dynamical oscillations. This formalism offers the possibility of a superposition of dynamical chaos on top of Landau type chaos. For this problem again numerical experiments are wanting.

E. and F. Non-linear dynamics and linear thermal equations in the presence of weak and strong coupling (NLW and NLS).

A simple procedure that takes care of weak dissipation consists in including formal linear viscosity terms in the model equations (III.1.8). The latter become

$$\ddot{q}_i + 2\lambda_i \dot{q}_i = - \partial/\partial q_i H \quad , \quad i = 1,2,\ldots,F \quad . \qquad III.3.5$$

The coefficients λ_i are the vibrational stability coefficients (cf. Papaloizou 1973) which obey the weak coupling condition $|\lambda_i| \ll 1/T_d$.

A mathematical analysis of these equations seems to be lacking, while extensive numerical integrations have been performed by Papaloizou (1973) under the assumption that the lowest mode is vibrationally unstable, the higher modes being stable ($\lambda_1 < 0$, $\lambda_j > 0$, $j = 2,3,\ldots,F$). These experiments were conducted with the purpose of analyzing whether the model equations can lead to an amplitude saturation or not. In Papaloizou's calculations the unstable mode alone is initially excited, which implies that energy sharing with higher modes is not efficient, except for modes which are strongly coupled by resonances with the fundamental mode. More recently we have attempted to study more realistic initial conditions, by randomly exciting a band of lower modes. If this band is very extended - which would occur if the excitation were due to thermal noise - the solution is chaotic at the outset, and a very strong energy exchange is taking place among the modes which eventually seems to bring about a statistically stationary state in which the average fluctuation amplitudes of the modes seem to be fixed by the formal viscosity coefficients (Perdang and Blacher 1984).

The alternative of an NLS approximation has not been considered so far.

IV. Conclusion

The main target of this review has been twofold. On the one hand I have laid stress on the observational fact that irregular variability among stars - and even among the most regular of the Regular Variables - is an overarching phenomenon most certainly not ascribable to observational error. It is the theorist's task, 'sub fide vel spe geometricantis naturae' (Giordano Bruno), to search for simple physical mechanisms accounting for the stellar asynchronism. On the other hand, I have supplied evidence that the stellar hydrodynamic equations conceal a surprisingly rich variety of fluctuating bounded solutions of different physical origins, which are mathematically distinct from mono-periodic or multi-periodic oscillations. It is the observer's duty to provide more precise and less lacunary velocity and light curves to enable the theorist to perform a meaningful comparison between observationally irregular variability and the different species of theoretically traced chaos. Surely, at this early stage of the theoretical development it is premature to believe that we have captured the truth regarding the origin of the observed disordered stellar variability. Besides, it is wise to keep in mind that 'truth is never pure and rarely simple'.

I have not touched upon a further aspect of chaos relevant to stellar oscillations, namely the stellar counterpart of quantum chaos (Berry 1983). In brief, stellar quantum chaos is a Hamiltonian chaos associated with non-integrable ray equations of the geometric acoustics approximation to the linear stellar oscillation equations. This type of chaos manifests itself in the statistics of the distribution of the linear asymptotic oscillation eigenvalues. Its astroseismological interest lies in its providing observational information on the symmetry of the star on different lengthscales (cf. Perdang 1984a for an exploratory analysis). It is likely that this concept will eventually turn out to become as important in stars as ordinary quantum chaos is in molecules.

Acknowledgements

It is a pleasure to thank M. V. Berry, S. Blacher, J. R. Buchler, G. Contopoulos, W. Däppen, M. Gabriel, L. Galgani, D. O. Gough, P. Ledoux, P. Renson, H. Robe, N. Simon and E. A. Spiegel for illuminating conversations. This paper was prepared while the author was staying at the Department of Physics, University of Florida, Gainesville. He wishes to thank the members of this institution for the hospitality extended to him. He gratefully acknowledges financial support through a NATO grant.

REFERENCES

Aizenman M. L., 1980, in 'Nonradial and Nonlinear Stellar Pulsation', 76, ed: Hill H. A., Dziembowski W. A., Springer Berlin.
Argelander F., 1848, Astron. Nachr. **26**, 369.
Argelander F., 1849, Astron. Nachr. **28**, 33.
Arneodo A., Coullet P. H., Spiegel E. A., Tresser C., 1983, 'Asymptotic Chaos' (preprint).
Arnold V., 1976, 'Méthodes mathématiques de la mécanique classique' ed. Mir, Moscow.
Arnold V. I., 1984, 'Catastrophe Theory', Springer, Berlin.
Auvergne M., Baglin A., 1984, 'A dynamical instability as a driving for stellar oscillations' (preprint).
Bailey S. I., 1899, Astrophys. J. **10**, 255.
Bailey S. I., 1902, Harvard Annals, **38**.
Baker N. H., Moore D. W., Spiegel, E. A., 1966, Astron. J. **71**, 845.
Balasz-Detre J., Detre L., 1962, Kleinere Veröff. Bamberg n° **34**.
Balona L. A., Dean J. F., Stobie R. S., 1981, Month. Not. Roy. Astron. Soc. **194**, 125.
Barbanis B., 1966, Astron. J. **71**, 415.
Bates R. H. T., 1982, Phys. Reports **90**, 203.

Becker F., 1924, 'Der veränderliche Stern ζ Geminorum',
    Berlin.
Belserene E. P., 1973 in 'Variable Stars in Globular Clusters
    and in Related Systems', 105-112, ed.: Fernie J. D.,
    Reidel Dordrecht.
Bemporad A., 1921, Mem. Soc. Astron. Italiana **1**, 229.
Berry M., 1983 in 'Chaotic Behavior of Deterministic Systems',
    171-271, Les Houches XXXVI, NATO ASI, ed. Ioss G.,
    Helleman R. H. G., Stora R. F., North-Holland, Amsterdam.
Birkhoff G. D., 1927, 'Dynamical Systems', Ann. Math. Soc.,
    Providence, Rhode Island.
Blacher S., 1981 (unpublished).
Blacher S., Perdang J., 1981, Month. Not. Roy. Astron. Soc.
    196, 109P.
Blacher S., Perdang J., 1981a, Physica **3D**, 512.
Blazhko S. N., 1926, Ann. de l'Observ. Astron. Moscou, série
    2, vol. **8**, livre 2, n° 2.
Bos R. J., Hill H. A., 1983, Solar Physics **83**, 89.
Breger M., 1969, Astrophys. J. Suppl. **19**, 79.
Breger M., 1979, Pub. Astron. Soc. Pacific **91**, 5.
Breger M., 1980, Space Science Reviews, **27**, 431.
Breger M., 1981, Astrophys. J. **249**, 666.
Broglia P., Guerrero G., 1972, Astron. Astrophys. **18**, 201.
Buchler J. R., 1984 (these proceedings).
Buchler J. R., Goupil M. J., 1984 Astrophysical J. **279**, 394.
Buchler J. R., Regev O., 1982, Astrophys. J. **263**, 312.
Buchler J. R., Regev O., 1983, Astron. Astrophys. **123**, 331.
Burki G., Mayor M., 1980, Space Science Reviews, **27**, 429.
Cacciari C., 1984, Astron. J. **89**, 231.
Cahn J. H., 1980, Space Science Reviews, **27**, 457.
Campbell W. W., 1901, Astrophys. J. **13**, 94.
Campbell L., Jacchia L., 1946, 'The Story of Variable Stars',
    Blackiston, Philadelphia and Toronto.
Chandrasekhar S., 1969, 'Ellipsoidal Figures of Equilibrium',
    Yale University, New Haven.
Cherewick T. A., Young A., 1975, Pub. Astron. Soc. Pacific,
    87, 311.
Chirikov B. V., 1979, Phys. Reports **52**, 263.
Christensen-Dalsgaard J., 1982 in 'Pulsations in classical and
    cataclysmic variable stars' 99-116, ed. Cox J. P., Hansen
    C. J., Joint Institute for Laboratory Astrophysics,
    Boulder, Colorado.
Christensen-Dalsgaard J., Gough D. O., 1982 Month Not. RoyAst
    ron. Soc. **198**, 141.
Christy R. F., 1964, Rev. Mod. Phys. **36**, 555.
Christy R. F., 1967, Methods in Computational Physics, 7, 191.
Claverie A., Isaak G. R., McLeod C. P., Van der Raay H. B.,
    Roca Cortes T., 1981, Nature **293**, 443.
Constantinescu D. M., Michel L., Radicati L. A., 1979, J. de
    Physique, **40**, 147.

Contopoulos G., 1960, Zsch. f. Astrophys. **49**, 273.
Contopoulos G., 1963, Astron. J. **68**, 763.
Contopoulos G., Magnenat P., Martinet L., 1982, Physica **6D**, 126.
Coullet P. H., Spiegel E. A., 1983, SIAM J. Appl. Math., **43**, 776.
Coutts C. M., 1973 in 'Variable Stars in Globular Clusters and Related Systems', 145-149, ed.: J. D. Fermie, Reidel, Dordrecht.
Cox A. N., 1982 in 'Pulsations in Classical and Cataclysmic Variable Stars', 157-169, ed.: Cox J. P., Hansen C. J., Joint Institute for Laboratory Astrophysics, Boulder.
Cox J. P., 1976, Ann. Rev. Astron. Astrophys. **14**, 247.
Cox J. P., 1980 'Theory of Stellar Pulsations', Princeton Univ. Press.
Cox J. P., 1982, Nature, **299**, 402.
Cragg T. A., 1983, J. Am. Assoc. Variable Star Observers, **12**, 20.
Curtiss, 1905, Lick. Obs. Bull. **3**, 168.
Däppen W., 1985, these proceedings.
Däppen W., Perdang J., 1984, Mem. Soc. Astr. Italiana **55**, 299.
Däppen W., Perdang J., 1984a, 'Non-linear Stellar Oscillations. Non-radial Mode-Interactions' (preprint).
Davis M. S., 1982, J. Am. Assoc. Variable Star Observers, **11**, 27.
Demers S., Fernie I. D., 1966, Astrophys. J. **144**, 440.
Detre L., 1966, Inf. Bull. Var. Stars, n° 152.
Detre L., Szeidl B., 1973 in 'Variable Stars in Globular Clusters and in Related Systems, 31-34, ed. Fernie J. D., Reidel, Dordrecht.
Deubner F. L., 1975, Astron. Astrophys. **44**, 371.
Deubner F. L., Ulrich R. K., Rhodes E. J., 1979, Astron. Astrophys. **72**, 177.
Deupree R. G., 1976, Los Alamos Report LA-6383.
Deupree R. G., 1977, Astrophys. J. **211**, 509 and **214**, 502.
Dziembowski W., 1980 in 'Nonradial and Nonlinear Stellar Pulsation', 22-33, ed. Hill H. A., Dziembowski W. A., Springer, Berlin.
Dziembowski W., 1982, Acta Astron. **32**, 147.
Dziembowski W., Kovacz G., 1984, Month. Not. Roy. Astron. Soc. **206**, 497.
Eddington A. S., 1918, Month. Not. Roy. Astron. Soc. **79**, 2.
Eddington A. S., 1919, Month. Not. Roy. Astron. Soc. **79**, 177.
Eddington A. S., Plakidis J., 1929, Monthly Notices Roy. Astr. Soc. **90**, 65.
Eggen O. J., 1979, Astrophys. J. Suppl. **41**, 413.
Eggen O. J., 1983, Astron. J. **88**, 361.
Faulkner D. J., Shobbrook R. R., 1979, Astrophys. J. **232**, 197.
Fernie J. D., 1979, Astrophys. J. **231**, 841.

Fernie J. D., 1981, Astrophys. J. **243**, 576.
Fernie J. D., 1983, Astrophys. J. **265**, 999.
Fischer P. L., 1969, Ann. der Universitäts - Sternwarte Wien, **28**, 139.
Fischer P. L., 1969, in 'Non-Periodic Phenomena in Variable Stars', 331-338, edit.: L. Detre.
Fontaine G., McGraw J. T., Coleman L., Lacombe P., Patterson, J., Vauclair G., 1980, Astrophys. J. **239**, 898.
Gabriel M., 1984, private comm.
Gelly B., Fossat E., 1984 in 'Proceedings of the 25th Liege International Astrophysical Colloquim'.
Gilman C., 1978, Sky and Telescope, **57**, 400.
Gough D. D., 1984, private comm.
Grec G., Fossat E., Pomerantz M. A., 1980, Nature, **288**, 541.
Hagen J. G., 1921, 'Die veränderlichen Sterne. I. Band. Geschichtlich-technischer Teil', Herder, Freiburg im Breisgau.
Henon M., 1983 in 'Chaotic Behavior of Deterministic Systems', p. 53-170, Les Houches, XXXVI, NATO ASI, ed.: Ioss G., Helleman R.H.G., Stora R., North-Holland, Amsterdam.
Henon M., Heiles C., 1964, Astron. J., **69**, 73.
Hertzsprung E., 1919, Astron. Nachr. **210**, 17.
Hertzsprung E., 1926, Bull. Astron. Inst. Netherlands 3, 115.
Hill H. A., Stebbins R. T., Brown T. M., 1976 in 'Atomic masses and fundamental constants', edit.: Sanders J. H., Wapstra A. H., Plenum NY.
Hodson S. W., Stellingwerf R. F., Cox A. N., 1979, Astrophys. J. **229**, 642.
Hoffmeister C., 1923, Astron. Nachr. **218**, 326.
Hoffmeister E., 1967, Zsch. Astrophys. **65**, 194.
Huth H., 1967, Mitt. veränd. Sterne, Band 4, Heft 3.
Hurst H. E., 1955, Proc. of the Institution of Civil Engineers Part I, 519-577.
Jarzebowski T., Jerzykiewicz M., Lecontel J. M., Musielok B., 1979, Acta Astron. **29**, 517.
Kepler S. O., 1984, Astrophys. J. (in press).
Kepler S. L., Robinson E. L., Nather R. E., 1982 in 'Pulsations in classical and cataclysmic variable stars', 73, ed: Cox J. P., Hansen C., Joint Institute for Laboratory Astrophysics, Boulder, Colorado.
Kepler S. O., Robinson E. L., Nather E., 1983, Astrophys. J. **271**, 744.
Klapp J., Goupil M. J., Buchler J. R., 1984 Astrophys. J. (submitted).
Kolmogorov A. N., 1957, (reprinted as Appendix D in R. Abraham 1967 'Foundations of Mechanics', New York).
Kraft R. P., 1972 in 'The Evolution of Population II Stars', 69, ed.: Davis Philip A. G., Dudley Obs. Rep. **4**, Albany.
Krogdahl W. S., 1955, Astrophys. J. **122**, 43.

Kukarkin B. V., 1954, 'Erforschung der Struktur und Entwicklung der Sternsysteme of der Grundlage des Studiums veränderlicher Sterne', Akademie-Verlag, Berlin.
Kukarkin B. V., Rastorgouev A. S., 1973 in 'Variable Stars in Globular Clusters and in Related Systems', 180-184, ed.: Fernie J. D., Reidel, Dordrecht.
Kukarkin B. V., Kholopov P. N., Efremov Yu. N., Kukarkina N.P., Kurochin N. E., Medvedeva G. I., Perova N. B., Fedorovich V. P., Frolov M. S., 1969 'General Catalogue of Variable Stars', 3rd edition, Nauka, Moscow.
Kukarkin B. V., Kholopov P. N., Artukhina N. M., Fedorovich V. P., Frolov M. S., Goranskij V. P., Gorynya N. A., Karitskaya E. A., Kireeva N. N., Kukarkina N. P., Kurochkin N. E., Medvedeva G. I., Perova N. G., Ponomareva G. A., Samus' N. N., Shugarov S. Yu., 1982, 'New General Catalogue of Suspected Variable Stars', 1982, Nauka, Moscow.
Kwee K. K., 1967, Bull. Astron. Inst. Neth. Suppl. 2, 97.
Lacy C. H., 1973, Astron. J. **78**, 90.
Landau L., Lifshitz E. M., 1959, 'Fluid Mechanics', Pergamon Press, Oxford.
Ledoux P., 1951, Astrophys. J. **114**, 373.
Ledoux P., 1958, Hdb. d. Phys. **51**, 605.
Ledoux P., 1962, Bull. Acad. Roy. Belgique, Cl. Sci. 5e série, **48**, 240.
Ledoux P., Walraven Th., 1958, Hdb. d. Phys. **51**, 353.
Leighton R. B., 1963, Ann. Rev. Astron. Astrophys. **1**, 19.
Lesh J. R., Aizenman M. L., 1975 in 'Multiple Periodic Variable Stars', 11-32, IAU Coll. **29**, ed.: Fitch W. S., Reidel Dordrecht.
Lesh J. R., Aizenman M. L., 1978, Ann. Rev. Astron. Astrophys. **16**, 215.
Lloyd Evans T., 1983, The Observatory, **103**, 276.
Lloyd C., Pike C. D., 1984, The Observatory, **104**, 9.
Lomb N. R., 1978, Monthly Not. Roy. Astron. Soc. **185**, 325.
Lorenz E. N., 1963, J. Atmosph. Sci. **20**, 130.
Ludendorff H., 1919, Astron. Nachr. **209**, 217.
Ludendorff H., 1928, Hdb. d. Astrophys. **6**, 49.
Lynden-Bell D., Katz J., 1981, Proc. Roy. Soc. London A **378**, 179.
Madore B. E., Stobie R. S., Van den Bergh S., 1978, Month. Not. Roy. Astron. Soc., **183**, 13.
Makarenko E. N., 1972, Sky and Telescope, **44**, 225.
Mandelbrot B. B., 1977, 'Fractals, Form, Chance and Dimension', Freeman, San Francisco.
Mantegazza L., 1983, Astron. Astrophys. Suppl. **54**, 379.
Martin W. C., 1938, Ann. Sterrewacht Leiden, **17**, Part 2, 1-166.
Martin C., Plummer H. C., 1915, Month. Not. Roy. Astron. Soc. **75**, 566.
Martinet L., Magnenat P., 1981, Astron. Astrophys. **96**, 68.

McGraw J. T., Robinson E. L., 1976, Astrophys. J. Lett. **205**, 155.
Mengel J. G., 1973 in 'Variable Stars in Globular Clusters and in Related Systems', 214-230, ed.: Fernie J. D., Reidel Dordrecht.
Merrill P. W., 1938, 'The Nature of Variable Stars', Macmillan, New York.
Michelson A. A., Pease F. G., 1921, Astrophys. J., **53**, 249.
Morguleff N., Oskanian V., Rutily B., Terzan A., 1975 in 'Variable Stars and Stellar Evolution', 255-256, ed.: V. E. Sherwood, L. Plaut, Reidel, Dordrecht.
Morguleff N., Rutily B., Terzan A., 1976, Astron. Astrophys. **52**, 129.
Moser J., 1962, Nachr. Akad. Wiss. Göttingen, Math-phys Klasse 1.
Murdin P., Allen D., 1979, 'Catalogue of the Universe', Cambridge University Press.
Newcomb S., 1901, Astrophys. J. **13**, 1.
Nikolov N., Tsvetkov T. S., 1972, Astrophys. Space Sci. **16**, 445.
Niva G. D., 1979, Astrophys. J. Lett. **232**, L43.
Odell A., 1980, Astrophys. J. **236**, 536.
Oosterhoff P. T., 1941, Ann. Sterrewacht Leiden **17**, Part 4, 1-49.
Oosterhoff P. T., 1957, Bull. Astron. Inst. Neth., **13**, 317, 320.
Osborn W., 1969, Astron. J., **74**, 108.
Ostlie D. A., Cox A. N., Cahn J. H., 1982 in 'Pulsations in Classical and Cataclysmic Variable Stars', 297, ed.: Cox J. P., Hansen C. J., Joint Institute for Laboratory Astrophysics, Boulder.
Papaloizou J. C. B., 1973, Month. Not. Roy. Astron. Soc. **162**, 143, 169.
Payne-Gaposchkin C., 1954, 'Variable Stars and Galactic Structure', The Athlone Press, University of London.
Payne-Gaposchkin C., Gaposchkin S., 1938, 'Variable Stars', Harvard Observatory Monograph 5, Cambridge, Mass.
Percy J. R., 1980, Space Science Reviews, **27**, 313.
Percy J. R., Evans N. R., 1980, Space Science Reviews, **27**, 425.
Percy J. R., Welch D. L., 1981, Pub. Astron. Soc. Pacific **93**, 367.
Perdang J., 1977, 'Lecture Notes on Stellar Stability, Part Two', University of Padua GAT **95**.
Perdang J., 1979, 'Stellar Oscillations: The Asymptotic Approach', Lecture Notes, Troisieme cycle interuniversitaire en Astronomie et Astrophysique, FNRS Brussels.
Perdang J., 1983, Solar Physics, **82**, 297.
Perdang J., 1984, 'Dynamical Chaos and the Two-time Method', (preprint).

Perdang J., 1984a, 'Quantum Chaos and Moiré Patterns' (preprint).
Perdang J., Blacher S., 1982, Astron. Astrophys. **112**, 35.
Perdang J., Blacher S., 1984, Astron. Astrophys, **136**, 263.
Perdang J., Blacher S., 1984a, Month. Not. R. Astron. Soc. **209**, 905.
Perdang J., Blacher S., 1984b, 'Beyond the Stellar Limit Mass' (preprint).
Plakidis S., 1932, Month. Not. Roy. Astron. Soc. **92**, 460.
Plummer H. C., 1913, Month. Not. Roy. Astron. Soc. **73**, 661.
Poincaré H., 1890, Acta Math. **13**, 67.
Preston G. W., 1964, Ann. Rev. Astron. Astrophys. **2**, 23.
Preston G. W., Krzeminski W., Smak J., Williams J. A., 1963, Astrophys. J., **137**, 401.
Richens P. J., Berry M. V., 1981, Physica **2D**, 495.
Rosino L., 1951, Astrophys. J. **113**, 60.
Rosino L., 1973 in 'Variable Stars in Globular Clusters and in Related Systems', 51-67, ed.: J. D. Fernie, Reidel, Dordrecht.
Rössler O. E., 1976, Phys. Letters **57A**, 397.
Russell H.N., 1902, Astrophys. J. **15**, 252.
Russev R. M., 1975 in 'Variable Stars and Stellar Evolution', 563-565, ed.: Sherwood, Plaut.
Sandford R. F., 1928, Astrophys. J., **68**, 408.
Sandig H. U., 1948, Astron. Nachr. **276**, 247.
Schaltenbrand R., Tammann G., 1971, Astron. Astrophys. Suppl. **4**, 265.
Scherrer P. H., 1982 in 'Pulsations in classical and cataclysmic variable stars', 83-98, ed.: Cox J. P., Hansen C., Joint Institute for Laboratory Astrophysics, Boulder, Colorado.
Scherrer P. H., Wilcox J. M., Severny A. B., Kotov V. A., Tsap T. T., 1980, Astrophys. J. **237**, L97.
Schneller H., 1950, Astron. Nachr., **279**, 71.
Schwarzschild M., Härm R., 1970, Astrophys. J. **160**, 341.
Severny A. B., Kotov V. A., Tsap T. T., 1976, Nature, **259**, 87.
Severny A. B., Kotov V. A., Tsap T. T., 1979, Astron. Zhurn. **56**, 1137.
Shapley H., 1914, Astrophys. J. **40**, 448.
Shobbrook R. R., 1979, Month. Not. Roy. Astron. Soc. **189**, 571.
Siegel C. L., 1954, Math. Ann. **128**, 144.
Simon N. R., 1979, Astron. Astrophys. **75**, 140.
Simon N. R., 1980, Astrophys. J. **237**, 175.
Simon N. R., Schmidt E. G., 1976, Astrophys. J. **205**, 162.
Smith M. A., 1978, Astrophys. J. **224**, 927.
Sperra S.W., 1910, Astron. Nachr. **184**n 241.
Sterne T. E., 1934, Harvard College Observatory, Circular **386**, 1-36 (I), Circular **387**, 1-23 (II).
Sterne T. E., Campbell L., 1937, Ann. Harvard Coll. Obs. **105**, 459.

Stobie R. S., 1977, Month. Not. Roy. Astron. Soc. **180**, 631.
Stobie R. S., 1980, Space Science Reviews, **27**, 401.
Stobie R. S., Shobbrook R. R., 1976, Monthly Not. Roy. Astron. Soc. **174**, 401.
Stobie R. S., Shobbrook R. R., Pickup D. A., 1977, Month. Not. Roy. Astron. Soc. **179**, 389.
Stover R. J., Hesser J. E., Lasker B. M., Nather R. E., Robinson E. L., 1980, Astrophys. J. **240**, 865.
Sweigart A. V., Demarque P., 1973 in 'Variable Stars in Globular Clusters and in Related Systems', 221-228, ed.: Fernie J. D., Reidel, Dordrecht.
Sweigart A. V., Renzini A., 1979, Astron. Astrophys. **71**, 66.
Szabados L., 1980, Mitt. Sternw. der Ungarischen Akad. Wissensch. **76**, 1.
Szabados L., 1981, Mitt. Sternw. der Ungarischen Akad. Wissensch. **77**, 1.
Szabados L., 1983, Astrophys. Space Sci. **96**, 185.
Szeidl B., 1965, Mitt. Sternw. der Ungarischen Akad. Wissensch. **58**, 74.
Takeuti M., Aikawa T., 1980, Month. Not. Roy. Astron. Soc. **192**, 697.
Takeuti M., Aikawa T., 1981, Sci. Rep. Tokoku Univ. **2**, 3.
Takeuti M., Petersen J. O., 1983, Astron. Astrophys. **117**, 352.
Tassoul J. L., 1978, 'Theory of Rotating Stars', Princeton University Press, Princeton N.Y.
Tassoul J. L., Tassoul M., 1984, Astrophys. J. **279**, 384. (see also previous papers of the series).
Traub W. A., Mariska J. T., Carleton N. P., 1978, Astrophys. J. **223**, 583.
Tsesevich V. P., 1969, 'RR Lyrae stars', Israel Program for Sci. Translations, Jerusalem.
Valtier J. C., Baglin A., Auvergne M., 1979, Astron. Astrophys. **73**, 329.
Vandakurov Yu V., 1979, Soviet Astron. **23**, 421.
Vandakurov Yu V., 1981, Soviet Astron. Lett. **7**, 128.
Vauclair G., Bonazzola S., 1981, Astrophys. J. **246**, 947.
Walker M. F., 1954, Astrophys. J. **120**, 58.
Walker G. H., Ford J., 1969, Phys. Rev., **188**, 416.
Wesselink A. J., 1939, Astrophys. J., **89**, 659.
Whitney C.A., 1984 in 'Proceedings of the 25th LiegeInternati onal Astrophysical Colloquium'.
Willson L. A., 1982 in 'Pulsations in Classical and Cataclysmic Variable Stars', 269, ed.: Cox J. P., Hansen C. J., Joint Institute for Laboratory Astrophysics, Boulder.
Winget D. E., 1981, Ph. D. Thesis, University of Rochester.
Winget D. E., Fontaine G., 1982 in 'Pulsations in classical and cataclysmic variable stars', 46-67, ed.: J. P. Cox, C. Hansen, Joint Institute for Laboratory Astrophysics, Boulder, Colorado.

Winget D. E., Robinson E. L., Nather R. E., Fontaine G., 1982,
    Astrophys. J. Letters **262**, L11.
Winget D. E., Van Horn H. M., Hansen C. J., 1981, Astrophys. J.
    Letters, **245**, L33.
Wizinovich P., Percy J. R., 1979, Pub. Astron. Soc. Pacific
    91, 53.
Woltjer J., 1935, Month. Not. Roy. Astron. Soc. **95**, 260.
Woltjer J., 1937, Bull. Astron. Inst. Neth. **8**, 193.
Woltjer J., 1943, Bull. Astron. Inst. Neth. **9**, 441.
Wood P. R., 1974, Astrophys. J. **190**, 609.
Wood P. R., 1975 in 'Multiple Periodic Variable Stars', 69-85,
    IAU Coll. **29**, ed.: Fitch W. S., Reidel, Dordrecht.
Woodard M., Hudson H. S., 1983, Nature **305**, 589.
Yamaska A., Gonzalez S. F., Peniche R., Peña J. H., 1983,
    Publ. Astron. Soc. Pacific **95**, 447.
Young A., Furenlid I., 1980, Space Science Rev. **27**, 329.
Zaitseva G. V., Lyutyi V. M., Efremov Yu N., 1973, Soviet
    Astron. **16**, 856.

FOOTNOTES

[1] A well studied statistical problem of a similar nature is
encountered in the flooding of rivers. Here the relevant
statistical variable is the river height H_i in the i th
cycle, rather than the cycle length. The standard
deviation of the river height over a period of n cycles
would then obey Eq. (II.1.6) if the assumption of
independence of the individual fluctuations were to
hold. In fact, observation shows that in the case of the
river Nile s_n conforms to a law of form (II.1.6'), with
a = .7 (cf Hurst 1955). The deviation of a from its
standard value 1/2 supplies a measure of the correlations
between the fluctuations within different cycles.

[2] The factor 2 accounts for the fact that each period
measurement consists in two observations, each observation
being affected by an error e.

[3] The semiregulars, together with the irregular variables, are
referred to in the older literature as μ Cephei variables.

[4] α Vir, or Spica, being a component of a binary system, an
estimate of its mass can be made; moreover, its radius has
been measured interferometrically; acording to Odell
(1980), the mass of the primary is 10.9 ± .9 $M_\odot$ and the
radius is 7.84 $R_\odot$; the envelope composition is X = .70, Z
= .03; the linear fundamental radial period of such a

model reproduces the observed period of .174 d of this star.

[5] In a classification scheme recently proposed by Eggen (1979, 1983), these variables are called Ultra Short Cepheids.

[6] Under the requirement of spherical symmetry, m(r) represents the mass contained in the sphere of centre coinciding with the symmetry centre of the star.

[7] If the smoothness conditions are violated, integrability may be lost even if F isolating integrals exist. Richens and Berry (1981) have discussed instances of billiards in odd-shaped space regions (cf the L-shaped billiard) in which this type of non-integrability occurs. Berry refers to the corresponding Hamiltonians as <u>pseudo-integrable</u> Hamiltonians (cf. Berry 1983).

[8] Technically, the harmonic oscillator Hamiltonian is 'degenerate' in the sense that $dtm\,(\partial^2/\partial J_i \partial J_j H_o)$ vanishes identically; the KAM theorem stipulates that a given solution of the unperturbed Hamiltonian system survives qualitatively in the perturbed problem provided that the unperturbed Hamiltonian is not degenerate, and the frequencies of the unperturbed problem are non-resonant to any order.

[9] An actual proof of this conjecture should rely on the asymptotic behavior of the linear eigenfunctions of high order k, which enable one to evaluate the coupling coefficients $V_{ijk}^{(3)}$,..., analytically.

[10] The symmetry groups D_{nh} and C_{nv}, n integer, have as symmetry elements an n-fold axis C_n (a configuration of symmetry D_{nh} or C_{nv} is invariant under a rotation of $2\pi/n$ around this axis); D_{nh} possesses binary axes Λ_2 normal to C_n (there are n such axes if n is odd, and n/2 binary axes for n even, as a consequence of the C_n symmetry); C_{nv} possesses 'vertical' symmetry planes σ_v containing the axis C_n (there are n such planes for n odd, and n/2 vertical planes for n even); for $n \to \infty$, the axis C_∞ becomes a symmetry axis for rotations of arbitrary angles θ.

[11] Besides the adiabatic stationary flows, we have of course also the possibility of non-adiabatically driven stationary circulations (cf. Tassoul 1978, 1984).

[12] If $S(\underline{r},t) = S_1$ and $S(\underline{r},t) = S_2$ ($S_1 < S_2$) are two entropy surfaces in the star, the mass contained in the shell defined by these surfaces is given by

$$m(S_1,S_2) \equiv (1/4\pi)\int_0^{2\pi} d\phi \int_0^{\pi} d\theta \, \sin\theta \int_{S_1}^{S_2} dS \, \mu(S,\theta,\phi;t); \quad \text{III.2.7}$$

the latter relation supplies an implicit definition of the stratification field μ.

[13] Wesselink actually tried to apply the idea of relaxation oscillations to the group of variables now known under the collective name of 'cataclysmic variables'; the current canonical model for the latter is a binary made up of a Roche lobe filling cool dwarf and an accreting white dwarf. Irregular variability is due to the mass exchange in the double star configuration.

[14] For radial stellar pulsations the latter condition is always fulfilled. For non-radial pulsations around a state of spherical equilibrium, two difficulties arise; the toroidal modes belong to class (b), and there is an infinity of them; therefore they can never be discarded from the expansion (cf the discussion of section III.2); in the framework of conventional stellar approximations, the onset of convective instabilities occurs via an infinitely degenerate critical point, all g-modes becoming unstable simultaneously; in the treatment of stellar convection, an infinity of modes in the Woltjer expansion would then be needed due to the infinite power of class (a).

COSMIC ARRHYTHMIAS

E. A. Spiegel

Cosmic matter is not stationary. Most of it is probably turbulent. Though we cannot solve even the simple fluid equations that describe such dynamics, many astrophysicists consider that we are progressing by constructing simple models of the phenomena observed. The possibility of making models that have predictive value for some aspects of the observations while ignoring certain of the implications is at the bottom of the success story of the physical sciences. The discussion in this volume is about a variant of such model making, that has some newness about it. We are examining models that are inspired, not so much by the detection of specific physical processes, but by the observation of qualitative kinds of spatial or temporal variability.
 The models discussed are mathematical in spirit. Often they are not physically derived, but are equations that can be written down without detailed knowledge of the physical process that causes the observed phenomenon. For example, in studying the temporal behavior of the solar oscillations, we can find an interesting model system once we conclude that the sun is overstable. It is not necessary for further exploration to know the explicit physical mechanism. The theory of systems subject to mild instabilities leads to nonlinear equations for the amplitude of the variations and the forms of the equations may be decided once the linear theory has been done. I explain how that works in this discussion. Solutions of some simple examples of amplitude equations show a rich temporal behavior that has interested people in several fields. This workshop was held in order to ask whether the recent developments in this approach may have value for astrophysics.
 The kinds of simple models I am talking about are especially helpful when the system being described is chaotic. So chaos is the catchword that has been used to focus this discussion. In fact, whenever the underlying system has a

complicated temporal evolution, the methods I describe here are worth trying. Naturally, we ought to have some idea of the nature and the number of the possible instabilities to go on, that may be implied by certain patterns in the observed temporal behavior. Examples of the astrophysical phenomena that seem worth examing in this light are the solar cycle, the rapid burster, and luminosity variations of quasars such as 3C273.

This discussion is meant to be introductory. It says a bit about chaos in §**1**. Normal form theory (the way to construct amplitude equations for mildly unstable situations) is taken up in §§**2-5**. Despite my best efforts, it may not be easy to skim that part but it should be possible to skip it and go directly to the final sections. In those, I describe possible examples of astrophysical chaos.

1. Introduction to Chaos

Long range prediction is not possible for a chaotic system because of what is called sensitivity to initial conditions. Let us look at a version of this property in a simple system without chaos. We can get sensitive systems by going to conditions close to bifurcation. Astronomers know about bifurcations from the theory of the equilibrium figures of a rotating, self-gravitating body. As the angular velocity of the body is increased through certain critical values, instabilities beset this system in a competition between gravitational and centrifugal forces. This is simplistically modeled by a particle sliding around inside a bowl rotating about a vertical axis (Lamb, 1907; Jeans, 1928) or, what comes to the same thing, a bead sliding around on a circular hoop rotating about a vertical diameter (Andronov and Chaikin, 1930). The equation of motion, with nondimensional time, is

$$\ddot{X} = \tfrac{1}{2}\alpha\sin 2X - \sin X - \nu\dot{X}. \qquad (1.1)$$

Here X is the angle from the downward plumb line, a is the radius of the hoop, Ω is the angular speed of the hoop, ν is a drag coefficient, and

$$\alpha^2 = a\Omega^2/g. \qquad (1.2)$$

For any α and ν, (1.1) has equibrium solutions at $X = 0$ and $X = \pm\pi$, the latter being always unstable. The equilibrium solution at $X = 0$ is stable for $\alpha < 1$ and is unstable for $\alpha > 1$. When α is increased through unity, two new equilibria bifurcate from $X = 0$ and move to increasing X. When $\alpha > 1$, for generic initial conditions, the bead does not go to the bottom of the hoop but to one side of it. To predict which side, we may use a computer. Figure 1 shows trajectories in the phase

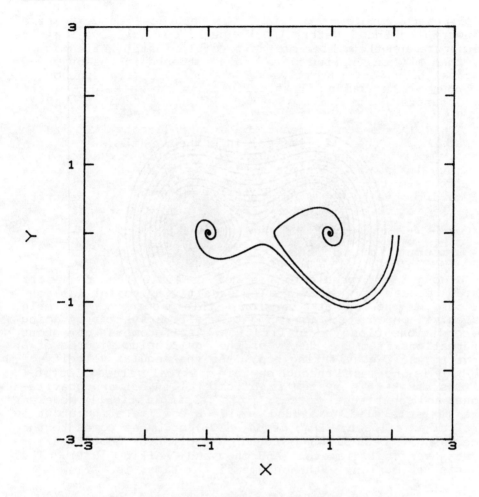

<u>Fig. 1</u>. Trajectories in the phase plane (X,Y), where Y = $\dot{X}$, obtained by solving (1.1) with α = 1.8 and ν = 0.5 for two initial conditions:
 (i) X_0 = 2.0, Y_0 = 0;   (ii) X_0 = 2.1, Y_0 = 0.
The orbits diverge near the saddle point at the origin.

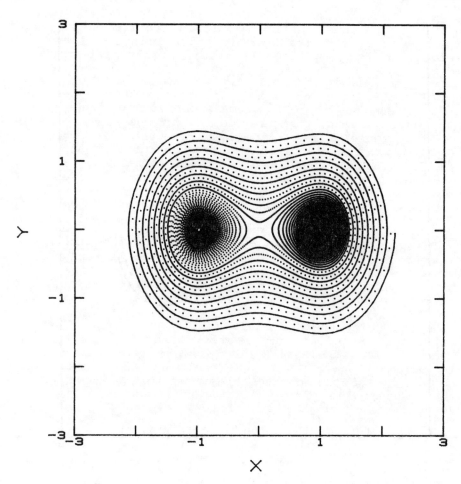

Fig. 2. Another pair of trajectories of (1.1) for $\alpha = 1.8$ and $\nu = 0.2$ for
 (i) $X_0 = 2.13$, $Y_0 = 0$; (ii) $X_0 = 2.20$, $Y_0 = 0$.

space $(X,\dot{X})$ found from a simulation of (1.1) for the parameter values indicated in the caption. The two phase trajectories emerge from neighboring phase points and remain close together for a while, then they rapidly part company and go to opposite sides of the hoop.

In Figure 2 we see the result of a similar pair of simulations, but this time, the drag ν has been reduced by a factor of 25. The two trajectories remain close for a longer time before their ways part, then they go to opposite sides of the hoop. This figure is meant to illustrate the point that if the two trajectories are close enough, the computer on which they were simulated could not distinguish them. As I make ν smaller, this problem grows increasingly acute and we arrive at a situation where, for a wide class of initial conditions, we are unable to predict the outcome of such a simple experiment. Such unpredictability for deterministic systems is one ingredient of chaos that, in this special case, has been localized in parameter space. The possibility of unpredictability has to be reckoned with in the study of such astrophysical phenomena as the sunspot cycle or pulsar timing noise. It may, for example, be impossible to predict with precision when the next intermission in solar activity will begin. But more on that in §6.

The inability to predict is really ours, but it comes about when the system is in a certain kind of state that I call pandemonium. The system described by (1.1) always decays to a stable equilibrium. We can keep the system from running down by forcing it externally. In this way, we get a sustained pandemonium: that is what many people call chaos.

We write (1.1) as the equivalent system

$$\dot{X} = Y$$
$$\dot{Y} = -\frac{\partial V}{\partial X} - \nu Y \qquad (1.3ab)$$

where V is the potential appropriate to (1.1). To simplify the discussion, we replace this potential by a quartic:

$$V = \tfrac{1}{4}X^4 - \tfrac{1}{2}aX^2 - ZX, \qquad (1.4)$$

where a is a new parameter. In (1.3b) we have included the external driving term, Z, whose time dependence may be specified explicitly, or given by an equation such as

$$\dot{Z} = -\varepsilon[Z + bX - cX^3], \qquad (1.5)$$

which couples Z to the state of the oscillator of (1.3). The system (1.3)-(1.5) may not seem much different from the original (1.1). Given the right parameter values we can have the same kinds of bifurcation that (1.1) exhibited, and with the corresponding equilibria. When the equilibria of (1.3)-(1.5)

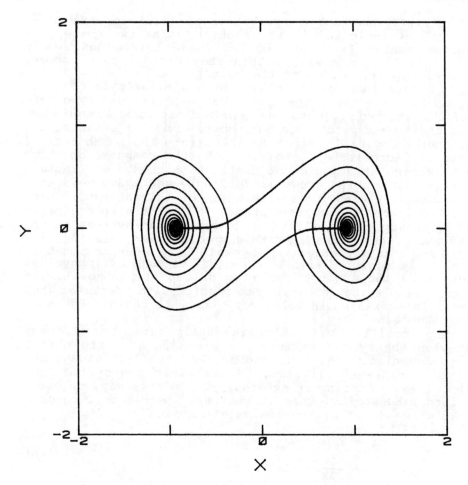

Fig. 3. A solution of (1.3)-(1.5) for:
 $a = .7$, $\nu = .1$, $b = 1$, $c = 1.2$, $\varepsilon = .02$.
A projection of the orbit into the X-Y plane is shown. The transients are suppressed.

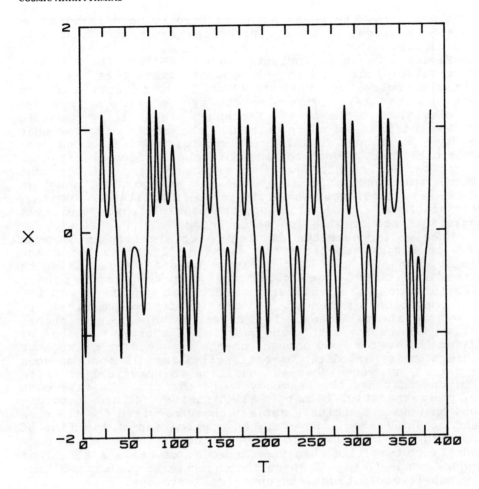

Fig. 4. X(t) from (1.3)-(1.5) for:
a = .6, b = .9, c = 1.1, ε = .1, ν = 0.
For long runs in time, the behavior continues as shown.

become unstable, the phase point is free to move in a three-dimensional space, and that introduces important changes in the dynamics.

Figure 3 shows a simulation of (1.3)-(1.5) for the parameter values indicated in the caption; transients have been given time to decay. What is shown is a projection of the orbit onto the X-Y plane; when the system crosses from one side of Fig. 3 to the other, Z changes sign. It is possible from the original output to determine the number of times that the phase point goes around each fixed point before going over to the other one, though this cannot be done from the figure. For the parameters of Fig. 3, this number turns out constant with the modest computational means at my disposal. Though we are close to conditions where the periodic orbits are numerous and unstable (Shil'nikov, 1968), the possible chaos that this suggests is too fragile for me to detect.

When we increase the damping to obtain results such as those in Fig. 4, aperiodicity is seen (see also Moore and Spiegel, 1966). This system may be said to be sensitive to initial conditions since a slight change in the starting point makes for a completely different orbit. But sensitivity to initial conditions carries with it sensitivity to numerical error and to external noise. In a case like this, better numerical methods cannot settle the question of whether you can, by running the system long enough, change the verdict on whether this system is periodic. Operationally, the distinction does not matter; a system that is sensitive to numerical error is often one that has the tendency to be chaotic. In this case too, the mathematics is not at all decisive. We are in conditions that do not strictly satisfy those required for the theorems of Shil'nikov. However, an application of his line of reasoning has been used to suggest that a robust chaos does occur in systems like this one (Arneodo, et. al., 1985). That is what we see in Fig. 5 where the chaos is so pronounced that many people would trust the numerical evidence.

Dynamics like that in Fig. 5 would be called quasiperiodic by astronomers who study stellar or solar variability, though mathematicians use this word in a different way. The system keeps doing essentially the same thing over and over, but does it slightly differently each time, as in the solar cycle. Many astronomical objects show this kind of aperiodicity, and the point of this discussion is that we can begin to think seriously about making mathematical models of such behavior. In a sense, we have always done this, if less systematically, since simple physical models in effect are mathematical. What is perhaps different now is that not everyone is ready to count a chaotic oscillator as a simple physical object, even if its equations are simple.

Because of the sensitivity to roundoff and other errors, we cannot be sure that a numerical trajectory like that in Fig. 5 represents a true solution in detail. Nevertheless,

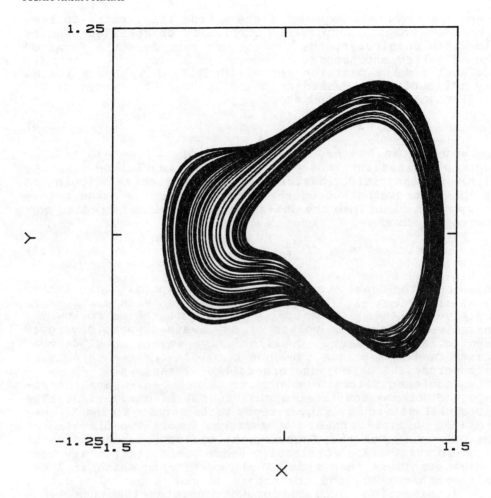

Fig. 5. Like Fig. 3, a projection of an orbit.
This is a strange attractor of (1.3)-(1.5), for:
a = .5, b = 1.3, c = 1.3, ε = .01, ν = .01.

there are important aspects of the motion illustrated in Fig. 5 that are robust, despite the difficulty of detailed prediction. In particular, the simulations reproduce the forms of the so-called attractors.

Let **x** be a position vector in (X,Y,Z) space, which we may think of as a Euclidean phase space. The system (1.3)-(1.5) may be abbreviated as

$$\dot{\mathbf{x}} = \mathbf{u} \qquad (1.6)$$

where $\mathbf{u}(\mathbf{x})$ can be read from (1.3)-(1.5). If we put a dense swarm of points into the phase space, we can follow it using (1.6); for the fluid dynamicist, this is the usual calculation of the particle path from the Eulerian velocity. The particles we introduce into the phase space are neither created nor destroyed, so that,

$$\mathbf{\nabla} \cdot \mathbf{u} = -\frac{1}{\rho}\frac{d\rho}{dt} \qquad (1.7)$$

where ρ is the density of the swarm. But explicit calculation shows that, for (1.3)-(1.5), $\mathbf{\nabla} \cdot \mathbf{u} = -(\varepsilon + \nu)$. When ε and ν are positive, ρ tends exponentially to infinity as we follow the particles. Hence the volume of any swarm of particles goes exponentially to zero. Qualitatively, we may conclude that swarms of phase points condense onto one or more objects of zero volume. Such objects are called attractors.

Stable equilibrium points and limit cycles are attractors. But more complicated objects may be attractors. The motion illustrated in Fig. 5 seems to lie on some kind of surace in the phase space. A surface, being two-dimensional does have zero volume. However, the trajectories do not seem to cover the surface completely, hence the attractor may have a dimension less than that of the surface in which it lies. On the other hand, the trajectory is not a periodic orbit, and its dimension could be somewhat greater than that of a limit cycle. That is why an attractor like that sketched in Fig. 5 is often called strange: it appears to have a fractal dimension (Mandelbrot, 1983).

In motion on a strange attractor, we have sustained pandemonium since two points starting close to each other will on average move apart exponentially in time. There are a few introductory books describing chaos and strange attractors (Schuster, 1984; Hao, 1984; Bergé, Pomeau, Vidal, 1984). They emphasize dissipative chaos, such as I have been discussing. But also they discuss Hamiltonian systems, where the quantity $\mathbf{\nabla} \cdot \mathbf{u}$ vanishes. Incompressibility of the phase flow is often taken as the definition of nondissipativity; there are no attractors in that case. There can still be chaos, called conservative or Hamiltonian chaos, or stochasticity. In plasma physics and stellar dynamics, stochasticity is paramount;

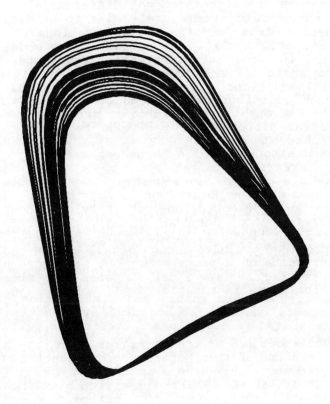

Fig. 6. A reconstruction of an attractor from X(t) by the use of delay coordinates for the same parameter values as for Fig. 5.

in astrophysics, dissipative systems and strange attractors may become interesting. But this is only a reading guide for the astromathematician. As Faraday put it, "I am fully aware that names are one thing and science another."

In the most elementary discussions of chaos, the mathematical systems are algebraic, not differential equations as here. The algebraic sytems are easier to handle. They can often be derived from the differential equations (Fowler, 1984; Arneodo, et. al., 1985). The derivation of model equations from the original partial differential equations, whether by physical or mathematical arguments, usually yield ordinary differential equations. I next examine the sytematic derivation of such equations for systems just beyond the onset of instability. We ought to be able to go directly from the nonlinear partial differential equations to the algebraic systems, but that has not been done in a systematic way as yet.

We need also to be able to assess a set of data and see whether its erratic component is due to deterministic effects. Suppose you are given a time sequence for a variable $X(t)$. If the phase space of the phenemonon is not too large, and there is an attractor, you could look for it by considering a space with coordinates $X, \dot{X}, \ddot{X}$, and so on. The number of dimensions is increased, until you see a structure that looks like an attractor, provided you have enough data. Ruelle has suggested a way to effect such a program without having to numerically differentiate $X(t)$. Make a set of pairs of numbers $X(t)$ and $X(t+\tau)$. Plot these in the plane. If you see something that looks like an attractor, proceed to anlyze it. Otherwise go on and look for structure in a plot of the set of triples, $X(t), X(t+\tau), X(t+2\tau)$. The rule of thumb is that τ should be chosen to be about 1/5 of the characteristic time seen in the data. The meaning of this unpublished suggestion of Ruelle is much analyzed and it has been used to great effect for laboratory data (Wolf, et al, 1985). For astronomical observations, where there are perforce gaps, this method promises to be especially effective. It is illustrated in Fig. 6.

2. Amplitude Equations

Suppose we study the development of instability in an astronomical body. Assume that we know the form of the equations that govern physical processes in this body. In astrophysics, we often are not sure about the values of parameters that describe the objects of interest or how to introduce boundary conditions. But let us assume that the parameters have at least been identified and that there are boundary conditions that can be written down.

Even if the equations are known, they are usually too hard to solve in general, and we look for simple solutions. Typically, we begin by seeking steady solutions, but even these are difficult to obtain without further simplification. We may impose axisymmetry or spherical symmetry or we may require the solution to be static. The steady solutions found in this way usually represent the first attempts to describe astronomical bodies. When such description fails, we sometimes rationalize the failure by showing that the known steady solutions are unstable. We may even go further and try to find the new solutions whose existence is indicated by the instability. Bifurcation theory offers a systematic way to seek such solutions by a reduction of the equations that is possible when the original state is mildly unstable. In this account, I shall not discuss how far the results may be pushed.

The model is described by a set of physical variables such as temperature, density, velocity, all as functions of position x and time t. We shall measure them with respect to their values in a selected steady solution. We denote by $U_I(x,t)$, $I=1,2,\ldots,N$, the values of these N variables minus the values in the steady solution. We imagine equations for these variables of the form

$$\partial_t \mathbf{U} = \mathcal{M}\mathbf{U} + \mathcal{N}(\mathbf{U}), \tag{2.1}$$

where $\mathbf{U}(x,t) = (U_1, U_2, \ldots, U_N)$, $\mathcal{M}$ is a linear operator and $\mathcal{N}$ is a strictly nonlinear operator. It may take some work to put the equations into this form, and in some cases it may not be the best thing to do. But it is useful to have a specific form of equation in mind for a general discussion and (2.1) is typical and contains the standard difficulties that come up.

The purpose in subtracting the steady solution from the state variables is to make $\mathbf{U} = 0$ a solution of (2.1). Since we have assumed that steady solutions have some relevance to the description of the object, we are typically in a situation where neither of the operators $\mathcal{M}$ or $\mathcal{N}$ contains t or ∂_t. The operators will generally depend on the parameters of the object, $\boldsymbol{\lambda} = (\lambda_1, \lambda_2, \ldots)$. Typical parameters are mass, angular momentum or chemical composition. For stars, these quantities may be considered as parameters when we are studying dynamical processes that go more quickly than the normally slow evolutionary changes in the steady state. Otherwise the approach described here has to be adapted for bifurcations from a nonstationary state (Lebovitz and Shaar, 1975).

Now consider the associated linear problem

$$\partial_t \mathbf{W} = \mathcal{M}\mathbf{W} \tag{2.2}$$

whose solutions are very useful in the discussion of the nonlinear solutions of (2.1) near to the onset of instability. Since we assume that $\mathcal{M}$ does not depend on t or ∂_t, the solutions of (2.2) are separable and we seek them in the form

$$\mathbf{W}(x,t) = e^{st}\boldsymbol{\Phi}(x). \tag{2.3}$$

This resembles the way we look for stationary solutions in quantum mechanics. Of course, if we had a Hamiltonian operator $\mathcal{H} = i\mathcal{M}$, (2.3) would be Schrödinger's equation, $\mathcal{M}$ would be anti-Hermitian and s (= iE) would be purely imaginary. These special properties of $\mathcal{M}$ are not typical of stability theory, but such a familiar example can guide our intuition.

After separation, (2.2) becomes

$$\mathcal{M}\Phi = s\Phi. \tag{2.4}$$

When we also introduce the linearized boundary conditions, we may find the proper value equation

$$\mathcal{F}(s;\lambda;\mathbf{k}) = 0. \tag{2.5}$$

The vector **k** is an indicator of the spatial strucure of the solution such as the index of a Legendre function or a wave vector of a Fourier component.

An important technical point arises when we have degeneracy, that is when there are multiple roots for s. In quantum mechanics, we have Hermitian operators so that, when there is a multiplicity ℓ, there are ℓ proper vectors with the same value of s. In that case, there is no problem in finding a complete set of proper vectors. When $\mathcal{M}$ is not selfadjoint, there may be fewer than ℓ proper vectors associated to a proper value with multiplicity ℓ. We then need some more vectors to complete a basis in the function space of **U**. These are called generalized proper vectors (Friedman, 1956), and their role is described in §5. When we indicate an expansion in proper vectors, it is understood that the generalized vectors are to be included in the basis.

It is convenient to classify proper vectors according to the value of s with which they are associated. Let

$$s = \eta + i\omega. \tag{2.6}$$

Vectors with small $|\eta|$ are called slow modes and those with large $|\eta|$ are called fast modes. We shall assume that there are no fast modes with $\eta > 0$. The fast, stable modes are sometimes called slave modes, following Haken (1977). The distinction between a slow and a fast mode is made easily when there is a gap in the spectrum of η values.

Let the fast modes be called f_j and the slow modes be called ϕ_i. We assume that there are a finite number n of slow modes and allow an infinite number of fast modes. When the spectrum of s is not continuous, an expansion of the state vector of (2.1) in terms of these modes has the form

$$\mathbf{U}(\mathbf{x},t) = \sum_{i=1}^{n} \alpha_i(t)\phi_i(\mathbf{x}) + \sum_j \beta_j(t)f_j(\mathbf{x}). \tag{2.7}$$

There are two senses in which the f_i and ϕ_j are vectors.

First, they are state vectors in the function space, like wave functions in quantum mechanics. Second they each have N components corresponding to the N functions of **U**. The relative amplitudes of these components are fixed as part of the modal structure. In (2.7), the amplitudes with which the modes contribute to **U** are the components α_i and β_j, which depend on t.

When we introduce (2.7) into (2.1), we obtain ordinary differential equations. If **α** and **β** stand for all the α_i and β_j, these equations have the forms

$$\dot{\boldsymbol{\alpha}} = \mathbf{M}\boldsymbol{\alpha} + \mathbf{F}(\boldsymbol{\alpha},\boldsymbol{\beta}) \tag{2.8}$$

$$\dot{\boldsymbol{\beta}} = \mathbf{K}\boldsymbol{\beta} + \mathbf{G}(\boldsymbol{\alpha},\boldsymbol{\beta}) \tag{2.9}$$

where **M** and **K** are matrices that operate on the vectors **α** and **β**, and **F** and **G** are nonlinear functions. That is, **F** and **G** and their first partial derivatives all vanish when **α** and **β** vanish. (If we do not expand in normal modes, there may be linear couplings between (2.8) and (2.9).)

The proper values of **K** have large negative real parts, so that, without the intervention of **G**, **β** would decay rapidly to zero. We may use this property to solve (2.9) in successive approximations by rewriting it as

$$\boldsymbol{\beta} = \mathbf{K}^{-1}[\dot{\boldsymbol{\beta}} - \mathbf{G}(\boldsymbol{\alpha},\boldsymbol{\beta})]. \tag{2.10}$$

In the leading approximation, we have **β** = **0**; this is sometimes called a Galerkin approximation. To get the next approximation, we put **β** = **0** in the right side of (2.10) and obtain the adiabatic approximation:

$$\boldsymbol{\beta} \simeq -\mathbf{K}^{-1}\mathbf{G}(\boldsymbol{\alpha},\mathbf{0}). \tag{2.11}$$

Then we introduce this approximation on the right of (2.10) and obtain yet a better approximation, and so on, to a series of approximations that gives us **β** in terms of **α**.

This series of approximations gives us, at each level, **β** as a function of **α**:

$$\boldsymbol{\beta} = \mathbf{B}(\boldsymbol{\alpha}). \tag{2.12}$$

When we introduce this into (2.8) we get the amplitude equation

$$\dot{\boldsymbol{\alpha}} = \mathbf{M}\boldsymbol{\alpha} + \boldsymbol{\Gamma}(\boldsymbol{\alpha}) \tag{2.13}$$

where

$$\boldsymbol{\Gamma}(\boldsymbol{\alpha}) = \mathbf{F}(\boldsymbol{\alpha},\mathbf{B}(\boldsymbol{\alpha})). \tag{2.14}$$

The reduction to (2.13) is simple in principle, but it

may be complicated to work out Γ. We need to know not only a certain number of modes of linear theory but, to perform the projections leading to (2.8) and (2.9), we have to know the modes of the adjoint linear problem as well. In cases where direct simulation on (2.1) is not a simple matter, it is worth the effort to derive (2.13), for the simulations on it are much easier than for (2.1) where we would have to resolve the spatial problem at each time step. There is an even simpler thing to do that may also be worthwhile.

It is possible to decide, on the basis of the the linear theory, what the form of the amplitude equation will be. We can then begin to examine solutions of the forms that (2.13) may take and discuss their solutions for various parameter values. This is something like studying fluid dynamics without specifying in advance what the value of the viscosity is. To unify the study of different problems with similar amplitude equations, it is well to have standard or normal forms for these equations. The next few sections are devoted to the derivation of such standard amplitude equations from the linear theory of the original problem. The relation of these normal forms of the equation to (2.13) comes about through transformations on (2.13). The phase point of the system (2.1) moves in a subspace of α-β space, and we would not have expected the α_i to be the most convenient variables to use as coordinates for this subspace. In §3, we transform to more suitable coordinates to find simple forms for the amplitude equations.

3. Normal Forms

To derive amplitude equations, we assumed that a clear separation between fast and slow modes could be made. This is not always possible, and there are many delicate situations that are topics for research but are unsuitable for inclusion here. We want to look at an ideal case that does not require special considerations.

Suppose that for a particular choice of the parameter, λ_0, the spectrum of s (see (2.4)) contains only values with $\eta = 0$ and $\eta < 0$ ($\eta = \mathbf{Re}\,s$). We have a case with a clear distinction between the two kinds of modes. Then the result implied in (2.12) can be made into a mathematical statement called the center manifold theorem (Carr, 1981). Though this is a special choice of spectrum, once we have an amplitude equation for λ_0, we can perturb it to get the equation for λ in the neighborhood of λ_0.

The modes (and generalized modes) with $\eta = 0$ span a subspace that is invariant under $\mathcal{M}$. I call it critical space (Arneodo et al, 1985), but it is also what is called center space for the linear theory (Guckenheimer and Holmes, 1983). The modes with $\eta < 0$ span another invariant subspace, the stable space. This breakup into a critical space and a stable space is like the decomposition made in applied mathematics of a linear space into a null space and a range (Friedman, 1956). The dimension of the null space is known as its nullity; let us call the dimension of critical space the criticality, κ.

The critical and stable modes are the fast and slow modes of the last section. We can make the same decomposition of **U** as in (2.7). The number of components in the amplitude vector **α** is κ. The discussion of §2 shows that the system moves quickly in the state space, with coordinates α_i and β_j, onto the subspace given by (2.12). Once the system is in this center manifold (Carr, 1981), the α_i are not necessarily the best choice of coordinates to describe the motion of the system point. To simplify (2.13) we can try transforming coordinates. Since we have already built the right linear theory into the amplitude equation, we want to keep the linear part of (2.13) intact. So we try the transformation

$$\boldsymbol{\alpha} = \mathbf{A} + \boldsymbol{\Psi}(\mathbf{A}) \tag{3.1}$$

where the transformation function **Ψ** is strictly nonlinear.

We want (2.13) to take an especially nice form that we write for now as

$$\dot{\mathbf{A}} = \mathbf{M}\mathbf{A} + \mathbf{g}(\mathbf{A}) \tag{3.2}$$

where a standard **g** is to be selected for its simplicity. When we substitute (3.1) and (3.2) into (2.13), we get

$$\mathcal{L}_{\mathbf{M}} \cdot \boldsymbol{\Psi} = \mathbf{T} - \mathbf{g} \tag{3.3}$$

where

$$\mathcal{L}_{\mathbf{M}} = M_{ij}A_j \partial_{A_i} - \mathbf{M} \tag{3.4}$$

is a linear operator and

$$\mathbf{T} = \boldsymbol{\Gamma}\bigl(\mathbf{A}+\boldsymbol{\Psi}(\mathbf{A})\bigr) - \mathbf{g}\cdot\partial_{\mathbf{A}}\boldsymbol{\Psi}. \tag{3.5}$$

In (3.4), we use the summation convention on repeated indices.

We seem to have one equation, (3.3), for the two unknowns, **g** and **Ψ**. But **g** is not an unknown; we want to pick it ourselves. The single unknown is **Ψ**; we have to solve (3.3) for it to get the **g** we want. Since (3.3) does not produce regular solutions for every **g**, we have to decide what the restrictions are before we make our choice. Once we know what the constraints on **g** are, we can choose the simplest **g** that is allowed. Loosely speaking, that is the idea of normal form theory (Arnold, 1983); it does not give a unique choice.

The forms of g for which the operator $\mathcal{L}_M$ can be inverted are decided by the operator itself, hence by M. That is why the linear problem decides a normal form for the amplitude equation. The important implication is that once the linear problem is solved, we can anticipate the kind of nonlinear time dependence that will occur near marginality, that is, near to λ_0. It seems well worth the effort to master a theory that does all this, especially as many of the manipulations are like those familiar from elementary quantum mechanics.

We have to perform two expansions. First, we expand λ around the critical point λ_0. This is the conventional parameter expansion of singular pertubation theory. Among other things, it permits us to find canonical forms for M in terms of the parameters (Coullet and Spiegel, 1983). Then, to find the normal form at λ_0, we expand g and Ψ in Taylor series in the components of A. When we gather terms of degree D in A_i, (3.3) is changed only mildly in appearance to

$$\mathcal{L}_M \cdot \Psi_D = T_D + g_D, \qquad (3.6)$$

where the subscripts mean that all the terms of degree D have been gathered together (the parameter expansion has not yet been done). The difference from (3.3) is that in (3.6) T_D is known at every stage in terms of solutions of lower degree. Hence (3.6) is a linear inhomogeneous equation, whereas (3.3) is nonlinear.

The condition that (3.6) is soluble is that vectors $z^{(i)}$ that are orthogonal to the left side of (3.6), should also be orthogonal to the right side. To make this meaningful, we need to define a scalar product for the space in which these vectors live. They depend on A and they are in a space spanned by monomials in the components of A. There is an analogy to the mathematics used in the elementary quantum mechanics of spinning particles (see §5), but there is no Pauli principle to contend with..

Once a scalar product is given, an adjoint operator, $\mathcal{L}_M^\dagger$, can be defined. The statement that $z^{(i)}$ is orthogonal to the left side of (3.6) is equivalent to

$$\mathcal{L}_M^\dagger z^{(i)} = 0. \qquad (3.7)$$

From (3.7) we can find the $z^{(i)}$ from a knowledge of only the linear operator. The constraint on g_D is that it should not be orthogonal to any of the $z^{(i)}$. This is not a unique prescription, and there is a gauge freedom. We may add to g a gauge that is any linear combination of nonnull proper vectors of the adjoint operator. Different choices of gauge give different Ψ, but the observed outcome is unaffected. We will see what the standard gauge choices have been in the example in §5. First, let us examine an important example of bifurcation theory.

4. Overstability

Supppose we have a sequence of hydrostatic solutions of the equations of stellar structure. Each member of the sequence is characterized by its luminosity, λ. In the linear stability theory for these models the modes have time dependence exp(st). We assume that all the modes are damped except one pair, $\phi(x)$, $\bar{\phi}(x)$. For this pair, $s = \pm i\omega + \mu$. We assume that μ crosses through zero as λ goes through a critical value, λ_0. Bifurcation theory permits us to calculate the nonlinear development of this overstability for λ near to λ_0.

For this illustration we assume there is no degeneracy. The normal modes satisfy the equations

$$\mathcal{M}\phi = (\mu+i\omega)\phi, \quad \mathcal{M}\bar{\phi} = (\mu-i\omega)\bar{\phi}, \quad (4.1)$$

where the overbar means complex conjugate. Then we look for solutions of (2.1) of the form

$$\mathbf{U}(x,t) = \alpha\phi + \bar{\alpha}\bar{\phi} + \mathbf{W} \quad (4.2)$$

where $\mathbf{W}$ is a linear combination of all the other modes, which are stable. We concluded in §2 that the time dependence of $\mathbf{W}$ is given entirely in terms of α and $\bar{\alpha}$, provided that $|\mu|$ is small compared to the decay rate of any of the of the stable modes. The amplitude equation for α has the form

$$\dot{\alpha} = (\mu+i\omega)\alpha + \Gamma(\alpha,\bar{\alpha}), \quad (4.3)$$

where $\bar{\alpha}$ satisfies the complex conjugate of (4.3). In the general discussion of amplitude equations we wrote $\boldsymbol{\alpha}$ for the amplitude vector; here that corresponds to a vector with α and $\bar{\alpha}$ as components. If we use that form, in (2.13) we would have the matrix $\mathbf{M} = i\omega\begin{pmatrix}1 & 0\\ 0 & -1\end{pmatrix} + \mu\begin{pmatrix}1 & 0\\ 0 & 1\end{pmatrix}$.

Without trying to decide what Γ is like, we immediately move to simplify (4.3) by the nonlinear transformation

$$\alpha = A + \Psi(A,\bar{A}) \quad (4.4)$$

so that A satisfies

$$\dot{A} = (\mu+i\omega)A + g(A,\bar{A}). \quad (4.5)$$

The equation for Ψ is found by substitution into (4.3) to be

$$\mathcal{L}\Psi = T - g \quad (4.6)$$

where the linear operator is

$$\mathcal{L} = i\omega(A\partial_A - \bar{A}\partial_{\bar{A}}) + \mu(A\partial_A + \bar{A}\partial_{\bar{A}}) - (\mu+i\omega) \quad (4.7)$$

and the nonlinear term has been abbreviated by

$$T(A,\bar{A}) = \Gamma(A+\Psi,\bar{A}+\bar{\Psi}) - g\partial_A\Psi - \bar{g}\partial_{\bar{A}}\Psi. \qquad (4.8)$$

Let us suppose that the various functions of A and $\bar{A}$ that arise can be expanded in Taylor series, that is, as linear combinations of monomials that are conveniently abbreviated as

$$|D\,K\rangle = A^{(D-K)}\,\bar{A}^{K}. \qquad (4.9)$$

The $|D\,K\rangle$ form a basis on the amplitude space in which g, T and Ψ live. These basis vectors are the proper vectors of $\mathcal{L}$:

$$\mathcal{L}|D\,K\rangle = \Lambda_{DK}|D\,K\rangle \qquad (4.10)$$

where

$$\Lambda_{DK} = i\omega(D-2K-1) + \mu(D-1). \qquad (4.11)$$

Let us introduce the Taylor series

$$\Psi = \sum_{D=0}^{\infty}\sum_{K=0}^{D}\Psi_{DK}|D\,K\rangle \;,\quad T = \sum_{D=0}^{\infty}\sum_{K=0}^{D}T_{DK}|D\,K\rangle \;,$$

$$g = \sum_{D=0}^{\infty}\sum_{K=0}^{D}g_{DK}|D\,K\rangle . \qquad (4.12)$$

Then (4.6) becomes

$$\Lambda_{DK}\Psi_{DK} = T_{DK} - g_{DK}, \qquad (4.13)$$

for $D \geq 2$. From (4.8) we see that $T_{2K} = 0$. Therefore we can solve (4.13) when $D = 2$, once we choose g_{2K}. At every new D, thereafter, T_D is known, if we solve sequentially.

If $\mu \neq 0$, then $\Lambda_{DK} \neq 0$, and we are able to solve (4.13) for any choice of g_{DK}. We can even choose $g_{DK} = 0$, but of course there is a catch. The solution is then $\Psi_{DK} = T_{DK}/\Lambda_{DK}$, and some of the Λ_{DK} vanish when $\mu = 0$. Even though the solution with $g = 0$ is valid at a particular λ, it is not uniformly valid when we start changing λ in the neighborhood of λ_0. Yet, this is just where we would have hoped the theory was best.

The way to get a uniformly valid amplitude equation is to take care of the singularity at λ_0. From (4.11), we see that when $\mu = 0$, Λ_{DK} vanishes when $K = \frac{1}{2}(D-1)$. Since K is an integer, Λ_{DK} vanishes only for odd D. With $D = 2L+1$, we have

K = L and $\Lambda_{L+1,L} = 0$, for L = 1, 2, 3, ... To avoid the small denominator $\Lambda_{L+1,L}$ near λ_0, we require that $g_{L+1,L} = T_{L+1,L}$ for integral L. All the other g_{DK} may be set equal to zero. The amplitude equation in this case is

$$\dot{A} = (\mu + i\omega)A + k_1 |A|^2 A + k_2 |A|^4 A + \ldots \quad (4.14)$$

where $k_L = T_{L+1,L}$. If we let

$$A = Re^{i\theta}, \quad (4.15)$$

then the Landau-Hopf equation (4.14) becomes

$$\dot{R} = (\mu + \alpha_1 R^2 + \alpha_2 R^4 + \ldots)R$$
$$\dot{\theta} = \omega + \beta_1 R^2 + \beta_2 R^4 + \ldots \quad (4.16a,b)$$

where $k_L = \alpha_L + i\beta_L$. The interpretation is that the nonlinear terms in the amplitude equation have renormalized the frequency and the growth rate. If the coefficients have unit order of magnitude and α_1 and μ have opposite signs, there are steady nontrivial solutions with $R \simeq \sqrt{(\mu/\alpha_1)}$. In that case, it may not be worth carrying higher terms than the first nonlinear corrections. (There may be other steady solutions to (4.15b), but they are of doubtful validity.) The principal solution, the long time limit of all initial conditions for $\mu > 0$ and $\alpha_1 < 0$, is a uniform circular motion. The circle is an attractor called a limit cycle.

5. Multiple Instabilities

In the simplest bifurcations, a single mode becomes unstable as a parameter is varied. In overstability, two modes simultaneously become unstable, but they form a complex conjugate pair whose joint stability is governed by a single parameter. Overstability becomes more complicated when, by varying a second parameter, we are able to tune the frequency to zero. This is an example of the degeneracy that we examine next. In this case, at a point in parameter space where both growth rate and frequency vanish, bifurcation from equilibrium into a relaxation oscillation is possible.

Suppose that the linear theory of (2.1) has a set of ℓ independent control parameters $\boldsymbol{\lambda} = (\lambda_1, \lambda_2, \ldots, \lambda_\ell)$. As in the theory of overstability, we can find a transformation that linearizes the amplitude equation at a generic value of $\boldsymbol{\lambda}$. But if we approach an exceptional point $\boldsymbol{\lambda}_0$, where bifurcation occurs, the linearizing transformation diverges. It is better to seek out the points of bifurcation and deal with the associated singularities straightaway. This is a standard strategy, like that used in getting solutions of differential equations by expanding about the singular points.

Assume that, in some neighborhood of parameter space, the characteristic value equation (2.5), $\mathcal{F}(s;\lambda;\mathbf{k}) = 0$, has ℓ roots with $|\text{Re} s|$ small. These are the slow modes and all the other solutions lie well to the left of the imaginary axis in the s-plane. Let this neighborhood contain a point λ_0 such that when $\lambda \to \lambda_0$, $\mathcal{F}(s;\lambda;\mathbf{k})$ approaches $s^m \mathcal{G}(s;\lambda_0;\mathbf{k})$, where the roots of $\mathcal{G} = 0$ have distinctly negative real parts. The ratio

$$\mathbf{P} = \mathcal{F}/\mathcal{G}, \qquad (5.1)$$

for λ near λ_0, is well approximated by a polynomial of degree ℓ that figures in the theory (Coullet and Spiegel, 1983). The amplitude equation should not lose the knowledge of the growth rates of the important slow modes. If we arrange that

$$\det(\mathbf{M} - s\mathbf{I}) = \mathbf{P}(s;\lambda;\mathbf{k}), \qquad (5.2)$$

the linear theory of the amplitude equation will be that of the slow modes.

At λ_0, where $\mathbf{P}$ reduces to s^ℓ, we denote the linear operator by $\mathcal{M}_0$. An important technical point is illustrated by the situation where the equation

$$\mathcal{M}_0 \phi = 0 \qquad (5.3)$$

has only one solution even though $s = 0$ occurs with multiplicity $\ell > 1$. We need more vectors to complete a basis in the function space of $\mathbf{U}$ (Friedman, 1956).

We return to the linear problem

$$\partial_t \mathbf{U} = \mathcal{M} \mathbf{U}. \qquad (2.2)$$

For $\lambda = \lambda_0$, we look for a solution that is a linear combination of terms of the form $t^{\ell-k} \psi_k(\mathbf{x})$ where k varies from 1 to ℓ. The ψ_k are going to be the ℓ basis vectors associated to the characteristic value 0 at λ_0; they are called generalized null vectors. The linear combination works provided that

$$\mathcal{M}_0 \psi_k = \mu_{k-1} \psi_{k-1}, \quad k = 1, 2, \ldots, \ell, \qquad (5.4)$$

where, for agreement with (5.3), ψ_1 has to be ϕ and μ_0 has to vanish. The remaining $\ell-1$ constants, μ_k, are arbitrary since (2.2) is linear. Usually, the μ_k are taken to be unity, but I prefer to choose

$$\mu_k = [k(\ell - k)]^{1/2}. \qquad (5.5)$$

If we use the normal modes as a basis for **U** when $\lambda \neq \lambda_0$, the matrix representation of $\mathcal{M}$ is diagonal. In this example, when we go to λ_0, the slow modes collapse onto the single null vector ψ_1 and the diagonal representation breaks down, as at a coordinate singularity in relativity. It is well to select a basis consisting of functions that turn into the ψ_k as $\lambda \to \lambda_0$. These may involve the slow normal modes, but will not be identical to them. Then we may proceed as in the general discussion to derive the amplitude equations. If we do not wish to choose basis functions yet, we can still carry out the formal developments of §§2 and 3. Alternatively, we may look at the problem in another way (Coullet and Spiegel, 1983) that brings out its structure.

We are looking for a solution **U**(**x**,t) of equation (2.1). The general arguments suggest that, for this problem, when λ is near to λ_0, the time dependence of **U** is carried by a set of functions $\mathbf{A} = (A_1, A_2, \ldots, A_\ell)$. We make this explicit with the ansatz, à la Bogoliubov,

$$\mathbf{U}(\mathbf{x},t) = \mathbf{V}(\mathbf{x}, \mathbf{A}(t)) \tag{5.6}$$

where **A** satisfies the amplitude equation

$$\dot{\mathbf{A}} = \mathbf{M}\mathbf{A} + \mathbf{g}(\mathbf{A}). \tag{3.2}$$

We have to specify **M** and to choose **g**. Any **M** that satisfies (5.2) will serve and we want to pick the simplest allowed **g**.

With $\partial_t \mathbf{U} = \dot{\mathbf{A}} \cdot \partial_\mathbf{A} \mathbf{V}$, and on account of (3.2), (2.1) becomes

$$\mathcal{L}\mathbf{V} = \mathcal{N}(\mathbf{V}) - \mathbf{g}\cdot\partial_\mathbf{A} \tag{5.7}$$

where

$$\mathcal{L} = M_{ij} A_j \partial_{A_i} - \mathcal{M}, \tag{5.8}$$

with summation convention on repeated indices. In $\mathcal{L}$ there is an $\mathcal{M}$ where there was an **M** in (3.3) because we have not expanded in normal modes.

We can develop in Taylor series in the components of **A**. The linear terms give

$$\mathcal{L}\mathbf{V}_1 = \mathbf{0}. \tag{5.9}$$

Since $\mathbf{V}_1$ is linear in A, we write

$$\mathbf{V}_1 = A_i \phi_i. \tag{5.10}$$

This is a solution of (5.9) provided that

$$\mathcal{M}\phi_i = M_{ki}\phi_k. \tag{5.11}$$

Hence $\mathbf{M}^\dagger$ is the matrix representation of $\mathcal{M}$ on the slow modes

ϕ_i, as those who pursued the details of the discussion in §2 found.

At λ_0, $\mathbf{M}^\dagger$ becomes $\mathbf{M}_0^\dagger$, a matrix with zeros everywhere except on the diagonal just below the principal one, as we see from (5.4). To specify $\mathbf{M}$ and its transpose, we can use the standard forms given by Arnold (1983; Gilmore, 1981) or we can use perturbation theory on (5.11). The aim, in either case, is to get normal forms for the matrix representation of $\mathcal{M}$. When we expand in $(\lambda - \lambda_0)$, the leading order term is (5.11) evaluated at λ_0, that is, (4.4). In the next order,

$$\mathcal{M}_0 \delta\phi_i - [M_0]_{ji} \delta\phi_j = \delta\mathcal{M}_0 \psi_i - \delta[M_0]_{ji} \psi_j. \tag{5.12}$$

The operators on the left side are known and their adjoints are found by standard procedures such as partial integration and transposition. The homogeneous adjoint equation is

$$\left[\mathcal{M}_0^\dagger - \mathbf{M}_0\right] \psi^\dagger = \mathbf{0}. \tag{5.13}$$

Since $\psi^\dagger$ is orthogonal to the left side of (5.12), it is orthogonal to its right side too. This condition permits us to fix the allowed choices of $\delta\mathbf{M}$ and so we know $\mathbf{M}$ to $O(\delta\lambda^2)$. For example, for $\ell = 2$, $\mathbf{M}_0 = \begin{pmatrix} 0 & 1 \\ 0 & 0 \end{pmatrix}$ and we have the critical polynomial

$$s^2 - \beta s - \alpha = 0. \tag{5.14}$$

The form proposed by Arnold is

$$\mathbf{M} = \begin{pmatrix} 0 & 1 \\ \alpha & \beta \end{pmatrix} \tag{5.15}$$

Now we have a suitable set of slowly evolving functions, and we can make expansions in them. We get $\mathbf{V} = \mathbf{A} \cdot \boldsymbol{\phi} + \mathbf{Y} \cdot \boldsymbol{\phi} + \mathbf{B} \cdot \mathbf{f}$, where the coefficients $\mathbf{Y}$ and $\mathbf{B}$ are nonlinear functions of $\mathbf{A}$. We put this expansion into (5.7), to recover (3.3),

$$\mathcal{L}_\mathbf{M} \cdot \mathbf{Y} = \mathbf{T} - \mathbf{g}, \tag{3.3}$$

where the symbols are defined in (3.4) and (3.5). We can as well work with this equation as with (5.7). As for overstability, the dangerous place is at λ_0. So we expand about λ_0, and we examine only the leading order here.

With the recovery of (3.3), we rejoin the general development and the normal form theory (Arnold, 1983). To make the formalism resemble one that many will recognize we expand the functions in amplitude space in Taylor series. This is equivalent to using momomials in the components of $\mathbf{A}$ as basis vectors. We designate a basis function this way:

$$|\xi\rangle = N_\xi A_1^{\xi_1} A_2^{\xi_2} \ldots, \tag{5.16}$$

where $\xi = (\xi_1, \xi_2, \ldots, \xi_\ell)$ and the normalization constant N_ξ is arbitrary. We define a multiplicity parameter j by

$$\ell = 2j + 1. \tag{5.17}$$

In the case of overstability, we introduced the degree D. To give the reductions a more familiar appearance for those who have studied atomic physics, we use instead the quantity

$$J = jD = j(\xi_1 + \xi_2 + \ldots + \xi_\ell), \tag{5.18}$$

Naturally, we also introduce

$$M = \sum_{k=0}^{2j+1} (-j+k-1)\xi_k. \tag{5.19}$$

These definitions are calling on a resemblance of the normal form calculations to the determination of spin eigenfunctions in quantum mechanics (Arneodo, Coullet and Spiegel, 1984). An underlying reason for the similarity is that $\mathbf{M}_0$, as given by (5.4) and (5.5), is the matrix representation of the operator known as $\mathbf{J}_+$ in quantum mechanics. Accordingly, we let

$$\mathbf{M}_1 = \tfrac{1}{2}(\mathbf{M}_0 + \mathbf{M}_0^\dagger), \quad \mathbf{M}_2 = -\tfrac{i}{2}(\mathbf{M}_0 - \mathbf{M}_0^\dagger), \quad \mathbf{M}_3 = \tfrac{1}{2}[\mathbf{M}_0, \mathbf{M}_0^\dagger]. \tag{5.20}$$

We may verify that

$$[\mathbf{M}_i, \mathbf{M}_j] = i\varepsilon_{ijk}\mathbf{M}_k. \tag{5.21}$$

where i,j,k run through 1,2,3, and ε_{ijk} is the permutation symbol. The same algebra applies to $\mathcal{L}_{\mathbf{M}_0}$ because

$$\left[\mathcal{L}_\mathbf{K}, \mathcal{L}_\mathbf{L}\right] = \mathcal{L}_{[\mathbf{L},\mathbf{K}]}. \tag{5.22}$$

I have underlined the similarity to calculations found in books on quantum mechanics, so that those already familiar with that subject can see at once how such calculations go. Rather than give the general theory here, I shall describe an example. To get started, we perform the Taylor expansion and gather together all terms of like degree, or J. This gives us (3.6) essentially, which to leading order in $\lambda - \lambda_0$, is

$$\mathcal{L}_{\mathbf{M}_0} \cdot \mathbf{T}_J = \mathbf{T}_J - \mathbf{g}_J, \tag{5.23}$$

where the subscript naught has been left off the variables but put on $\mathbf{M}$ where it matters most. Let us write

$$\mathcal{L}_{\mathbf{M}_0} = \mathbf{D} - \mathbf{M}_0, \quad \mathbf{D} = [\mathbf{M}_0]_{ij} A_i \partial_{A_j}. \tag{5.24}$$

For $j = \tfrac{1}{2}$ ($\ell = 2$), we have $\mathbf{M}_0 = \bigl(\begin{smallmatrix}0&1\\0&0\end{smallmatrix}\bigr)$ and we shall refer to the two components of **A** as A and B. Then

$$\mathbf{D} = B \partial_A. \quad (5.25)$$

There are two components of ξ, so we can also rename the monomials: $|J,M\rangle = |\xi\rangle$. If we also adopt a convenient normalization, we have

$$|J,M\rangle = N A^{J-M} B^{J+M}, \quad N = [(J+M)!(J-M)!]^{-1/2}. \quad (5.26)$$

We see that

$$\mathbf{D}|J,M\rangle = [J(J+1)-M(M+1)]^{1/2} |J,M+1\rangle. \quad (5.27)$$

If, in (5.5), we identify j with J and k with J-M, we find that the normalization in (5.27) is μ_{J-M}. The matrix representation of **D** in this basis is therefore a block like $\mathbf{M}_0$ but for the appropriate J. This simple statement holds for $j = \frac{1}{2}$. (For higher j, the matrix representation of **D** is block diagonal with blocks $\mathbf{M}_0$ of various sizes along the principal diagonal.) The equivalence between the matrix operators in spin space and certain differential operators is well-known (Edmonds, 1957). The difference from quantum mechanics is that terms corresponding to antisymmetric functions do not arise.

The matrix representation of $\mathbf{D}^\dagger$ is the transpose of that of **D**, hence it is $\mathbf{M}_0^\dagger$ for j = J. Therefore

$$\mathbf{D}^\dagger |J,M\rangle = [J(J+1)-M(M-1)]^{1/2} |J,M-1\rangle. \quad (5.28)$$

We verify readily that this is the case if

$$\mathbf{D}^\dagger = A \partial_B. \quad (5.29)$$

We find that

$$\mathbf{D}^\dagger |J,-J\rangle = 0$$
$$\mathbf{D}^\dagger |J,-J+1\rangle = (2J)^{1/2} |J,J\rangle. \quad (5.30)$$

We have $\mathbf{M}_0^\dagger = \begin{pmatrix} 0 & 0 \\ 1 & 0 \end{pmatrix}$, so $\mathbf{M}_0^\dagger \begin{pmatrix} 1 \\ 0 \end{pmatrix} = \begin{pmatrix} 0 \\ 1 \end{pmatrix}$ and $\mathbf{M}_0^\dagger \begin{pmatrix} 0 \\ 1 \end{pmatrix} = 0$. So we readily conclude that

$$\mathcal{L}_{\mathbf{M}_0}^\dagger [A^{2J} \begin{pmatrix} 0 \\ 1 \end{pmatrix}] = 0.$$

$$\mathcal{L}_{\mathbf{M}_0}^\dagger [A^{2J} \begin{pmatrix} 1 \\ 0 \end{pmatrix} + BA^{2J-1} \begin{pmatrix} 0 \\ 1 \end{pmatrix}] = 0. \quad (5.31)$$

For each J, $\mathcal{L}_{\mathbf{M}_0}^\dagger$ has two null vectors and 2J-1 additional generalized null vectors. If we expand $\mathbf{g}_J$ in these vectors, only the components on the two null vectors are determined by the solvability conditions. We can set all the other components equal to zero. The amplitude equation at λ_0 is

$$\dot{A} = B + \sum_{J=1}^{\infty} k_{J2} A^{2J}$$
$$\dot{B} = \sum_{J=1}^{\infty} k_{J1} A^{2J} + \sum_{J=1}^{\infty} k_{J2} A^{2J-1} B, \qquad (5.32)$$

where the k_{JI} are constants that are determined by the solvability condition at each step of the calculation.

We can transform the nonlinear terms to a more conventional form. There is a gauge freedom in the choice of the nonlinear terms: we can add any linear combination of generalized null vectors to $\mathbf{g}_J$. Since

$$\mathcal{L}_{\mathbb{M}_0}^\dagger [A^2 \begin{pmatrix}1\\0\end{pmatrix}] = - A^2 \begin{pmatrix}0\\1\end{pmatrix}, \qquad (5.33)$$

if we add $-k_{J2} A^2 \begin{pmatrix}1\\0\end{pmatrix}$ to $\mathbf{g}_J$, we eliminate the nonlinearity from the first of equations (5.32). The normal form is then one first found by Bogdanov (1975; see also Arnold, 1983; Coullet and Spiegel, 1983). If we replace $\mathbb{M}_0$ by $\mathbb{M}$ from (5.15) this is

$$\dot{A} = B$$
$$\dot{B} = \alpha A + \beta B + f(A) + Bh(A), \qquad (5.34)$$

where f and h are functions whose Taylor expansions we can calculate to any desired order.

We can write (5.34) as a single equation for A:

$$\ddot{A} - [\beta + h(A)] \dot{A} - \alpha A - f(A) = 0. \qquad (5.35)$$

Since f and h are nonlinear functions, $f(0) = 0$. Hence, $A = 0$ is an equilibrium solution, or fixed point, of the flow in the (A,B) plane. When α and β are both negative, this is a stable fixed point, but when β increases through zero for α fixed and negative, the equilibrium becomes overstable, or vibrationally unstable. A new attractor forms that is a periodic orbit called a limit cycle. At first, this limit cycle is a nice oval, but as β is increased a corner forms, where the motion is slow. As a function of t, A becomes increasingly anharmonic and a relaxation oscillation develops. Normally, this is thought of as a highly nonlinear phenomenon. It occurs for large enough β. But the value of β for which the relaxation oscillation appears may be made as small as we like by a suitable choice of α. This is an aspect of bifurcation theory with several control several parameters (Arnold, 1984). If we go on to the the case where three parameters are involved, we may have chaos as close as we want to the onset of instability (Arneodo, et al, 1985). That is what is illustrated in the model of §1.

6. Qualitative Approaches

Parameter space is like a blue cheese. The veins are surfaces on which a single mode is marginally stable, if we consider a complex conjugate pair of modes as a single complex mode. Where two veins cross, two modes are simultaneously marginal and we have the example, discussed in §5, of competing instabilities (Spiegel, 1972). We may have any multiplicity of crossings and, in neighborhoods of these, the methods that I described in §§2-5 may be used to derive amplitude equations. Many of the equations discussed in chaos theory were not originally derived in that way, but rather by physical or intuitive arguments. One advantage of the formal methodology is that it can guide the qualitative arguments even when the full formalism is not used.

When only one mode, $\phi(x)$, is near to marginality, we can describe the solution by $A(t)\phi(x)$ plus terms of higher order in the amplitude, A. If initially A is infinitesimal, for a while at least, it will vary like exp(st) where s depends on the parameters, λ, and $\text{Re}\, s = 0$ on $\lambda = \lambda_0$. The equation for A is the Landau-Hopf equation $\dot{A} = sA +$ nonlinear terms in A. The nonlinear terms are A^2, A^3, and so on. There is no coordinate transformation that removes these nonlinearities, though some may have vanishing coefficients because of a symmetry in the problem. The actual calculation of the coefficents in the leading terms took some time in the theory of hydrodynamical instability (Drazin and Reid, 1981).

When two modes $\phi(x)$ and $\psi(x)$ are marginal at λ_0, we approximate the solution by $A(t)\phi(x)+B(t)\psi(x)$ plus nonlinear terms in A and B. We let $\mathbf{A} = (A,B)$ and write the amplitude equation $\dot{\mathbf{A}} = \mathbf{M}\mathbf{A} +$ nonlinear terms. The matrix representation $\mathbf{M}$ satisfies $\det(\mathbf{M}-s\mathbf{I}) = s^2-\alpha s-\beta \equiv P(s)$ and the amplitude equation can be written as $(D_t^2 - \alpha D_t - \beta)A =$ nonlinear terms. The nonlinear terms in this equation are A^2, $A\dot{A}$, $\dot{A}^2$, A^3, $A^2\dot{A}$, and so on. We have seen how to calculate their coefficients, but there are other approximations that may be of use. Very near to the marginal situation, the growth rates are very small and so are the coefficients α and β, which vanish at λ_0. This means that time derivatives are small, and with a suitable scaling, $|\dot{A}| \ll |A|$. We obtain asymptotic normal forms (Arneodo, et al, 1985), which can be surmised from the linear theory.

The amplitude equation (5.35) is only one of several examples of two competing instabilities (Coullet and Spiegel, 1983); it is the case where the critical polynomial does not factor in real terms. Then there is only one null vector of the linear operator at λ_0. When there are more competing instabilities, there are correspondingly more possibilities. The Lorenz (1963) system of equations is an example where the cubic critical polynomial can be written as $(s-\mu)(s^2-\alpha s-\beta)$, where the parameters are all real. Lorenz' equations were derived by an expansion in normal modes of the Boussinesq equations for two-dimensional convection with slippery boundaries. Truncation of this expansion to three modes gives a third-order system of differential equations. The Lorenz system has chaotic (numerical) solutions (Sparrow, 1982) and this is suggestive about fluid flows. The nature of the derivation is such that the system ought not to be applicable to the convection problem when the chaos appears. Yet, accurate numerical simulations on the full Boussinesq equations for this problem do reveal chaos (McLaughlin and Orszag, 1982).

In the case of doubly diffusive convection, such as magnetoconvection, thermohaline convection, and semiconvection (Spiegel, 1972)) another third order system was derived, this time by physical arguments (Moore and Spiegel, 1966; a special case is the model described in §1). This system also predicts chaos, especially near to the joint onset of thermoconvective instability (Defouw, 1970) with the usual two modes of doubly diffusive convection. On the basis of this model D.W. Moore and I suggested that doubly diffusive convection would be chaotic. The justification was weak here too and our prediction was not widely credited, not even by the group that confirmed it numerically (D.R. Moore, et al, 1983).

The moral is not new one. It is often true that crude derivations lead quickly to results that we find interesting, but we are not sure whether to accept them. We may be guided by such results, but must maintain our awareness that we are on dangerous ground. Landau's derivation of his amplitude equation was once not considered very rigorous by some, but with hindsight, we are not surprised to confirm that he saw clearly into the problem. Now we have a variety of techniques that lead us to acceptable amplitude equations (Coullet and Spiegel, 1983), but they have validity criteria that, when strictly applied, do not give us much scope for using the results. With asymptotic results it is often not bad to apply them liberally, but I think that it remains desirable to lean heavily on the qualitative approach for many astrophysical purposes, especially when it may be enriched by an understanding of normal form and center manifold theory. Let me describe some examples in which I have been interested.

Stellar Pulsation

When Moore and I devised our model, we were thinking of solar oscillations. With Baker, we went on to use the same kind of model to describe radial stellar pulsations (Baker, Moore and Spiegel, 1966). We used a one-zone model of stellar pulsation (Baker, 1966) that gave a third-order system of equations, with a behavior like that we have already seen in §1. But the new third-order system is nastier looking than the ones we have looked at; the term involving the third derivative is $A^3\ddot{A}$. We can simplify such a system by normal form theory but at the cost of omitting some of the real complications of stellar opacities. It is possible to use a combination of Padé regrouping with normal form theory to get closer to reality, but for the present I think that the method of hands-on physical arguments is best. It does not seem warranted to perform hard calculations like those needed to get the coefficients in the normal forms until we have made a better contact with observed phenomena.

There are several theoretical discussions leading to chaotic pulsation models (for example, Buchler and Perdang, 1979; Auvergne, Baglin and Morel, 1981). We next have to see whether a qualitative connection to observations exists. I believe that a promising class of stellar objects to examine for chaotic pulsation is the ZZ Ceti variable (see the papers in VanHorn and Weidemann (1979) for both theoretical and observational aspects). These are pulsating white dwarfs whose erratic and intermittent light curves seem to have the right mixture of order and disorder. Their periods of about ten minutes makes them good to study observationally and the presence of at least two instability mechanisms is encouraging on the theoretical side.

Other objects among the many stars in Perdang's article in this volume or in Auvergne's (1983) discussion may be chaotic. These ought to be scrutinized in turn. I would like to add the quasar 3C273 to these lists of candidates for chaotic behavior. This is a difficult object both observationally, on account of its long time scale, and because we know so little about its physical nature. Its luminosity varies on a time scale of about 13 years. Press (1978) has reviewed the debate over whether this object is periodic or not and proposes that its variation is an example of 1/f noise. The fact that it looks periodic to some and random to others leads me to think of a strange attractor. If there is a supermassive object at work, it hovers near to $\gamma = \frac{4}{3}$ and this is a favorable condition for slow modes, as students of pulsation theory realize. Rick White and I tried to make a third-order model, but the physics of these objects is in such doubt that we did not push the project to completion. Here is certainly a case for further phenomenological modelling.

Solar Cycle

The solar activity cycle is an example of a repetition of the same thing over and over again, but not in exactly the same way two times in a row. The motion on the strange attractor in Fig. 5 has just this property and it is natural to imagine that a strange attractor has something to do with the solar cycle (Spiegel, 1977). The paucity of sunspots during the time of Newton is further evidence supporting this notion, since intermittency is a familiar feature of chaos. In chaos theory, there are several qualities of intermittency (Pomeau, 1983), of which the most pronounced reminds one of turbulence. In flows at very high Reynolds numbers there are patches of intense turbulence interspersed with reasonably laminar flow. In the solar cycle, the occurence of intermissions, or minima as Eddy (1976) call them, reminds me of the strong intermittency of fully developed turbulence. The Maunder intermission is a near cessation of the solar activity cycle for about six or seven cycles.

Gudzenko and Chertoprud (1978, and references cited there) modelled the solar cyclic activity by a second-order system that produced limit cycles, with the idea that the aperiodicity is the result of a noise source external to the cyclic process. The limit cycle that they fit to the solar data is given in the commentary by Ruzmaikin (1981), who proposes the Lorenz attractor as a possible description of the solar dynamo cycle, since it has already been used to describe a disk dynamo (Malkus, 1972). The aperiodicity is there, but it is hard to find anything like the Maunder intermission. Fifth-order systems, such as the dynamo model described by Childress and Fautrell, 1982) can be richer in this kind of behavior, though another fifth-order model (Cattaneo, Jones, and Weiss, 1984) has shown no episodes like the Maunder intermission.

I would like to indicate how an understanding of bifurcaton theory permits you to express certain involved physical ideas about a phenomenon like the solar cycle rather simply. The model I want to describe actually grew out of discussions with D.W. Moore some years ago. Like everyone else who first comes on chaotic behavior, we were excited by the prospects for explaining a number of phenomena that seemed hard to comprehend, including of course the solar cycle. But when Eddy (1976) made it clear that there were real intermissions in solar activity, our simple model seemed too simple. We were led to invoke more competing processes, but were not satisfied with the physical processes we had chosen.

Later, Weiss and I (1980) proposed a simple vision of the solar activity cycle that is more defensible and that can give rise to similar dynamical equations. We were worried about the problem of building and storing strong magnetic fields in the solar convection zone. We proposed that the strong fields are formed immediately <u>below</u> the convective zone, from which they escape cyclically through an instability, probably involving magnetic buoyancy (Parker, 1979). The dynamics of the magnetic layer below the normal convective zone on this view provides an example of competing instabilities like those in doubly-diffusive convection. We can expect the instability to produce a Landau, Hopf or Bogdanov bifurcation (Childress and Spiegel, 1980). Suppose that it is the last and richest of the three so that the critical matrix is

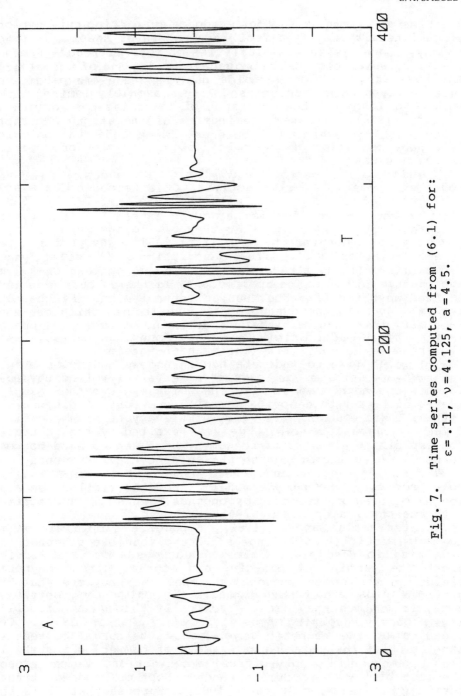

Fig. 7. Time series computed from (6.1) for: $\varepsilon = .11$, $\nu = 4.125$, $a = 4.5$.

$\mathbf{M}_0 = \begin{pmatrix} 0 & 1 \\ 0 & 0 \end{pmatrix}$ as in §5. Suppose also that this instability is fed by a simple dynamo process in the convection zone, describable by a Lorenz system: $\mathbf{M}_b = \begin{pmatrix} 0 & 0 & 0 \\ 0 & 0 & 0 \\ 0 & 0 & 0 \end{pmatrix}$. The five by five matrix of the full system will have two blocks $\begin{pmatrix} 0 & 1 \\ 0 & 0 \end{pmatrix}$ and one (0). I will not write the normal form for that case here but will save space by referring to a special case I have already studied (Spiegel, 1980):

$$\dot{A} = B$$
$$\dot{B} = -A^3 - 2AX + ZA - \epsilon\nu B$$
$$\dot{X} = Y \quad (6.1)$$
$$\dot{Y} = -X^3 - A^2 + ZX - \epsilon\nu Y$$
$$\dot{Z} = -\epsilon[Z + a(A^2 + X^2 - 1)].$$

Notice that if $A = B = 0$, we recover the third-order Lorenz system in a form favored by people doing asymptotics (Robbins, 1979, Marzec and Spiegel, 1980). For a wide range of parameter values, the system seems to go nearly into this Lorenzian subspace, just as a relaxation oscillator follows a slow manifold. The example of Fig. 7 shows mild chaos and pronounced intermissions of moderate length. Though it does not greatly resemble the solar cycle, it has some lessons to teach us.

When we measure the sunspot activity with the Wolf number (a measure of sunpot activity: Izenman, 1985), we use a positive definite quantity. The magnetic field, which is strongly implicated in the process, has polarity. That is why A rather than |A| is plotted versus time in the figure. Most of the time, A changes sign from cycle to cycle, but sometimes it does not. People tend to take it for granted that the solar magnetic polarity flips every eleven years and speak of the cyle time as twenty-two years. The model reminds us that we cannot draw such a conclusion just because we have seen a few reversals. In the model, the analogue of the eleven years is the time taken to go around a lobe of the attractor.

Why should we expect to be able to make useful models of such a simple kind? This is where the methods of analyzing data from the point of view of dynamical system theory show their power. Alan Wolf and I have been studying a set of data provided by J.A. Eddy that contains a list of daily Wolf numbers for the past century. There seem to be several time scales in the data and possibly some of them arise in noise (both solar and observational) extrinsic to the process. We think we may be seeing an attractor in the data when it is reconstructed in the same way that Fig. 6 of §1 was made from the output of (1.3)-(1.5). Methods recently developed for determining the dimensions of strange attractors (Wolf, et al, 1985), when applied to these data, suggest that the dimension of the suspected solar attractor is less than five. The analysis is still in progress, but the implication of this tentative result is that a five-dimensional phase space ought to be sufficient to describe much of the temporal dynamics. This makes modelling such as I have just described seem promising.

Problems of Resolution

In observing sunspots we have no serious spatial resolution problem. Even though we cannot see the other side of the sun, we can look at data in multiples of 28 days to compensate for this lack. The sun is exceptional in astrophysics, where we are usually troubled by problems of spatial resolution. Pulsars provide another special case. The radiation from them is confined to a narrow angle and it may come from a small part of the star. On the other hand we move through the beam quite rapidly because of the rotation of these neutron stars, so we see something of the spatial structure in this way. But the rotation period is not constant. The pulsars are spinning down and the rotation rate has superposed fluctuations known as timing noise (Shapiro and Teukolsky, 1983) whose determination is coupled to the spatio-temporal structure of the emission region. Despite the accuracy of these data, a study of the chaos in them is difficult since both the emission mechanism and the rotation rate could be chaotic. This example is a good one for chaoticists to think about as rotation especially should be accessible to simple modelling.

The model of §1 may be adapted for a first look at this question. Instead of fixing the angular velocity of the hoop, we impose a torque that depends on the rotation rate (Shapiro and Teukolsky, 1983). This time we do not treat the moment of inertia as constant, but allow it vary with the bead's motion on the hoop. This model already gives some interesting fluctuations in its rotation rate about the mean, but they will die down after a time. However, suppose that a pulsar pulsates (for example, because of overstability of its magnetoelastic crust when heated from below; Steiner, 1973). We can model this by an external forcing of the bead's position. In this way we can describe a chaotic spindown process. Further enrichment is possible with the inclusion of internal degrees of freedom and nonlinear coupling between core and mantle.

Another example where the problem of resolution presents itself is the rapid burster, an unusual bursting X-ray source (Lewin and Joss, 1981). One sees distinct bursts of widely different energies and separated by widely different times. There is much more detail, but this was already enough for Celnikier (1977) to propose that the sequence of energies seen in the bursts might be represented by a one-dimensional return map. Oddly enough, there seems to be no report of an attempt to examine the data to see whether the bursting sequence is deterministic. This may in part be due to the difficulty in obtaining data. Thanks to Walter Lewin and Herman Marshall, Alain Arneodo and I did obtain about five days worth of MIT data. Four of the days were sequential and the fifth was a month later.

Arneodo has studied the data in many ways and there are some preliminary things to be reported. If a one-dimensional return map is made of the time intervals between bursts for all the data, it does not look at all deterministic. But if the fifth day from a month later is taken out, the plot resembles a deterministic map. In fact, Arenodo has observed that a similarity to one seen in a chaotic biological system (Hayashi, Ishizuka and Hirakawa, 1983). However, the scatter is larger than we would like and since there exist many more data, it does not seem reasonable to go ahead without including them in the discussion. Nevertheless an interesting feature has already emerged.

The behavior on the fifth day of the data from MIT has a completey different look from those on the other four days. This is perfectly normal in a chaotic source with intermittency. But the plot of the data for this fifth day looks more like a scatter diagram than a deterministic return map. Alternation between deterministic and normal random behavior is not typical of chaotic behavior and one might speculate on many reasons for this. The one we favor is that the onset of nondeterministic behavior is a result of low spatial resolution. Arneodo and I conjecture that there are typically a few sources of bursts on the X-ray star. When there is only one source, we observe a nice return map, only slightly marred by external noise. When there are two or more sources going at once, we see the sort of thing that occurs on the isolated fifth day. We have modeled this, but until there are more data for comparison, we simply offer this as a paradigm for the astrophysical problem of low spatial resolution. The discussion in the next section gives a theroretical scenario that may bear on this question.

7. Modulational Chaos and Solar Waves

Chaos may arise from ordinary differential equations, but they must be of at least third order. To get such equations by formal expansion procedures we assumed that the system had three modes in nearly marginal conditions. The competition among these modes causes chaos arbitrarily closely to the joint onset of the three instabilities (Arneodo, et al, 1985; Arneodo, Coullet and Spiegel, 1985). The volume of parameter space in which we can expect to be near to such special conditions may be small unless we are in a turbulent state or the system is very large. In the latter case, we may get a continuum of solutions of the linear theory so that if one mode is nearly marginal, many are.

The simplest pertinent configuration arises often in astrophysical models; it is a slab of fluid that could represent a convective envelope, an accretion disk or a galactic disk. We idealize by supposing that the system is infinite in the x and y directions and finite in the z-direction. For the sake of discussion, assume that only the gravest modes in z may be unstable. Let these gravest modes have wave vector **k** in the x-y directions. As before, suppose that the modes have time dependence exp(st) and write $s = \sigma(\lambda,k)$ for the gravest modes, to emphasize the dependence on the parameters λ and the wave number $k = |\mathbf{k}|$. We assume that we may find λ_0 and k_0 such that **Re**σ and d**Re**σ/dk both vanish when $\lambda = \lambda_0$ and $|\mathbf{k}| = k_0$. For the other modes, with time dependence exp(st) we have **Re**$s < 0$. So for λ near to λ_0, the grave modes with $|\mathbf{k}|$ near to k_0 have small **Re**σ and all other modes are markedly damped in linear theory.

This is a basic problem in the theory of pattern formation (Wesfried and Zaleski, 1983). To avoid some of its complexities, we suppose that the problem is two-dimensional. For small $|\lambda - \lambda_0|$, we look for solutions of (2.1) of the form $A(x,t)\exp(ik_0 x - i\omega_0 t) + c.c. +$ nonlinear, where $\omega_0 = \mathbf{Im}\sigma(\lambda_0,k_0)$. In this special case we have allowed only a wave travelling to the right, though we cannot always get away with this. When it does work, we get a single equation for the amplitude, or modulation, function $A(x,t)$. In the treatments of this problem, the dependence of A on x and t is usually presumed to be very weak; this means that A really depends on $X = \varepsilon x$ and $T = \varepsilon t$ where $\varepsilon^2 = |\lambda - \lambda_0|$. The equation for A is the time-dependent Ginzburg-Landau equation:

$$\partial_t A = \alpha A + \beta \partial_x^2 A + \gamma |A|^2 A. \qquad (7.1)$$

For real σ, the constants α,β,γ are real and (7.1) is a non-linear diffusion equation (Newell and Whitehead, 1969; Segel, 1969) whose solutions become steady after a time. When there is overstability, the coefficients are complex (Newell, 1974) and we get time dependence. Numerical solutions with periodic boundary conditions, when the domain in x is large enough, are chaotic (Moon, Huerre and Redekopp, 1982). Under appropriate conditions, long-lived solitary waves develop and interact strongly (Bretherton and Spiegel, 1983).

The solutions in Fig. 8 show $|A|^2$ in spacetime; we do not see the actual waves but only the modulational amplitude. In observations we may see only the modulation also, and the solitary waves in the figure would be bursts. But if the resolution is high, we would see each structure of the figure as a series of bursts. All this depends on the parameters of the underlying physical system and the temporal resolution of the observations (assuming poor spatial resolution). A good example to study is provided by solar waves, for the high spatial resolution makes it easier to grasp what is happening.

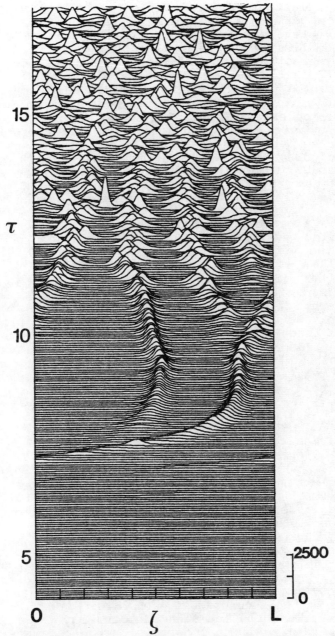

Fig. 8a. Solutions of (7.1) with $|A|^2$ evolving from white noise of low after Bretherton and Spiegel (1983). For $\alpha = 1$, $\gamma = i$, $\beta = 1+100i$ with L as domain size. $\zeta = x/L$ and scaled time is ordinate.

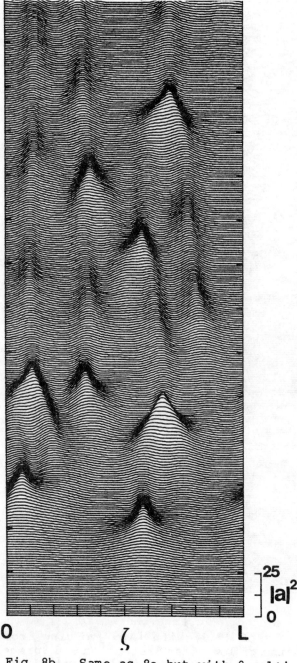

Fig. 8b. Same as 8a but with β = 1+i.

But does the theory apply?

The theoretical problem with the sun is that the outer part, which we see, is turbulent and we cannot say much that is quantitative without using a notion like eddy diffusivity. If we allow this usage, we can treat the convection zone as a quiescent region on which to make the kinds of analysis that we have been discussing. In the past, other aspect of the turbulence have been considered important. The hot tenuous corona that envelops the sun was once thought to be heated by the dissipation of compressible waves (Rutten and Cram, 1981) generated by the turbulence in the solar atmosphere. When the five-minute oscillations of the sun were discovered, it was considered that the predicted waves were seen (Evans and Michard, 1962) and this was puzzling.

The turbulence that was supposed to excite the oscillations is broad band. The oscillations have most of their energy in a fairly narrow band at about five minutes. The spatial coherence of the oscillations suggested a length scale for which the thermal time is about the same as the characteristic time of the oscillation. This made things worse because Stokes (Rayleigh, 1877) had found that the most effective thermal damping of sound occurs when the two times are comparable. I don't know if this puzzle is yet resolved, but after much struggling to get around the seeming dilemma posed by Stokes' work, Moore and I finally realized that it actually provided the explanation, if only the the damping is negative. We proposed (1966) that the acoustic modes were overstable and went on to the kind of model described in §1 and to chaos. The erratic oscillations we found seemed to reflect an ordered aperiodicity that somewhat resembled the observed local solar observations. We were not allowed to be complacent with this superficial agreement for very long. The observations soon were seen to show a pronounced spatio-temporal intermittency that was not accessible to our crude modelling.

In the case of the five-minute modes, there is an effective acoustic wave-guide that limits the vertical extent of the waves (Leibacher and Stein, 1981). This shapes the amplitude distribution of the modes and probably influences their stability characteristics. It also lends some plausiblity to the application of the wave modulation theory picture of the five-minute waves. However, there are several instability mechanisms that may operate in the sun (Unno, et al, 1978). Bretherton and I supposed that one predominates in making the suggestion that the theory implied by (7.1) applies to solar waves. The solutions in Fig. 8 lead to world lines for the solitary wave groups. These depend on the bulk parameters of the model. If we could observe the world lines of the solar wave packets, we could by comparing with the parameter in (7.1), we would have a new way to study properties of the outer solar layers. Observations reveal suitable objects for such a study (Deubner, Ulrich, and Rhodes, 1979).

8. Conclusion

Astrophysicists ought to take advantage of the approaches offered by the developments in dynamical system theory. If a system has an attractor, we can find out a lot about it even when there are temporal gaps in the data, as there usually are in astrophysical observations. The methods of reconstructing attractors from data and studying their properties may be the main techniques that astronomers can profitably borrow from the dynamical systems approach, even though observations rarely provide the quality of data that can be acquired in carefully conceived laboratory experiments.

The issues of data handling are discussed elsewhere in this volume. The question I raise is this: Why should we expect to see systems with low-dimensional attractors? The discussion of §K2-5 was concerned with finding conditions under which we may expect to find simple, low-dimensional dynamics. Laboratory experiments in chaos are designed to favor such conditions, but they cannot reproduce them. And even in the special conditions of a controlled experiment, why should one see Bénard cells despite the hurly burly of the constituent atoms? Even more to the present point: Why can we detect solar oscillations even though the turbulence in the sun must be fierce? Perhaps turbulence renormalizes the system so that it is effectively in the right state to allow simple ordered behavior to emerge.

In the case of laboratory experiments one is trying to determine properties, such as the dimension of an attractor, for a system **S** that is in interaction with an apparatus **A**. In astrophysics, we have observational noise too, but the instruments do not influence **S**. Nevertheless, there is often an equivalent ambient agent, **A**, that interferes with **S** in our studies. The external system **A** usually has a lot of modes that we don't want to think about, so we call it noise. We are dealing with a different problem than the one discussed in §§2-5. If **A** + **S** were all there were in the world, assuming $h = 0$, **A** + **S** would be deterministic. The description of **A** + **S** should probably be Hamiltonian. If we want to describe **A** and **S** separately, each should be ascribed some dissipation to compensate for the truncations. That is why we often imagine **S** and **A** separately as dissipative systems that may have attractors. In the simplest descriptions, we put in the mean interactions through these dissipative terms and look for a simple system, **S**. For example, we use the fluid equations to describe stellar pulsation. If we want to allow for turbulence, we modify the viscosity. If the fluctuating part of the interaction between **A** and **S** is small, we may hope to find an **S** with attractors of low dimension. That is how I interpret the attention given to steady and periodic systems in the past. It is simply that their attractors have dimensions zero and one, respectively.

We can say this with simple equations that may help us to think about these issues, especially if they are explored with numerical experiments. Suppose we make a series of observations on **S** that gives us observed values of the variables $x = (x_1, x_2, \ldots, x_n)$. Imagine that there is a dynamic of **S** that is described by the map

$$x' = f(x) + \varepsilon g(x, X), \qquad (S)$$

where $X = (X_1, X_2, \ldots, X_N)$ are the state variables of **A** and ε is a coupling parameter that measures the degree of influence of **A** on **S**. We suppose that $n \ll N$. When X is a prescribed process and $\varepsilon = 0$, (S) gives us an attractor whose dimension d is rather low. If we try to measure d when $\varepsilon \neq 0$, we would in practice find an increased value. Once $d(\varepsilon)$ gets large, we tend to give up on **S**. As to **A**, its dynamic is

$$X' = F(X) + \delta G(x, X). \qquad (A)$$

This says that **A** may feel the influence of **S**, however slight. We usually ignore this influence but it may be an important way to bring order into a highly turbulent system.

One way to proceed is to give **A** a statistical description and to study **S**. We can also begin to study large systems like **A** to try to extract the δ terms. In part, that is what turbulence theory is about. In the solar convection zone, we might think of **S** as the solar cycle as in §**6** or as the solar oscillations as in §**7**. In both cases the turbulence is **A**. We usually do not much worry about what the oscillations do to the turbulence, but here is a case where we might see some effect of the δ-term. Could the oscillations play a role in determining the granule size? This is the kind of question that astrophysics offers to the subject, besides prompting the images rendered by (A) and (S).

Nowadays people are beginning to see systems with dimensions between two and three as relatively simple. They have attractors like those in §**1**. But almost all the detections occur in specially designed experiments. In astrophysics, where the observations tend to lead the theory, we have no convincing evidence for chaos or strange attractors outside the solar system. Nevertheless, the ideas of dynamical system theory may prove useful in making models, as I tried to illustrate §**6**. But I think that the most important idea that we can take from the developments in dynamical system theory is its general outlook. The discussion of §**2** on center manifold theory shows how simply we can get a dimensional reduction of a large system. This important lesson should lead us to reexamine systems of low dimension and ask why we see them at all.

Why has the universe such a low dimension? Why do the old three-dimensional space of Newtonian dynamics and the four-dimensional Einstein-Minkowsi space-time work so well?

Even the enlarged five-dimensional Kaluza-Klein theory or the recent superstring models with ten or so dimensions do not seem to require very many dimensions. Is our very existence dependent on the fact that we are living in a center manifold of modest dimension? Should that be true, it is unlikely that we can discover the high dimensional phase flow that has squashed us down into our present, probably temporary, low-dimensional state. But the ideas of the general theory give us a way to think about such a situation.

Suppose that that we are in a flow of enormous dimension that went by a fixed point some fifteen billion years ago. In the neighborhood of that point, only a few slow modes exist. So we are now moving out away from that fixed point in a divergent flow temporarily confined to a low dimensional manifold. If we extrapolate our motion back without allowing for the possibility of higher dimensionality earlier on, we might well conclude that something quite drastic occured at the fixed point. If we think ahead, we may anticipate breaking out of the locally invariant manifold into the rigors of a higher dimensionality. Such dreams are difficult to have without the geometrical vision of dynamical system theory. It remains to be seen whether this outlook can be applied on a more useful scale in astrophysics.

Acknowledgement. The support of much of the work that informed this account came from N.S.F. grants to Columbia University. The current one is PHY80-23721. The manner of preparation of this and of many other manuscripts with which I have been involved owes much to the talents and patience of E.T. Scharlemann and J.C. Theys, to whom I am ever grateful.

References

Andronov, A.A., Vitt, E.A. and Khaikin, S.E., Theory of Oscillators, Pergamon (1966); original Russian version (1930), translation published by Princeton Press also, but omits Vitt.

Arneodo, A., Coullet, P.H., Spiegel, E.A., Turbulence and Chaotic Phenomena in Fluids; edited by T. Tatsumi, North-Holland (1984) 215-220.

Arneodo, A., Coullet, P.H., Spiegel, E.A., Geophys. & Astrophys. Fluid Dyn., 31 (1985) 1-48.

Arneodo, A., Coullet, P.H., Spiegel, E.A. and Tresser, C., Physica, 14D (1985) 327-347.

Arneodo, A., Coullet, P. and Tresser, C., J. Stat. Phys. 27 (1982) 171-182.

Arnold, V.I., Geometrical Methods in the Theory of Ordinary Differential Equations, Springer-Verlag, Grundlagen der mathematischen Wissenschaften 250 (1983).

Auvergne, M. in Instabilities Hydrdynamiques et Applications Astrophysiques, A. Baglin, ed., Soc. Francais des Specialités d'Astronomie (1983) 319-333.
Auvergne, M., Baglin, A. and Morel, P.-J., Astron. & Astrophys., 104 (1981) 47-56.
Baker, N.H., in Stellar Evolution, R.F. Stein and A.G.W. Cameron, eds., Plenum Press (1976).
Baker, N.H., Moore, D.W., and Spiegel, E.A., Astron. J., 9 (1966) Abstract.
Bergé, P., Pomeau, Y. and Vidal, C., L'Ordre dans le Chaos, Hermann, Paris (1984).
Bretherton, C.S. and Spiegel, E.A., Phys. Lett., 96A (1983) 152-156.
Buchler, R. and Perdang, J., Astrophys. J., 231 (1979) 524.
Defouw, R.J., Astrophys. J., 160 (1970) 659.
Carr, J., Applications of Centre Manifold Theory, Springer-Verlag, Applied Mathematical Sciences 35 (1981).
Celnikier, L.M., Astronomy and Astrophysics, 60 (1977) 421-422.
Childress, S. and Fautrell, Y., Geophys. & Astrophys. Fluid Dyn., 22 (1982) 235-279.
Childress, S. and Spiegel, E.A., in Variations of the Solar Constant, S. Sofia, ed. NASA Conference Pub. 2191 (1980) 273-292.
Coullet. P.C. and Spiegel, E.A., SIAM J. Appl. Math., 43 (1983) 775-819.
Deubner, F.-L., Ulrich, R.K. and Rhodes, E.J., Jr., Astron. Astrophys., 72 (1979) 177.
Drazin, P.G. and Reid, W.H. Hydrodynamic Stability, Camb. Univ. (1981).
Eddy, J.A., Science, 192 (1976) 1189.
Edmonds, A.R., Angular Momentum in Quantum Mechanics, Princeton University Press (1957) p. 26.
Evans, J.W. and Michard, R., Astrophys. J., 136 (1962) 493-506.
Fowler, A., Sudies in Applied Math. (1984) 215-233.
Friedman, B., Principles and Techniques of Applied Mathematics, John Wiley and Sons (1956).
Gilmore, R., Catastrophe Theory for Scientists and Engineers, Wiley & Sons (1981).
Guckenheimer, J. & Holmes, P., Nonlinear Oscilllations, Dynamical Systems, and Bifurcations of Vector Fields, Springer-Verlag, Applied Math. Sciences 42 (1981).
Gudzenko, L.I and Chertoprud, V.E., in The Kinematics of Simple Models in the Theory of Oscillations, N.G. Basov, ed., Proc. P.N. Lebedev Physics Institute, 90 (1976); English Version Published by Consultants Bureau of Plenum Press (New York, 1978) Chapt. VIII, 153-199.
Haken, H., Synergetics. An Introduction. Nonequilibrium Phase Transitions in Physics, Chemistry and Biology. Springer-Verlag (1977).

Hao, B.-L., Chaos, World Scientific Pub. Co. (1984).
Hayashi, H., Ishizuka, S., Hirakawa, K., Phys. Lett., 98A (1983) 474-476.
Izenman, A.J., Math. Intelligencer, 7 (1985) 27-33.
Jeans, J.H., Astronomy and Cosmogony, Camb. Univ. (1928) p. 193.
Lamb, H., Proc. Roy. Soc. A lxxx (1907) 170.
Lebovitz, N.R. and Schaar, R.J., Studies in applied Math., LIV (1975) 229-260.
Leibacher, J.W. and Stein, R.F. in the Sun as a Star, S. Jordan, ed., NASA SP-450 (1981) 263-287.
Lewin, W.G.H. and Joss, P.C., Space Sci. Rev., 29 (1981) 3-87.
Lorenz, E.N., Jour. Atmos. Sci., 20 (1963) 130-141.
Malkus, W.V.R., Trans. Am. Geophys. Union, 53 (1972) 617.
Mandelbrot, B.B., The Fractal Geometry of Nature, Freeman (1983).
Marzec, C.J. and Spiegel, E.A., SIAM J. App. Math., 38 (1980) 403-421.
McLaughlin J.B. and Orszag, S.A., J. Fluid Mech., 122 (1982) 123-142.
Moon, H.T., Huerre, P. and Redekopp, L.G., Phys. Rev. Lett., 49 (1982) 458.
Moore, D.R., Toomre, J., Knobloch, E. and Weiss, N.O., Nature, 303 (1983) 663-667.
Moore D.W. and Spiegel, E.A., Astrophys. J., 143 (1966) 871-887.
Newell, A.C., Envelope Equations, Lectures in Applied Math., 15 (1974) 157-163.
Newell, A.C. and Whitehead, J.A., J. Fluid Mech., 38 (1969) 279-303.
Parker, E.N., Cosmical Magnetic Fields, Oxford (1979).
Pomeau, Y., in Nonlinear Dynamics and Turbulence, G.I. Barenblatt, G. Iooss and D.D. Joseph, eds. Pitman (1983) 295-304.
Press, W.H., Comments on Astrophys., 7 (1978) 103-119.
Robbins, K.A., SIAM J. Applied Math., 36 (1979) 457-472.
Rutten, R.J. and Cram, L.E. in the Sun as a Star, S. Jordan, ed., NASA SP-450 (1981) 3-10.
Ruzmaikin, A.A., Comments on Astrophys., 9 (1981) 85-96.
Schuster, H.G., Deterministic Chaos, Physik-Verlag, Weinheim (1984).
Segel, L.A., J. Fluid Mech., 38 (1969) 203-224.
Shapiro, S.L. and Teukolsky, S.A., Black Holes, White Dwarfs and Neutron Stars, Wiley-Interscience (1983).
Shil'nikov, L.P., Math. Sbornik **6** (1968) 427-438.
Sparrow, C., The Lorenz Equations:Bifurcations, Chaos and Strange Attractors, Applied Math Sciences 41, Springer-Verlag (1982).
Spiegel, E.A., Convection in Stars II. Special Effects, Ann. Rev. Astron. & Astrophys. (1972) 261-304.

Spiegel, E.A., in Problems of Stellar Convection, E.A. Spiegel and J.-P. Zahn, eds., Lecture Notes in Physics 71, Springer-Verlag (1977) p.3.
Spiegel, E.A., Ann. N.Y. Acad. Sci., 357 (1980) 305-312.
Spiegel, E.A. and Weiss, N.O., Nature, 287 (1980) 616.
Steiner, J.M., Bull. Aust. Math. Soc., 9 (1973) 153-154; and Thesis, Math. Dept., Monash Univ. (1973).
Unno, W., Osaki, Y., Ando, H. and Shibahashi, H., Nonradial Oscillations of Stars, University of Tokyo Press, (1978).
Weiss, N.O., Cattaneo, F., Jones, C.A., Geophys. & Astrophys. Fluid Dyn. (1984).
Wesfried, J.E. and Zaleski, S., eds., Cellular Structures in Instabilities, Lecture Notes in Physics 210, Springer-Verlag (1984).
Wolf, A., Swift, J.B., Swinney, H.L. and Vastano, J.A., Physica D (1985) in press.

A PERTURBATIVE APPROACH TO STELLAR PULSATIONS

J. Robert Buchler

ABSTRACT

   An asymptotic perturbation formalism is adapted to the
problem of the nonlinear behavior of stellar pulsators. It
leads to amplitude equations which govern the temporal, regular
or chaotic modulations of the amplitudes of the excited
modes. These amplitude equations have been calculated and
solved for realistic population I (Classical) Cepheid models,
which have an internal resonance of the type $\Omega_2 \simeq 2\Omega_0$ and
exhibit a secondary bump on their light and velocity curves.
The agreement with the results of numerical hydrodynamic codes
is very good and is achieved with a hundredfold reduction in
computing time. The potential usefulness of the method for
other types of pulsation, in particular irregular pulsation, is
discussed.

INTRODUCTION

   Variable stars have been discovered relatively late,
considering that astronomy can be regarded as one of the oldest
professions. The first scientific observation of the
variability of o Ceti, also called Mira Ceti, seems to have
been made by Fabricius in 1596 (Ledoux and Walraven 1958), but
the systematic study of variable stars did not seriously get
started until two centuries later. This year actually marks
the bicentenary of the discovery of δ Cephei, which has given
its name to and is the prototype of the so-called Classical
Cepheids variables. These Classical Cepheids are the most
regular and probably the best understood of all variable stars.

   The study of variable stars is of interest to several
scientific disciplines, in particular, to the physicist because
of the interplay between various physical effects associated

with the pulsation and amplitude saturation mechanism, to the
astrophysicist, because of the constraints they put on stellar
models and on stellar evolution and, finally, to the astronomer
and cosmologist, because of the existence of period-luminosity
relations, which, in the case of the Cepheids, make these the
most reliable distance indicators.

The theoretical work on stellar pulsators so far has
fallen into two categories, namely linear stability analyses,
on one hand, and numerical hydrodynamics, on the other.
Extensive studies of the linearized hydrodynamic and heat flow
equations, which describe the behavior of the stars, have been
instrumental to our understanding of the mechanisms for the
pulsational instability and to delineating the regions of
instability in the H-R diagram. They have also yielded periods
in good agreement with observation. This linear approach,
however, can yield no information on the amplitude saturation
mechanism nor on the finite amplitude behavior in general.

The alternative route has been the brute force numerical
integration of the coupled hydrodynamic and heat flow
equations. In many, if not in most stellar pulsators the
thermal timescale is considerably shorter than the dynamic
timescale. As a result the cost of evolving a stellar model to
finite amplitude can be very high, especially if a fine mass-
zoning is required for good resolution. To our knowledge, the
problem of nonradial oscillations has never even been attempted
with the numerical approach. Even in the case of radial
pulsations the numerical hydrodynamic approach suffers from
additional drawbacks. Numerical stabilization techniques, like
artificial viscosity, introduce uncontrollable uncertainties
and distortions in the oscillations. Transient oscillations
may take a very long time to die out, which makes it very
difficult if not impossible to determine when or if a saturated
oscillation has been achieved. A similar difficulty arises
when the saturated motion is multi-periodic or chaotic. It is
only in the strictly periodic case that an astute iteration
scheme (Stellingwerf 1975) can with certainty zero-in on the
stable finite amplitude behavior. For these reasons, the
direct numerical approach is seen to be of very limited use as
a tool for surveying the finite amplitude behavior of stellar
models.

In order to more effectively deal with the search for
globally stable nonlinear behavior, on the one hand, and to
devise a computationally tractable approximation method, on the
other hand, perturbation methods can be taken advantage of in
the frequent cases where there are small parameters in the
problem. The aim is to reduce the system of partial
differential equations governing the hydrodynamic and thermal

behavior of a star to a small set of ordinary differential equations for the amplitudes of the excited modes, equations which involve only the (slow) thermal time. Thus, the gain in computing time is twofold. In addition, the structure of these amplitude equations allows one to explore and map in some detail the possible amplitude limiting behavior as a function of the stellar model parameters, something which could not possibly be obtained with a numerical hydrocode.

The underlying assumption of essentially all perturbative approaches to stellar pulsations up to date (e.g., Castor in J. P. Cox 1980, Buchler, Yueh and Perdang 1977, Buchler 1978, Regev and Buchler 1981, Simon 1981, but see Buchler and Pesnell 1984, Dziembowski and Kovacs 1984, Vandakurov 1981, Takeuti and Aikawa 1981) is one of quasi-adiabaticity, namely that the ratio $\varepsilon(m)$ of the dynamic timescale to the local thermal timescale is much smaller than unity throughout the whole star. (The question of how one might properly define such a local parameter is discussed in Buchler and Pesnell 1985). The obvious advantage of a quasi-adiabatic perturbation approach lies in the use of the nicely behaved, real adiabatic eigenvectors as a basis for the expansion. The quasi-adiabatic approximation, however, necessarily breaks down in the outermost layers of the star where the specific internal energy is very small and the thermal readjustment time can become considerably shorter than the dynamic time. In many stars the mass in these outer nonadiabatic layers is quite small and the quasi-adiabatic approximation is useful, at least qualitatively. In the application to a realistic stellar model, however, a cut-off must be introduced at the transition region, which is vaguely defined as the mass-shell where $\varepsilon(m_{tr}) \simeq 1$. This cutoff has been found to cause an intolerable quantitative uncertainty in the results (Pesnell and Buchler 1984). We note, nevertheless, that the quasi-adiabatic approximation yields amplitude equations which have the correct structure, so that they lend themselves to analytical considerations; the difficulty lies in getting the correct coefficients of these equations. In order to alleviate the problems introduced by this somewhat adhoc cutoff but keep the advantages of a basis of real eigenvectors, Buchler and Regev (1982) proposed to include in the formalism the nonadiabatic region beyond m_{tr} by treating it in the highly nonadiabatic limit, i.e., by assuming that $\varepsilon(m) \gg 1$ for $m > m_{tr}$. The eigenvectors obtained in this way are quite similar to the linear nonadiabatic ones. Unfortunately, the abrupt transition from quasi-adiabaticity on the inner side of the transition region to high nonadiabaticity on the outer side causes some numerical problems, at least in the case of classical Cepheid and β-Cephei models which have been looked at in detail by Pesnell et al. (1985).

In view of the practical difficulties with the quasi-adiabatic approach mentioned above, Buchler and Goupil, (hereafter BG84) devised a radically different nonadiabatic formalism. That formalism does does not use a local perturbation parameter $\varepsilon(m)$, but rather, uses the global parameter r, which is the ratio of the linear oscillation frequency of the modes of interest to their linear growth rates. It is now irrelevant that $\varepsilon(m)$ is of order unity in the transition region and gets very large in the surface layers as long as r is small. The price one has to pay for the removal of the deficiencies of the previous approaches is the use of the dual basis of complex linear nonadiabatic (LNA) eigenvectors (e.g., Castor 1971) as opposed to the real adiabatic eigenvectors used in the quasi-adiabatic approaches.

We shall discuss here the nonadiabatic perturbation formalism developed by BG84 and discuss its application to some realistic stellar models. The formalism leads to amplitude equations (**AE**) of the same type as those derived by other authors with similar or different methods, generally in the context of convection in fluids. A very nice historical summary and comparison of the different approaches can be found in the introduction of Coullet and Spiegel (1983). The method of Coullet and Spiegel is somewhat more general and powerful than ours. Since our emphasis is on the amplitude equations themselves and on their applicability (and application) to stellar pulsations, rather than on mathematical methods, we shall not discuss this other work here, but refer the reader directly to the source.

The Amplitude Equations

The basic philosophy of the perturbation method is first to assume that a dichotomy of the linear modes is possible into a strongly damped group, on the one hand, and into a marginally unstable (or marginally stable) group, on the other. Phrased more mathemathically, one assumes the existence of a slow or center manifold (e.g., Holmes 1981). The second assumption is that the nonlinearities remain weak during the pulsation. Amplitude saturation then results from a balance between the assumed gentle instability and the low order nonlinearities. This balance can lead to a rich variety of behavior ranging from pulsations with constant amplitudes, to singly or multiply periodic pulsations, to highly irregular (chaotic) pulsations.

For convenience of notation we restrict ourselves to radial pulsations. In principle, the formalism can readily be generalized to nonradial pulsations although as we shall later point out, this poses practical difficulties.

The equations governing the behavior of a stellar model are given in a Lagrangean description by (e.g., Cox and Giuli 1968)

$$\frac{d^2}{dt^2} R(m) = - 4\pi R^2 \frac{\partial}{\partial m} p(v,s) - \frac{Gm}{R^2} \equiv - g[R(m), s(m)], \qquad (1)$$

$$\frac{d}{dt} s(m) = - \frac{1}{T} \frac{\partial}{\partial m} (4\pi R^2)^2 \frac{c}{3\kappa(v,s)} \frac{\partial}{\partial m} aT^4 \equiv h[R(m), s(m)], \qquad (2)$$

where v is the specific volume and the other symbols have their usual meaning. We have assumed here that the heat transport occurs through radiative conduction. We have also disregarded convection, turbulent viscosity, rotation and magnetic fields. Equations (1,2) have to be supplemented with the mass conservation equation

$$v = \frac{4\pi}{3} \frac{\partial}{\partial m} R^3, \qquad (3)$$

an equation of state $p = p(v,s)$ and an expression for the opacity $\kappa = \kappa(v,s)$. The right hand sides of equations (1,2) are thus expressible in terms of R(m), s(m) and their spatial derivatives.

The functions g(m) and h(m) are now expanded about the hydrostatic and thermal equilibrium state with

$$(g[R_e(m), s_e(m)] = 0), \qquad (h[R_e(m), s_e(m)] = 0).$$

Let $\delta R(m) = R(m) - R_e(m)$ and $\delta s(m) = s(m) - s_e(m)$. The function g(m) then assumes the form

$$g = - \left(\frac{4Gm}{R^3} \delta R - 4\pi R^2 \frac{\partial}{\partial m} \delta p\right) - \left(-4\pi R^2 \frac{\partial}{\partial m} \delta^2 p - \frac{2Gm}{R^4} (\delta R)^2 - 8\pi R \delta R \frac{\partial}{\partial m} \delta p\right) + \ldots, \qquad (4)$$

with

$$\delta p = \left(\frac{\partial p}{\partial v}\right)_s \delta v + \left(\frac{\partial p}{\partial s}\right)_v \delta s, \qquad (5)$$

$$\delta^2 p = \frac{\partial p}{\partial v} \delta^2 v + \frac{1}{2} \frac{\partial^2 p}{\partial v^2} (\delta v)^2 + \frac{\partial^2 p}{\partial v \partial s} \delta v \delta s + \frac{1}{2} \frac{\partial^2 p}{\partial s^2} (\delta s)^2. \qquad (6)$$

With the help of eq. (3), δv, $\delta^2 v, \ldots$ can be expressed in terms of the δR and derivatives thereof. The g is thus expressible in the form

$$g = (\mathbf{g}_R \delta R + \mathbf{g}_s \delta S) + (\frac{1}{2} \mathbf{g}_{RR} \delta R \delta R + \mathbf{g}_{Rs} \delta R \delta s + \frac{1}{2} \mathbf{g}_{ss} \delta s \delta s) \ldots \qquad (7)$$

Mutatis mutandis we get a similar expansion for h.

At this stage it is advantageous to introduce another variable $\delta V \equiv d(\delta R)/dt$ and to define a 'vector' $z \equiv (\delta R, \delta V, \delta s)$. Equations (1,2) can then be written in a compact notation as a set of three coupled partial differential equations, which are first order in time

$$\frac{d}{dt} z = A z + N_2 zz + N_3 zzz + \ldots, \tag{8}$$

where

$$A = \begin{pmatrix} 0 & 1 & 0 \\ -g_R & 0 & -g_s \\ h_R & 0 & h_s \end{pmatrix} \tag{9}$$

is the linear nonadiabatic (**LNA**) operator for radial stellar pulsations (Castor 1971). N_2, N_3, etc. are the quadratic, cubic, etc. nonlinearities in the combined oscillator and the heat flow equations. The form of these equations (8) and (9) is obviously very general and transcends the context of radial stellar pulsations.

The operators g_R and h_s can both be made self-adjoint and, with appropriate boundary conditions, constitute separate Sturm-Liouville problems, corresponding to adiabatic oscillations and thermal diffusion modes (Pesnell and Buchler 1984), respectively. The LNA operator **A**, however, is not self-adjoint and gives rise to a dual basis set of left ('bra') and right ('ket') eigenvectors with complex eigenvalues $\sigma \equiv i\Omega + \kappa$:

$$\sigma_\alpha |\alpha\rangle = A |\alpha\rangle , \tag{10}$$

$$\sigma_\alpha \langle\alpha'| = \langle\alpha'| A , \tag{11}$$

where we have introduced an $\exp(\sigma t)$ time dependence in the modes. We have adorned the bras with a prime to denote the fact that $\langle\alpha'| \neq$ transpose of $|\alpha\rangle$. The bras and kets satisfy the orthogonality condition

$$\langle\alpha'|\beta\rangle = \delta_{\alpha\beta}. \tag{12}$$

Because the operator **A** is real, the spectrum of eigenvalues splits into two complex conjugate mirror branches and a real branch. The latter represents the 'secular' and 'thermal diffusion' modes. The spectral representation of the operator **A** is thus

$$A = \Sigma_\nu (\sigma_\nu |\nu\rangle\langle\nu'| + c.c.) + \Sigma_k \sigma_k |k\rangle\langle k'|. \tag{13}$$

In the following we shall use Greek indices for the (complex) vibrational and Roman indices for the (real) thermal modes and assume that the sum includes the complex conjugates (c.c.) where appropriate.

As mentioned above, the gist of our perturbation approach is first, to split the modes into two categories, one for which $\kappa \equiv \mathrm{Re}\sigma$ is small compared to $\mathrm{Im}\sigma$ (the excited, marginally stable or unstable modes) and the other for which $\mathrm{Re}\sigma$ is large, and, of course, assumed negative (the slave modes). Physically, the slave modes are assumed to be so strongly damped that their transients decay essentially instantly; they do however affect and distort the motion which is governed by the excited modes. We shall denote by r the ratio of the growth- or decay-rate, $\kappa \equiv \mathrm{Re}\sigma$, to the oscillation frequency, $\Omega \equiv \mathrm{Im}\sigma$, of the excited modes, and assume r to be small. Second, we consider the oscillation to be <u>weakly nonlinear</u>, i.e., we introduce the scaling $z' = \lambda z$ and, as is common, instantly drop the prime.

Illustration of the Perturbation Formalism

We shall now illustrate the perturbation formalism with an application to the case of two excited resonant vibrational modes ($2\Omega_\alpha \simeq \Omega_\beta$) and p excited thermal modes. We define the resonance frequency

$$\Omega = (2\Omega_\alpha + \Omega_\beta)/4, \tag{14}$$

and the off-resonance parameter

$$\Delta = (2\Omega_\alpha - \Omega_\beta)/(2\Omega). \tag{15}$$

In this case we thus have three small parameters, the linear stability parameter, r, the anharmonicity parameter, λ, and the off-resonance parameter Δ.

The LNA operator **A** will be written in a form which exhibits the small parameters r, λ and Δ explicitly

$$\mathbf{A} = \mathbf{A}_u + r\mathbf{A}_p + \Delta\mathbf{A}_d, \tag{16}$$

where

$$\mathbf{A}_u = i\Omega|\alpha\rangle\langle\alpha'| + 2i\Omega|\beta\rangle\langle\beta'| + \sum_{\text{slaves}} \sigma_\nu|\nu\rangle\langle\nu'|, \tag{17}$$

$$r\mathbf{A}_p = \kappa_\alpha|\alpha\rangle\langle\alpha'| + \kappa_\beta|\beta\rangle\langle\beta'| + \sum_{k=1,p} \kappa_k|k\rangle\langle k'|, \tag{18}$$

$$\Delta A_d = \frac{i\Omega\Delta}{2} |\alpha\rangle\langle\alpha'| - i\Omega\Delta|\beta\rangle\langle\beta'|. \tag{19}$$

Equation (8) thus assumes the form

$$\frac{d}{dt}|z\rangle = A_u|z\rangle + r\,A_p|z\rangle + \Delta\,A_d|z\rangle$$
$$+ \lambda\,N_2|zz\rangle + \lambda^2 N_3|zzz\rangle + \ldots, \tag{20}$$

where $N^{(2)}|zz\rangle$ denotes the application of the quadratic operator onto the two kets.

At this stage, we need to decide on an ordering of the small parameters. We shall choose here $O(r) = O(\lambda) = O(\Delta)$. We note in anticipation that this choice gives nontrivial amplitude equations with the lowest nonlinearities. In physical terms, we restrict ourselves to stellar models which hold for these orderings.

From here on, any one of a number of asymptotic perturbation methods (e.g., Nayfeh and Mook 1979), can be used. We have chosen (BG84) the method of 'multiple scales'; first, because it is straightforward and systematic and, second, because it yields a time-dependence for the amplitudes rather than merely fixed points as, for example, the Lindstedt-Poincaré method does.

In summary, the method consists of defining an asymptotic series

$$|z\rangle = |z_0\rangle + \lambda|z_1\rangle + \lambda^2|z_2\rangle + \ldots, \tag{21}$$

and multiple times

$$t_0 = t, \quad t_1 = \lambda t, \quad t_2 = \lambda^2 t, \quad \ldots, \tag{22}$$

which are considered independent variables so that the time derivative becomes

$$\frac{d}{dt} = \frac{d}{dt_0} + \lambda\frac{d}{dt_1} + \lambda^2\frac{d}{dt_2} + \ldots. \tag{23}$$

These quantities are introduced into eq. (20) and the terms of the successive powers of λ are gathered. In lowest order, this yields

$$\frac{d}{dt_0}|z_0\rangle - A_u|z_0\rangle = 0, \tag{24}$$

with the solution corresponding to the excited modes of interest

A PERTURBATIVE APPROACH TO STELLAR PULSATIONS

$$|z_0\rangle = \frac{1}{2} a_\alpha |\alpha\rangle \exp(i\Omega t_0) + \frac{1}{2} a_\beta |\beta\rangle \exp(2i\Omega t_0) + \text{c.c.}$$

$$+ \sum_{k=1,p} b_k |k\rangle . \qquad (25)$$

The amplitudes, a, for the vibrational modes are complex and the amplitudes, b, for the thermal modes are real. All amplitudes are considered functions of t_1, t_2, ...

The functional dependence of the amplitudes on t_1 is determined by the next order in λ. This equation is of the form

$$\frac{d}{dt_0}|z_1\rangle - \mathbf{A}_u|z_1\rangle = -\frac{d}{dt_1}|z_0\rangle + \mathbf{A}_p|z_0\rangle + \mathbf{A}_d|z_0\rangle + \mathbf{N}_2|z_0 z_0\rangle \equiv |R(t_0)\rangle. \qquad (26)$$

Solutions exist only if the following <u>compatibility conditions</u> are satisfied

$$\oint \exp(-i\Omega t_0) \langle \alpha' | R(t_0)\rangle dt_0 = 0, \qquad (27)$$

$$\oint \exp(-2i\Omega t_0) \langle \beta' | R(t_0)\rangle dt_0 = 0, \qquad (28)$$

$$\oint \langle k' | R(t_0)\rangle dt_0 = 0, \quad k = 1,\ldots,p. \qquad (29)$$

These conditions yield the <u>amplitude equations</u> (**AE**)

$$\frac{da_\alpha}{dt_1} = \kappa_\alpha a_\alpha + \frac{1}{2} i\Delta\Omega a_\alpha + \frac{1}{2} \langle \alpha' | \mathbf{N}_2 | \overline{\alpha^* \beta} \rangle a_\alpha^* a_\beta + \sum_k \langle \alpha' | \mathbf{N}_2 | \overline{\alpha k} \rangle a_\alpha b_k, \qquad (30)$$

$$\frac{da_\beta}{dt_1} = \kappa_\beta a_\beta - i\Delta\Omega a_\beta + \frac{1}{2} \langle \beta' | \mathbf{N}_2 | \overline{\alpha\alpha} \rangle a_\alpha^2 + \sum_k \langle \beta' | \mathbf{N}_2 | \overline{\beta k} \rangle a_\beta b_k, \qquad (31)$$

$$\frac{db_k}{dt_1} = \kappa_k b_k + \frac{1}{4} \langle k' | \mathbf{N}_2 | \overline{\alpha\alpha^*} \rangle |a_\alpha|^2 + \frac{1}{4} \langle k' | \mathbf{N}_2 | \overline{\beta\beta^*} \rangle |a_\beta|^2$$

$$+ \sum_{\ell m} \langle k' | \mathbf{N}_2 | \overline{\ell m} \rangle b_\ell b_m , \qquad (32)$$

where we have defined $|\overline{\alpha\beta}\rangle \equiv |\alpha\beta\rangle + |\beta\alpha\rangle$. In going from eqs. (27-29) to (30-32), we have assumed the eigenvectors to be normalized to unity as in eq. (12).

The successive terms in the expansions may be regrouped and the solution written in the form of Coullet and Spiegel (1983):

$$|z\rangle = |z_0\rangle + \lambda |z_1\rangle + \lambda^2 |z_2\rangle + \ldots \qquad (33a)$$

$$\equiv \frac{1}{2} \Sigma F_\mu(\{\tilde{a}_\nu\})|\mu\rangle + |W\rangle, \qquad (33b)$$

where the sum runs over the excited modes and where $|W\rangle$ denotes a vector in the space orthogonal to the excited modes. Introducing the 'complete' amplitudes $a_\nu = a_\nu \exp(i\Omega_\nu t)$ we can obtain the 'reconstituted' amplitude equations (in the parlance of Spiegel, 1981) by summing

$$\frac{d}{dt}\tilde{a} = \frac{\partial}{\partial t_0}\tilde{a}_0 + \lambda \frac{\partial}{\partial t_1}\tilde{a}_1 + \lambda^2 \frac{\partial}{\partial t_2}\tilde{a}_2 + \ldots \quad .$$

These amplitude equations then agree with those obtained with Coullet and Spiegel's formalism. The equations are given in the Tables.

We should point out that while the decomposition (33b) is unique, the amplitude equations and concomitantly the 'nonlinear' amplitudes F_μ are not unique. In the next section we shall see an example of how terms can be reshuffled.

Lowest Order Amplitude Equations

In the tables we classify the various AE according to the order of the nonlinearity which gives rise to nontrivial amplitude equations. Such a classification is convenient and useful because it gives an indication of the programming effort required in an application to a realistic system. In Table 1 we have summarized the three cases giving rise to quadratic nonlinearities. Under (T1a) we exhibit the amplitude equations for the case of a single nonresonant oscillatory mode α interacting with p thermal modes, again with the same ordering of the small parameters $O(\lambda) = O(r)$. By 'nonresonant' we mean that there exists no relation of the form

$$\left|\Sigma_\nu \ell_\nu \text{Im}\sigma_\nu\right| < \varepsilon\ell \left|\text{Im}\sigma_\alpha\right|, \quad \ell_\nu \text{ integer}$$

with $\ell = |\Sigma \ell_\nu|$, the order of the resonance, equal to three. The case treated above thus corresponds to a third order resonance. The AE (T1a) clearly has as a subcase the interaction of only thermal modes.

Another case involving only quadratic nonlinearities

arises for the 3rd order resonance condition of the type $\Omega_\alpha + \Omega_\beta \simeq \Omega_\gamma$. Defining here

$$\Omega = \Omega_\alpha + \Omega_\beta, \qquad (34a)$$

$$\Delta = (\Omega_\gamma - \Omega_\alpha - \Omega_\beta)/\Omega, \qquad (34b)$$

with the same ordering as in the previous cases, one obtains the set of amplitude equations (T1c).

In Table 2, we exhibit only one of the cubic AE which corresponds to the case of nonresonant oscillatory modes and which has the monomode oscillation as a subcase.

In order for amplitude equations to be able to yield chaotic solutions in some range of parameter values they need to be sufficiently coupled and to be of sufficiently high order. It is well known that two coupled first order equations can only give rise to regular attractors. Some equations, like the AE of (T2) with $\nu = 2$ or (T1) with $p = 1$ constitute really only two coupled equations, which one sees best by replacing the complex amplitudes a_α by their moduli, A_α, and phases, ϕ_α. Indeed T2 yields AE of the type

$$\frac{dA_\alpha}{dt} = \kappa_\alpha A_\alpha + (\mathrm{Re}T_\alpha)A_\alpha^3 + (\mathrm{Re}S_\alpha)A_\alpha A_\beta^2 \qquad (35a)$$

$$\frac{dB_\alpha}{dt} = \kappa_\beta A_\beta + (\mathrm{Re}T_\beta)A_\beta^3 + (\mathrm{Re}S_\beta)A_\beta A_\alpha^2 \qquad (35b)$$

$$\frac{d\phi_\alpha}{dt} = (\mathrm{Im}T_\alpha)A_\alpha^2 + (\mathrm{Im}S_\alpha)A_\beta^2 \qquad (35c)$$

$$\frac{d\phi_\beta}{dt} = (\mathrm{Im}T_\beta)A_\beta^2 + (\mathrm{Im}S_\beta)A_\alpha^2 \qquad (35d)$$

and

$$\frac{db_k}{dt} = \kappa_k b_k + U_\alpha A_\alpha^2 + V_k b_k^2 \qquad (36a)$$

$$\frac{dA_\alpha}{dt} = \kappa_\alpha A_\alpha + (\mathrm{Re}P_k)A_\alpha b_k \qquad (36b)$$

$$\frac{d\phi_\alpha}{dt} = (\mathrm{Im}P_k)b_k \qquad (36c)$$

respectively. The quantities T_ν, S_ν, U_α, V_α, P_k represent the quadratic coefficients in eqs. (T2) and (T1a). In both cases the first two equations are independent of the phase(s).

It is possible to generalize our amplitude equations somewhat. So far the growthrates κ_ν appear only linearly and it would be worthwhile to try to introduce higher order corrections in the κ_ν. A close inspection of the inverse operators appearing in the expression for $|z\rangle$ (BG64, e.g. eqs. 26-27) suggests that the $i\Omega_\nu$ by replaced everywhere by σ_ν or σ_ν^* where appropriate. Fortunately, this heuristic generalization can be made readily rigorous within the framework of Coullet and Spiegel's formalism. The latter has the advantage of treating the small parameters λ and κ_ν as independent. Indeed, they first compute amplitude equations when there exists a center manifold, ie., when the κ_ν are strictly zero, and then they use another perturbation expansion for small κ_ν. This procedure involves a great deal of algebra in the general case. Here, because their operator $\mathbf{M}$ is diagonal in the perturbations κ_ν, it is easy to see that the latter can be absorbed into their resolvent $\mathbf{L}$. The κ_ν thus only appear in connection with the unperturbed eigenvalues $i\Omega_\nu$ as σ_ν, respectively $-\sigma_\nu^*$.

The cubic amplitude equations (Table 2 or eqs. 35a-c) have divergencies in the presence of a third order resonance because of the occurrence of small denominators of the form $(2\Omega_\alpha - \Omega_\beta)$. For the same reason, the first correction $|z_1\rangle$ to the solution is also divergent. We have already pointed out that our amplitude equations are not unique and one can take advantage of that freedom to eliminate the divergencies (Buchler and Kovács 1985). In the case of a resonance of the type $2\Omega_\alpha - \Omega_\beta$ the apposite transformation is

$$\tilde{a}' = \tilde{a} + \frac{\langle\alpha|\mathbf{N}_2|\overline{\alpha^*\beta}\rangle}{\sigma_\beta + \sigma_\alpha^* - \sigma_\alpha} \tilde{a}^* \tilde{b} \quad, \tag{37}$$

$$\tilde{b}' = \tilde{b} + \frac{\langle\beta|\mathbf{N}_2|\alpha\alpha\rangle}{2\sigma_\alpha - \sigma_\beta} \tilde{a}^2 \quad. \tag{38}$$

It may not come as a surprise that the transformed amplitude equations are identical with the 'resonant' amplitude equations (T1b) at least up to the third order in λ. The cubic amplitude equations are thus only valid out of resonance, whereas the resonant amplitude equations are valid in, as well as out of resonance, provided of course, that there is no additional nearby resonance to spoil the game. In the formalism of Coullet and Spiegel (1983), the choice between the two types of amplitude equations arises from compatibility conditions of the form

$$i(\Omega_\beta - 2\Omega_\alpha)v_1 + g_1 = \langle\alpha|\mathbf{N}_2|\overline{\alpha^*\beta}\rangle \quad, \tag{39}$$

$$i(2\Omega_\alpha - \Omega_\beta)V_2 + g_2 = \langle\beta|\mathbf{N}_2|\alpha\alpha\rangle \quad , \tag{40}$$

where the V_i represent quadratic terms in the nonlinear amplitudes and the g_i represent quadratic terms in the amplitude equations. In resonance, the only possible choice is $g_i = 0$ with $V_i \neq 0$ arbitrary (thus, conveniently chosen = 0), yielding the resonant amplitude equations. There is an alternative choice out of resonance of $V_i = 0$ and $g_i \neq 0$, giving rise to the cubic amplitude equations.

We would also like to point out that the reconstituted amplitude equations of the Table are more general than the original amplitude equations of the perturbation expansion and contain the latter as subcases, depending on the ordering of the small parameters. For example, when the appropriate scalings, $0(\lambda) = 0(r) = 0(\Delta)$, are introduced into the amplitude equations (T1b), together with an asymptotic expansion for the amplitudes and with multiple timescales, then one readily recovers the amplitude equations (30-32). On the other hand, when the scaling $0(\lambda^2) = 0(r)$ and $0(\Delta) = 0(1)$ is introduced, one recovers the cubic amplitude equations

$$\frac{\partial}{\partial t_2} a_\nu = \kappa_\nu a_\nu + \Sigma_\mu T_{\mu\nu} a_\nu |a_\mu|^2 \quad . \tag{41}$$

Of course, the latter could have been obtained more easily from the reconstituted cubic amplitude equations (T2) directly.

We have already mentioned that the resonant amplitude equations (36) have enough structure to allow for chaotic solutions, whereas the nonresonant equations (35) have not, although both can be derived from the same reconstituted amplitude equation (T1b). Even in the case of fixed points, the physical saturation mechanism is very different in both limits. This can be best visualized if one asumes an adiabatic evaluation of the nonlinear terms, ie., one computes the nonlinear 'matrix elements' $\langle\nu|\mathbf{N}_2|\nu\nu'\rangle$ and $\langle\nu|\mathbf{N}_3|\nu'\nu''\nu'''\rangle$ with the adiabatic eigenvectors; thus, these are purely imaginary (Klapp, Goupil and Buchler 1985, hereafter, KGB). Equations (35a,b) show that the nonresonant amplitude equations have no bounded fixed points. It is thus the nonlinear dissipation which must balance the linear driving to effect a saturation. On the other hand, in the resonant case, saturation can occur in this adiabatic approximation. The physical reason is simply that the nonlinear energy transfer to the second linearly stable resonant mode acts as an effective nonlinear dissipation for the first mode.

Table 1

Quadratic Amplitude Equations

a) <u>nonresonant oscillating mode coupled to nonoscillatory modes</u>

$$\frac{d}{dt} \tilde{a}_\alpha = (i\Omega_\alpha + \kappa_\alpha)\tilde{a}_\alpha + \sum_k \langle \alpha' | \mathbf{N}_2 | \overline{\alpha k} \rangle \tilde{a}_\alpha b_k$$

$$\frac{d}{dt} b_k = \kappa_k b_k + \frac{1}{4} \langle k' | \mathbf{N}_2 | \overline{\alpha \alpha^*} \rangle |a_\alpha|^2 + \sum_{\ell m} \langle k' | \mathbf{N}_2 | \overline{\ell m} \rangle b_\ell b_m$$

$$k = 1,\ldots,p$$

b) <u>third order resonant oscillatory modes</u> $(2\Omega_\alpha \simeq \Omega_\beta)$ <u>coupled to nonoscillatory modes</u>

$$\frac{d}{dt} \tilde{a}_\alpha = (i\Omega_\alpha + \kappa_\alpha)\tilde{a}_\alpha + \frac{1}{2}\langle \alpha' | \mathbf{N}_2 | \overline{\alpha^* \beta} \rangle \tilde{a}_\alpha^* \tilde{a}_\beta + \sum_k \langle \alpha' | \mathbf{N}_2 | \overline{\alpha k} \rangle \tilde{a}_\alpha \tilde{b}_k$$

$$\frac{d}{dt} \tilde{a}_\beta = (i\Omega_\beta + \kappa_\beta) \tilde{a}_\beta + \frac{1}{2}\langle \beta' | \mathbf{N}_2 | \overline{\alpha\alpha} \rangle \tilde{a}_\alpha^2 + \sum_k \langle \beta' | \mathbf{N}_2 | \overline{\beta k} \rangle \tilde{a}_\beta b_k$$

$$\frac{d}{dt} b_k = \kappa_k b_k + \frac{1}{4}\langle k' | \mathbf{N}_2 | \overline{\alpha\alpha^*} \rangle |\tilde{a}_\alpha|^2 + \frac{1}{4}\langle k' | \mathbf{N}_2 | \overline{\beta\beta^*} \rangle |\tilde{a}_\beta|^2 +$$

$$+ \sum_{\ell m} \langle k' | \mathbf{N}_2 | \ell m \rangle b_\ell b_m \qquad k = 1,\ldots,p$$

c) <u>Third order resonant oscillatory modes</u> $(\Omega_\alpha + \Omega_\beta \simeq \Omega_\gamma)$ <u>coupled to nonoscillatory modes</u>

$$\frac{d}{dt} \tilde{a}_\alpha = (i\Omega_\alpha + \kappa_\alpha)\tilde{a}_\alpha + \frac{1}{2} \langle \alpha' | \mathbf{N}_2 | \overline{\beta^* \gamma} \rangle \tilde{a}_\beta^* \tilde{a}_\gamma + \sum_k \langle \alpha' | \mathbf{N}_2 | \overline{\alpha k} \rangle \tilde{a}_\alpha b_k$$

$$\frac{d}{dt} \tilde{a}_\beta = (i\Omega_\beta + \kappa_\beta)\tilde{a}_\beta + \frac{1}{2} \langle \beta' | \mathbf{N}_2 | \overline{\alpha^* \gamma} \rangle \tilde{a}_\alpha^* \tilde{a}_\gamma + \sum_k \langle \beta' | \mathbf{N}_2 | \overline{\beta k} \rangle \tilde{a}_\beta b_k$$

$$\frac{d}{dt} \tilde{a}_\gamma = (i\Omega_\gamma + \kappa_\gamma)\tilde{a}_\gamma + \frac{1}{2} \langle \gamma | \mathbf{N}_2 | \overline{\alpha\beta} \rangle \tilde{a}_\alpha \tilde{a}_\beta + \sum_k \langle \gamma' | \mathbf{N}_2 | \overline{\gamma k} \rangle \tilde{a}_\gamma b_k$$

$$\frac{d}{dt} b_k = \kappa_k b_k + \frac{1}{4} \langle k'|\mathbf{N}_2|\overline{\alpha\alpha}*\rangle |\tilde{a}_\alpha|^2 + \frac{1}{4} \langle k'|\mathbf{N}_2|\overline{\beta\beta}*\rangle |\tilde{a}_\beta|^2$$

$$+ \frac{1}{4} \langle k'|\mathbf{N}_2|\overline{\gamma\gamma}*\rangle |\tilde{a}_\gamma|^2 + \sum_{\ell m} \langle k'|\mathbf{N}_2|\overline{\ell m}\rangle b_\ell b_m \quad k = 1,\ldots,p$$

Table 2

Cubic Amplitude Equations

Nonresonant Oscillatory Modes

$$\frac{d}{dt} \tilde{a}_\nu = (i\Omega_\nu + \kappa_\nu)\tilde{a}_\nu + \sum_\mu T_{\nu\mu} \tilde{a}_\nu |\tilde{a}_\mu|^2$$

the $T_{\nu\mu}$, which now involve both $\mathbf{N}_2$ and $\mathbf{N}_3$ are lengthy to write and we refer to BG84 (eqs. 72-83).

Application To Stellar Pulsations

As a testcase for the applicability of AE to the modelling of stellar pulsations, we have chosen the so-called Bump Cepheids (e.g., A. N. Cox 1980), which exhibit a secondary maximum (bump) on their light and velocity curves. On the basis of a comparison of numerical hydrodynamic models with the period ratios obtained with a LNA analysis, it was first suggested by Simon and Schmidt (1976) that the bumps have their origin in the resonant interaction between the fundamental radial mode and the second overtone when $2\Omega_0 \simeq \Omega_2$. However, at that time no formalism was available to model and explore this hypothesis.

The AE apposite to this resonant case are given by our eqs. (30) and (31). The fundamental mode is noted by α or 0 and the resonant second overtone by β or 2. We ignore here the possible concomitant excitation of nonoscillatory modes. By introducing moduli and phases,
$a_\alpha = A \exp(i\theta_\alpha)$, $a_\beta = B \exp(i\theta_\beta)$, $\Gamma = \theta_\beta - 2\theta_\alpha$,

it is possible to reduce the two eqs. (30) and (31) for the complex amplitudes to a set of only three equations for the real variables A, B and Γ (KGB 1984)

$$\frac{dA}{dt} = \kappa_\alpha A + (R_\alpha \cos\Gamma - I_\alpha \sin\Gamma)\, AB, \tag{42}$$

$$\frac{dB}{dt} = \kappa_\beta B + (R_\beta \cos\Gamma + I_\beta \sin\Gamma)\, A^2, \tag{43}$$

$$\frac{d\Gamma}{dt} = -2\Omega\Delta + (I_\beta \cos\Gamma - R_\beta \sin\Gamma)\, A^2/B \\ - 2(I_\alpha \cos\Gamma + R_\alpha \sin\Gamma)\, B, \tag{44}$$

where we have set

$$\langle \alpha' | \mathbf{N}_2 | \overline{\alpha*\beta} \rangle = 2(R_\alpha + iI_\alpha),$$
$$\langle \beta' | \mathbf{N}_2 | \overline{\alpha\alpha} \rangle = 2(R_\beta + iI_\beta).$$

We now denote by α the fundamental mode and by β the second overtone.

The AE (42-44) depend on 7 parameters $(\Delta, \kappa_\alpha, \kappa_\beta, R_\alpha, I_\alpha, R_\beta, I_\beta)$ and are too complicated to allow for a detailed study of their solutions in this space. It is possible, however, to discuss the general types of asymptotic behavior, which are: (1) fixed points, corresponding to singly periodic, constant amplitude oscillations of the stellar model; (2) limit cycles, corresponding to periodic energy transfer between the two modes (periodic amplitude modulations); (3) strange attractors, leading to bounded, but chaotically varying amplitudes and phases; (4) no bounded attractors indicating that no balance is possible between the linear driving and the quadratic nonlinearities and that higher order nonlinearities must be considered, or that the system is too nonlinear to be handled with this formalism. Similar equations occur in plasma physics and have been shown to give rise to chaotic oscillations (Wersinger et al. 1980).

For the stellar models of interest here, $\kappa_\alpha > 0$ (linearly unstable fundamental) and $\kappa_\beta < 0$ (linearly stable resonant overtone) with $\kappa_\alpha + \kappa_\beta < 0$. It is easy to see that this latter condition guarantees that the divergence of the trajectories in phase space is always negative, which is a necessary condition for the existence of attractors. It does not guarantee that they are bounded, however. The line ($A = B = 0$), which corresponds to the state of no pulsation is thus always a line of unstable fixed points.

In physical terms, attractors (amplitude saturation) can then arise as follows: The fundamental mode being vibrationally unstable (overstable) saps thermal energy from the star and converts it into pulsational (mechanical) energy. At sufficiently large amplitude, this energy gets shared with the second mode through the quadratic nonlinearities. However, the second mode, being damped, restores energy to the thermal reservoir. An attractor results from a balance of these effects.

The fixed points of the AE are obtained by setting the right hand sides of equations (42-44) equal to zero, which leads to a quadratic equation for cotan Γ. As long as the discriminant is positive, there are therefore two solutions for Γ (mod π). The amplitudes are then given in terms of these solutions by

$$A^2 = \kappa_\alpha \kappa_\beta (R_\alpha \cos\Gamma - I_\alpha \sin\Gamma)^{-1} (R_\beta \cos\Gamma + I_\beta \sin\Gamma)^{-1}, \quad (45)$$

$$B = -\kappa_\alpha (R_\alpha \cos\Gamma - I_\alpha \sin\Gamma)^{-1}.$$

The mod π ambiguity is resolved by requiring B to be positive. Two fixed points (in addition to the trivial A = B = 0) exist thus, if expression (45) is also positive. These fixed points can be attractors only if their 3 linear stability roots all have negative real parts. For the stellar models which we have investigated, we find that the fixed points always have one real stability root and a complex conjugate pair of roots. One of the fixed points is always found to be unstable with both fixed points merging and then disappearing when the discriminant vanishes. Another interesting case, a Hopf bifurcation, arises when the real part of the complex conjugate pair of roots vanishes for the stable fixed point.

It is instructive to describe the behavior of the solutions in the neighborhood of the resonance, $P_2/P_0 \simeq 1/2$ or $\Delta \simeq 0$. We recall (eq. 25) that the velocity profile can be written as

$$\delta V(m,t) = \text{Re}\left[\Omega A(t) |\alpha(m)\rangle_R \exp(i\Omega t) + 2\Omega B(t) |\beta(m)\rangle_R \exp(2i\Omega t + i\Gamma - i\pi/2)\right]. \quad (46)$$

The eigenvectors are customarily normalized so that the radial component is unity at the surface, $|\delta R(M)\rangle = 1$. The symmetrical velocity curve thus occurs for $\Gamma = -\pi/2$. It can be shown from the AE that this does not happen exactly at $\Delta = 0$, but at

$$\Delta = \Delta_s \equiv (\kappa_\beta R_\beta / I_\beta - 2\kappa_\alpha R_\alpha / I_\alpha)/(2\Omega).$$

When the quadratic coefficients in eqs. (42-44) are

computed with adiabatic eigenvectors, they become purely immaginary (e.g., Buchler and Regev 1982, Dziembowski 1982). The deviation from exact resonance is thus, a measure of the nonadiabaticity of the nonlinear terms. In our Cepheid models, the ratios (R_ν/I_ν) are always fairly small ($\simeq$ 10%) so that the deviation is not very large.

To the extent that one can consider the κ_ν, R_ν, I_ν constant in the neighborhood of the resonance, one can show, for a sequence of models, that the behavior of Γ as a function of Δ is that of a arccot. Since the nonadiabatic effects and thus, the shift, are not exactly the same for all sequences of the stellar models, one expects a broadening of this arccot curve. In addition, the coefficients in the AE vary as one traverses the resonance. One predicts then that all stellar pulsators, for which the AE (40-42) apply, should fall on an arccot curve which has a narrow width near the resonance, and which becomes increasingly fuzzy on either side of the resonance. This prediction is indeed satisfied for both observational data and numerical hydrodynamic stellar models for $\Delta\Omega{>}0$; for $\Delta\Omega{>}0$ some complications arise because the stable and unstable fixed points merge and eqs. (42-44) do not have any bounded fixed points beyond.

On the other hand, for the amplitudes, the AE predict a typical cosec Γ behavior:

$$B \simeq \frac{\kappa_\alpha}{I_\alpha} \frac{1}{\sin \Gamma} \quad \text{and} \quad A^2 \simeq -\frac{\kappa_\alpha \kappa_\beta}{I_\alpha I_\beta} \frac{1}{\sin^2 \Gamma}, \qquad (47)$$

whereas the ratio B/A is constant. Thus, the reproduction of the behavior of this ratio as inferred from observation, constitutes a stringent test of the models.

The construction of the numerical models and the evaluation of the coefficients in the AE is discussed in detail in KGB85. We only mention that an equation of state and opacity law, taking into account hydrogen, helium and a representative metal ionization, has been used. Convection is inefficient (Christy 1975) and has been disregarded.

Figure 1 shows the results of KGB for an 8 solar mass population I Cepheid models (X = 0.66, Y = 0.30, Z = 0.04) in a Herzsprung-Russell (luminosity versus effective temperature) diagram. The lines marked κ_ν denote the boundary below which mode ν is linearly unstable. The line for κ_2 falls well below the range of the diagram. Also shown are lines of constant period ratio P_2/P_0. Qualitatively similar diagrams are obtained for different masses and compositions.

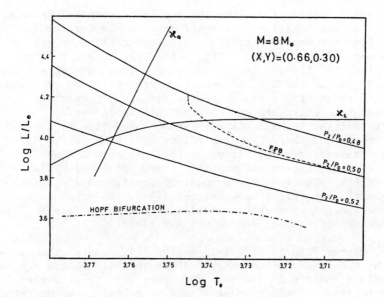

Figure 1

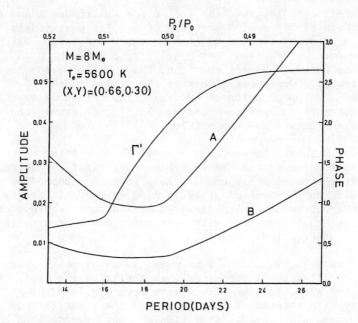

Figure 2

The dash-dotted line marks the Hopf bifurcation. We have investigated the behavior of 4 solar mass models for log T_e = 3.7 and find the Hopf bifurcation to be subcritical (e.g., Marsden and McCracken 1976). This means that before the Hopf bicurcation, the fixed point is surrounded by an unstable limit cycle and that at the Hopf bicurcation, the solution can jump to a limit cycle with a finite radius (in A, B, Γ space) if such a solution exists beyond the unstable limit cycle. In that alternative, one has the coexistence of a fixed point and a limit cycle and thus, the possibility of a hysteresis. Another interesting possibility, is the collision of such a limit cycle with the second unstable fixed point which could result in the formation of a homoclinic orbit, and a resultant strange attractor (e.g., Guckenheimer 1984). Unfortunately, the investigation of these possibilities necessitates the consideration of higher order nonlinearities.

The dashed line in figure 1, marked FPB, denotes the locus where the two fixed points merge. Above that line there are no fixed points and the AE (30 and 31) do not seem to give rise to bounded behavior. Higher order AE may be necessary to describe the behavior of the models in that region.

Figure 2 exhibits the behavior of the amplitudes A and B and the phase Γ'=Γ + π of a sequence of 8 solar mass models along a constant T_e (= 5600 K) line in the H-R diagram, which indeed have the properties just described, namely the approximate arccot behavior for Γ and the cosec behavior for the amplitudes. One notes also that the minima in the amplitudes and the crossing of π/2 for Γ' do not occur exactly at the resonance as mentioned above. Similar diagrams occur for different masses and composition.

Figure 3 shows the surface velocity as a function of time for the various models of the sequence. One clearly observes the peregrination of the secondary bump from the descending (right) side to the ascending (left) side as P_2/P_0 increases through the resonance, (the so-called Hertzsprung progression) which is in rough agreement with both numerical hydrodynamic models and observation.

In figure 4a, we show the temporal behavior of the velocity profiles for a number of mass-shells in the star. For illustrative purposes, the curves have been displaced vertically with respect to each other and the inner (bottom) curves have been enhanced. This figure should be compared to a similar figure obtained from numerical hydrodynamic calculations , figure 4b, which has been adapted from Christy (1968) for a similar, but not identical model. (The plotted mass zones are not the same, nor are the scalings for the inner

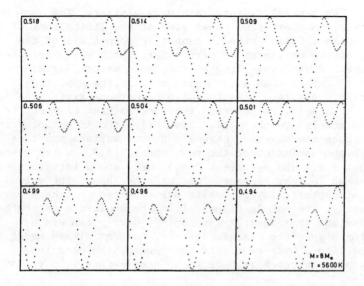

Figure 3

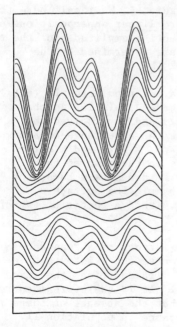

Figure 4a

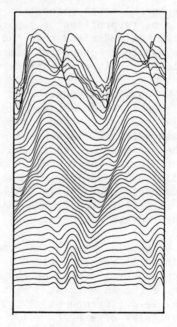

Figure 4b

zones.) The general features are very similar. The comparison, however, also shows some of the limitations of the AE formalism. Clearly the AE cannot reproduce all of the physics of the envelope; in particular, shock waves are beyond its scope. However, as long as the shock waves play an incidental rather than determining role in the amplitude saturation mechanism, this is not a bothersome limitation if surveys of models, rather than the construction of specific and detailed models are the goal. Similarly, the AE cannot pretend to reproduce the observed strongly skewed surface profiles, although the inclusion of the next order, $|z_1\rangle$, which contains a $\cos(3\omega t + \phi_3)$ and a $\cos(4\omega t + \phi_4)$ term, would introduce some improvement in that respect. On the other hand, the very dilute atmospheric layers, which are expected to behave very nonlinearly, are not instrumental in the amplitude saturation mechanism and therefore, again, do not overly concern us. If a model for a specific star is desired, one can always model its behavior with a hydrocode. In that sense, the numerical hydrodynamic approach is seen to be complementary to the AE formalism.

Figure 5 shows separately the fundamental and second overtone LNA modes which make up figure 4. We recall here that the velocity profiles of figure 4 are <u>linear</u> superpositions of the two LNA modes and that the respective amplitudes of the two modes and their phase relationship are determined by the <u>nonlinear</u> amplitude equations (40-42).

We have also constructed models for population II (metal poor) Cepheids (paradigmatized by BL Her), similar to the models of King, Cox and Hodson (1981) and have solved the corresponding AE. Again, we observe the appearance of a bump in the vicinity of the resonance ($P_2/P_0 \simeq 1/2$) and a Herzsprung progression of that bump. As in the case of the population I Cepheids, whenever our AE predicts a bump, the hydrodynamic models also exhibit a bump. The surface velocities inferred from our AE are within a factor of two of the hydrodynamic models.

We recall here that the mere existence of a resonance condition is not sufficient to guarantee the existence of a Herzsprung progression of the bumps. In addition, the fixed point of the AE must also be stable. It is therefore incorrect to say (e.g., Carson et al. 1981) that the absence of bumps in the vicinity of a $P_1/P_0 \simeq 1/2$ resonance in population II Cepheids, e.g., constitutes an argument against the resonance origin of the bumps. On the care of the $P_1/P_0 \simeq 1/2$ resonance, the necessary condition for the existence of attractors, namely $\kappa_0 + \kappa_1 < 0$ is not satisfied (at least not when convection is neglected).

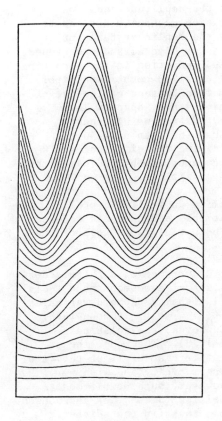
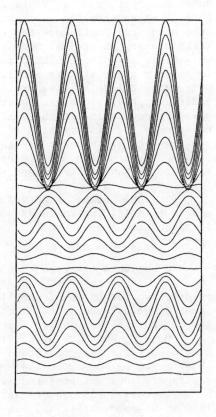

Figure 5a   Figure 5b

The correspondence of our results with those results of numerical hydrocodes suggests that the amplitude equation formalism captures the basic saturation mechanism and can be used with some confidence to study the nonlinear behavior of large classes of stellar models. As we have already mentioned, we achieve an enormous (hundredfold) reduction in the computing cost for a given stellar model. A detailed comparison with hydrodynamic models, however, is marred because of the use of different physical input (equation of state, opacity law, boundary conditions at the surface): differencing schemes, coarseness and distribution of mesh points, to name just a few. An effort is under way to minimize the differences between our approach and a specific numerical hydrodynamic code so as to make a tighter comparison possible.

The application of this formalism to realistic stellar models is still in its infancy. At this stage we are reduced to speculations about the further potential usefulness of the approach.

It has been observed that the occurrence of a third order resonance of the type $\Omega_\alpha + \Omega_\beta \simeq \Omega_\gamma$ is quite common among radial pulsators. For example, some low luminosity Cepheid variables (Beat Cepheids) obey the relation $\Omega_0 + \Omega_1 \simeq \Omega_3$ (Simon 1978) and for some AI Vel stars (Dwarf Cepheids) the relation $\Omega_0 + \Omega_1 \simeq \Omega_4$ seems to hold (Simon 1979). The modelling of Beat Cepheids still constitute a puzzle: about 11 of them are observed to pulsate with a double periodicity and with apparently constant amplitudes. However, such double-mode pulsation has defied numerical modelling (Simon, Cox and Hodson 1980). Dziembowski (1982) has shown that (in the adiabatic limit at least) the AE have no stable fixed point.

Equation (T1a), or a generalization of T2 with the inclusion of thermal modes, is also of interest in radial stellar pulsators. In the real branch of LNA modes, there exist some (the 'secular' modes) whose eigenvalues can be of the same order and even smaller than the growth rate(s) of the excited oscillatory mode(s). It is conceivable that the interaction of the oscillatory mode with these thermal modes gives rise to a regular or irregular modulation of the oscillations. An intriguing regular modulation of an essentially monoperiodic Cepheid has been observed (HD7803, e.g., Burki et al. 1983). We are planning to test such a possibility with realistic stellar models.

Finally, we want to also briefly mention nonradial pulsators. It is well known (e.g., Ledoux and Walraven 1958) that the adiabatic problem has a large number of poloidal modes which asymptotically split into two branches; the p-modes,

which have an accumulation point at infinity for large quantum
numbers and the g-modes, which, on the other hand, have an
accumulation point at the origin. In addition, in the absence
of rotation and magnetic fields, there is a $(2\ell + 1)$ fold
degeneracy in m. The possibilities of resonances and of the
simultaneous excitation and interaction of a large number of
modes thus exist, together with the increased likelihood of
strange attractors. In the case of the interaction of a bunch
of contiguous modes, the large number of coupled AE make the
latter impractical to use. In that case, it might be
preferable to introduce a continuous mode index and transform
the AE into coupled integro-differential equations; if the
kernel describing the interactions is sufficiently peaked it
may then be possible to reduce the integral operator to a sum
of derivatives, in which case the AE become a set of coupled
partial differential equations in time and in mode number.
However even then serious difficulties remain for the nonradial
case. Firstly, the numerical work is much more involved, even
for the LNA problem; secondly, and more seriously, the growth-
and decay-rates are physically less certain as they generally
involve the small scale structure of the stellar model and the
uncertainties of time-dependent convection.

Another problem which needs to be broached is that of the
'modal selection'. The AE which we have considered so far
apply to a given nonevolving model. This is a good approximation for studying the existence of attractors, since only the
envelope partakes in the pulsation, whereas the nuclear
reactions in the core effect an evolution which occurs on a
very long timescale. In general, several attractors coexist
and it is then of interest to explain how the observed variable
stars are distributed among these possible states of
pulsation. In the case of a hysteresis such a study can
provide a strong constraint on stellar evolution; for example,
stars traversing the instability strip from the right may find
themselves in a different attractor from those entering the
strip from the left. Along the same lines, one has the problem
of understanding the evolution of a pulsating stellar model
through an internal resonance, in particular, the nature of the
oscillation during the resonance and the selection of the
attractor as it emerges from the resonance. Since the AE
constitute asymptotic expansions an approach based on matching
these asymptotic expansions seems to be the appropriate route
of attack and is currently being pursued. (A similar, but much
simpler problem related to the roll during reentry of a
spacecraft has been considered by Ablowitz et al. 1973).

The AE (for the amplitudes a,b) were derived under the
assumption that the amplitude modulations, whether regular or
irregular, occur on the long timescale. The reconstructed AE

for the amplitudes $\tilde{a}, \tilde{b}$) can, in principle, give rise to rapid variability, i.e., on a dynamic timescale, but it is unlikely that they then provide an accurate representation of the behavior of the original system. It is therefore the relatively slow regular or irregular stellar variability which could be describable by the above types of amplitude equations. One may speculate that the rapid, irregular, but small fluctuations of the amplitudes and phases which are observed in some stars occur because the system is very weakly dissipative only and is already chaotic in the conservative (Hamiltonian) limit (cf. Perdang 1985). It seems likely to us that the Semi-Regular and especially the Irregular Variables, on the other extreme, behave in a strongly nonlinear way in which dissipation also plays a dominant role. Buchler and Regev (1982) have proposed a physical mechanism and constructed an idealized model which exhibits such irregular variability. It is interesting that the dynamics of this model is akin to that of the well known Baker-Moore-Spiegel (1968) model oscillator, which is one of the first dissipative systems in which a strange attractor has been found.

We have seen that the asymptotic perturbation formalism gives rise to AE, the solution of which seems to describe very well the behavior of some specific stellar models (the Classical Bump Cepheids and the Pop. II Cepheids). As has been mentioned, these AE also admit chaotic solutions (e.g., Wersinger et al. 1980), but not in the range of parameter values corresponding to Bump Cepheids. We are actively engaged in the exploration of a variety of different stellar models and the search for irregular behavior.

REFERENCES

Ablowitz, M. J., Funk, B. and Newell, A. C., 1973, Stnd. Appl. Math. L11, 51.
Baker, N., Moore, D. and Spiegel, E. A., 1966, Astrophys. J. 71, 845.
Buchler, J. R., 1978, Astrophys. J. 220, 629.
Buchler, J. R. and Goupil, M. J., 1984, Astrophys. J. 279, 394.
Buchler, J.R. and Kovacs, G., 1985, (in preparation).
Buchler, J. R. and Pesnell, W. D., 1984, Astrophys. J. (submitted).
Buchler, J. R. and Regev, O., 1982a, A & A 114, 188.
Buchler, J. R. and Regev, O., 1982b, Astrophys. J. 263, 312 (see also article in this volume).
Buchler, J. R., Yueh, W. R. and Perdang, J., 1977, Astrophys. J. 214, 510.
Burki, G., Mayor, M. and Benz, W., 1982, A & A 109, 258.
Carson, T. R., Stothers, R. and Vemury, S. K., 1981, Astrophys.

J. 244, 230.
Castor, J. I., 1971, Astrophys. J., 166, 109.
Christy, F. R., 1968, Quart. J. R. Astr. Soc. 9, 13.
Christy, F. R., 1975, in Cepheid Modelling, ed. Fischel & Sparks.
Coullet, P. and Spiegel, E. A., 1983, SIAM J. Appl. Math. 43, 776.
Cox, A. N., 1980, Ann. Rev.Astron. Astrophys. 18, 15.
Cox, J. P., 1978, in Current Problems in Stellar Pulsation Instabilities, eds. D. Fischel, J. R. Lesh and W. M. Sparks.
Cox, J. P. and Giuli, R. T., 1968, Principles of Stellar Structure, Gordon & Breach.
Dziembowski, W., 1982, Acta Astron. 32, 147.
Dziembowski, W. and Kovacs, G., 1984, Monthly Not. Roy. Astron. Soc. 206, 497.
Guckenheimer, J., 1984, SIAM J. Math Anal. 15, 1.
Hodson, S. W., Cox, A. N. and King, D. S., 1982, Astrophys. J. 253, 260.
Holmes, P. J., 1981, Physica 2D, 449.
King, D. S., Cox, A. N. and Hodson, S. W., 1981, Astrophys. J. 244, 242.
Klapp, J., Goupil, M. J. and Buchler, J. R., 1984, Astrophys. J. (submitted).
Ledoux, P. and Walraven, T. H., 1958, Handbuch der Physik, vol. LI.
Marsden, J. E. and McCracken, M., 1976, The Hopf Bifurcation and Its Applications, Appl. Math. Sci. 19, Springer Verlag.
Nayfeh, A. H. and Mook, D. T., 1979, Nonlinear Oscillations, J. Wiley Publ.
Perdang, J., 1985, see article in this volume.
Pesnell, W. D. and Buchler, J. R., 1985, Astrophys. J. (submitted).
Regev, O. and Buchler, J. R., 1981, Astrophys. J. 250, 769.
Simon, N. R., 1979a, A & A 74, 30.
Simon, N. R., 1979b, A & A 75, 140.
Simon, N. R., 1981, Astrophys. J. 247, 594.
Simon, N. R., Cox, A. N. and Hodson, S. W., 1980, Astrophys. J. 237, 550.
Simon, N. R. and Schmidt, E. G., 1976, Astrophys. J. 205, 996.
Simon, N. R. and Theays, T. J., 1981, Astrophys. J. 265, 996.
Spiegel, E. A., 1981, Lecture Notes, Summer Study Program in Geophysical Fluid Dynamics, Woods Hole Oceanographic Institution, WHO-81-102.
Stellingwerf, R. F., 1975, Astrophys. J. 195, 441.
Takeuti, M. and Aikawa, T., 1981, Sci. Rep. Tohoku Univ., 8th Ser. 2, No. 3.
Vandakurov, Yu., 1981, Sov. Astron. Lett. 7, 128.
Wersinger, J. M., Finn, J. M. and Ott, E., 1980, Phys. Fluids 23, 1142.

CHAOS AND NOISE

T. Geisel

Abstract

   An observed irregular signal may be due to chaos, noise, or both. This paper reviews methods for reconstructing a strange attractor from an observed signal if the underlying dynamics is chaotic. The influence of external noise acting on chaotic systems is illustrated in some examples. The statistical properties of intermittent chaos are reviewed, which might turn out to be particularly relevant in astrophysical problems.

1. Introduction

   In physics one often observes signals that seem to be irregular. This is true in particular for astrophysical observations. The astrophysical layman is puzzled by the variety of such observations, which are not only confined to what is known as irregular variables. There are e.g., X-ray sources which pulse semiregularly, and even some of the so-called regular variables do not appear to be entirely regular. We are led to the following questions: 1. Is it possible to define the regularlity or irregularity of the signal? 2. Since the signal is our main source of information about the object, can the irregularity of the signal tell us abouts its origins? 3. How many coupled variables are needed to model the irregular dynamics?

   In recent years techniques have been developed in the context of chaotic phenomena that are suitable to such a situation, and may help to answer the above questions. In particular, given only an irregular signal, it is possible to distinguish whether its origin is low dimensional chaos or noise. In the first case it is even possible to reconstruct the strange attractor, the ID-card of the chaotic behavior from the irregular signal. In Sect. 2 these techniques will be reviewed from a practical point of view.

A signal may not only be due to chaos or noise, it may also be due to chaos plus noise. The influence of external noise on a chaotic system is illustrated in Sect. 3. A phenomenon which is frequently encountered in astrophysics is an intermittent signal, e.g., from X-ray bursters. Intermittency may also arise as a typical phenomenon in chaotic systems. This is reviewed in Sect. 4, since it might turn out to account for astrophysical intermittent signals, although other explanations are possible.

2. Phase Space Reconstruction

One must clearly distinguish chaotic motions from random noise. When erratic motions are due to thermal fluctuations in physical systems, a large number (10^{23}) of particles and degrees of freedom are involved, which make the measurements look noisy. In Brownian motion, e.g., the erratic motion of a heavy pollen particle is due to collisions with the surrounding molecules of a liquid, and thus reflects the irregular dynamics of the 10^{23} molecules of that liquid. In contrast to this random noise, we understand chaos as an erratic motion which is generated by a (nonlinear) deterministic system with few degrees of freedom in the absence of external noise. Some of the simplest examples are found in periodically driven anharmonic oscillators [1].

For the sake of simplicity, consider first an undriven harmonic oscillator, whose phase space is two dimensional, consisting of position x and velocity v.

$$\dot{x} = v$$
$$\dot{v} = -\gamma v - \omega_o^2 x \qquad (2.1)$$

In the presence of damping γ, the motion settles down at $x = 0$, $v = 0$ asymptotically. Any initial condition x_o, v_o in the plane leads to this point, the attractor. The dimension is thus reduced from two to zero. If this damped harmonic oscillator is driven periodically the asymptotic motion is a harmonic oscillation with the frequency of the driving. Any initial condition leads to this limit cycle, which forms a closed curve in phase space. The dimension is thus reduced to one. Generally, the presence of dissipation in a dynamical system causes a contraction of phase volume, such that the asymptotic motion is confined to a space of lower dimension.

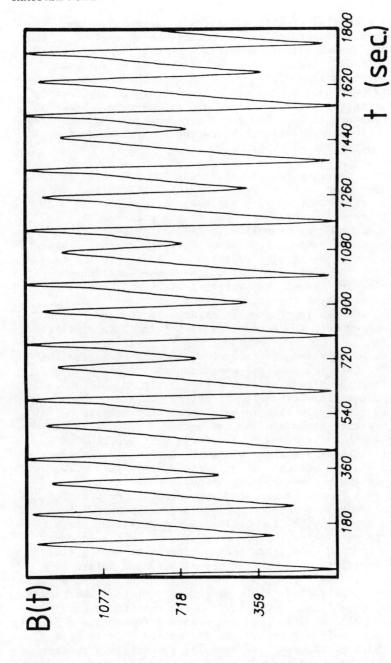

Fig. 1 Signal observed in an irregularly oscillating chemical reaction. The concentration $B(t)$ of one of the chemical constituents (bromide ions) was measured as a function of time.

If the dimension of this subspace is larger than two, chaotic motion is possible [2-4]. Besides being aperiodic, the motion on the chaotic attractor has a sensitive dependence on initial conditions, i.e., motions starting from neighboring initial conditions separate exponentially in the course of time. Low dimensional attractors having this property must be strange looking topological objects and are therefore known as strange attractors [2-4]. The sensitive dependence practically prevents long-term predictions, since the initial conditions are usually not known exactly.

Strange attractors have a particular importance in systems where the dimension of phase space is high or infinite. This is the case, e.g., in fluids, since they are governed by partial differential equations. Due to the presence of dissipation, the motion in multi-dimensional phase space may settle down on a strange attractor of very low dimension which acts like a low dimensional trap. The asymptotic motion is then completely determined within a low-dimensional subspace. If the topology of the strange attractor is known, then the asymptotic dynamics of the entire system are known. This explains the central importance of strange attractors for the chaotic dynamics even if the systems are of infinite dimension.

In view of this importance, one would like to derive the strange attractor from physical experiments. Often, however, only one irregular signal or time series has been observed. This is also a typical situation in astrophysics. Nevertheless, it is possible in many cases to reconstruct the strange attractor from a single time series [5]. One of the procedures has been proposed by the Santa Cruz Dynamical Systems Collective [6]. It consists in choosing the observed signal $x(t)$ and its successive derivatives $\dot{x}(t)$, $\ddot{x}(t)$, ..., as coordinates spanning the subspace of phase space to be reconstructed. It has been tested successfully for a set of differential equations due to Rössler [6,7]. It cannot be applied, however, if the time series is not detailed enough to determine the derivatives. Another procedure, which is more appropriate in such cases, has been suggested by Ruelle (in private communications): Use the signal $x(t)$ and delayed signals $x(t+T)$, $x(t+2T)$, ... as coordinates spanning the phase space to be reconstructed. This procedure will be illustrated below in the example of a chaotic chemical reaction. From a mathematical point of view, the existence of embeddings of differentiable manifolds into R^n is guaranteed by Whitney's embedding theorem. Takens has shown [8] that it is a generic property that the above procedures yield an embedding.

Random noise as described above results from the irregular dynamics of 10^{23} interacting particles. The above procedures

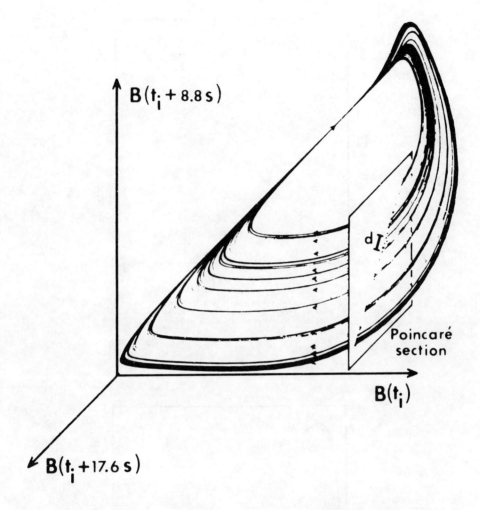

Fig. 2 Phase space reconstruction of a strange attractor from an irregular signal B(t). The tentative phase space is spanned by the signal and its delayed versions.

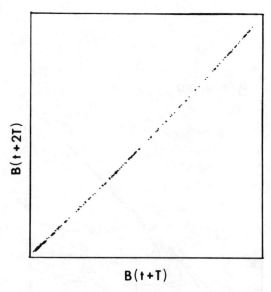

Fig. 3 a) Cross-section (Poincaré section) of an attractor as indicated by the plane in Fig. 2. This Poincaré section was derived from the data of Fig. 1.

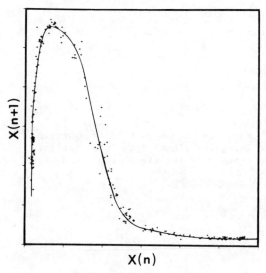

b) Return map obtained by recording the value of $B(t+T) = X$ for successive intersectons n and n+1 with the Poincaré section and plotting them as abscissa $X(n)$ and ordinate $X(n+1)$ [9].

are able to distinguish a chaotic signal from noisy signals. This is because phase space reconstruction will not work in low dimension if the dynamics is determined by 10^{23} degrees of freedom.

In order to demonstrate its practical applicability, I will illustrate the technique of phase space reconstruction in an example. The first application and most extensive studies were made by Roux et al. for the dynamics of chemical reactions of the Belousov-Zhabotinskii type [9]. The experiments are done in a stirred chemical reactor, into which different chemicals are injected in certain concentrations. The reaction involves more than 20 individual reactions and as many chemical constituents. The phase space therefore consists of at least 20 variables. Only one of those variables, however, the concentration of bromide ions B(t), was measured as a function of time. Fig. 1 shows an example which exhibits apparently aperiodic oscillations of the bromide ion concentration. Other aperiodic signals have been observed. The phase space reconstruction of one such signal is shown in Fig. 2. An orbit was constructed as a position vector as a function of time in a 3-dimensional coordinate system with the components B(t), B(t+8.8sec), B(t+17.6sec). The value chosen as a time delay is not crucial for the method work. The dimension of phase space must be increased successively until additional structure ceases to appear. In the present example, this was the case in 3 dimensions. The question of how many variables are needed to model the dynamics is thus answered. The strange attractor is reconstructed and noise can be ruled out as the major cause of the irregularity of the signal.

It is useful to construct a Poincaré section as is indicated by the plane in Fig. 2. A point is marked in the plane whenever the orbit intersects it. Fig. 3a shows a Poincaré section obtained in this way for the example of Fig. 1. The plane can be shifted to other positions along the attractor and thereby give a view of the attractor through a number of cross-sections. The points in Fig. 3a indicate that the cross-section is almost a one dimensional set. (For topological reasons it cannot be exactly one dimensional.) When this is the case we expect that the dynamics can be described to a good approximation by a one dimensional return map. Such a return map has been obtained from Fig. 3a and is shown in Fig. 3b. Since in Fig. 3a the coordinate B(t+2T) is correlated to a large extent with the value of B(t+T), only the latter is retained and is now denoted by X. In the course of time, an orbit successively intersects the plane. If the system is deterministic and the phase space construction was done correctly, the future evolution of the orbit is determined by its position in the plane at any intersection. In particular, X(n), the value of B(t+T) at the n-th intersection, should determine its successor X(n+1). In Fig. 3b X(n+1) was plotted against X(n). Apart from some scatter, the data points uniquely define a function, the return map.

Fig. 3b does not only demonstrate that the system is deterministic and that the phase space reconstruction has been done correctly, it also represents a central item for the dynamical evolution. As long as we are not interested in the motion of the orbit between two intersections, the return map incorporates the dynamics for all later times and determines its statistical properties. In going from Fig. 1 to Fig. 3b, the dynamics of a complex system of more than 20 coupled variables has thus been boiled down to its essentials, a simple return map. If the cross-section of the attractor in Fig. 3a had not been nearly one dimensional, a return map could have been constructed in two dimensions only. Algorithms, which can be applied to determine the dimensionality of an attractor directly from an irregular signal, are reviewed in the articles by Grassberger [10] and Guckenheimer [11]. The technique of phase space reconstruction has also been applied successfully to other systems, in hydrodynamic turbulence e.g. to baroclinc flow [12] and Couette-Taylor flow [13].

3. Chaos Plus Noise

Many dynamical systems in nature are subject to a certain amount of external noise. How does the noise influence the dynamical behavior? How can one distinguish periodic from chaotic behavior in such a case? In order to understand the influence of noise we will shortly review some phenomena in the noise-free case.

Once we know that a system can be described by a map similar to the one in Fig. 3b, we may ask what happens if the control parameters of the system change and the map deforms in one or another way. It is well known by now that changing the steepness of the map causes the period-doubling scenario to take place. Let us concentrate on this scenario to illustrate the role of noise. As long as we are interested in universal properties, we can consider classes of maps and need not know the explicit form of a map. For a review of these properties we may simply consider one representative of a class. A prominent example for the period-doubling scenario is the logistic map

$$x_{t+1} = r(x_t - x_t^2), \qquad (3.1)$$

where r is a parameter controlling the steepness of the map.

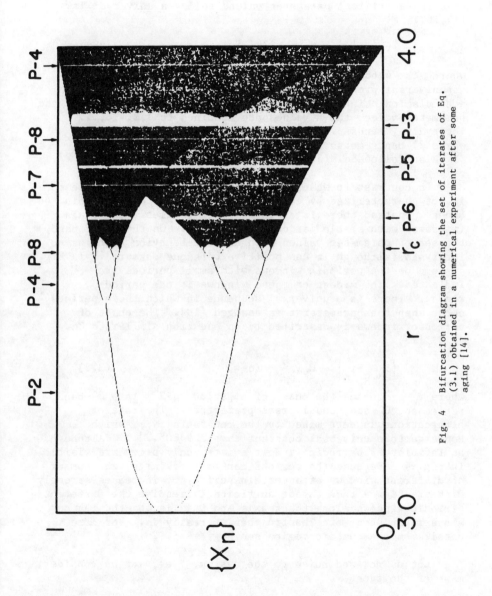

Fig. 4 Bifurcation diagram showing the set of iterates of Eq. (3.1) obtained in a numerical experiment after some aging [14].

Fig. 4 recalls the bifurcation diagram in the noise-free case [14]. For 1000 values of r in the interval [3,4], 700 iterated values of x were plotted after an initial 500 iterations. There are critical values r_k where a period 2^k appears and period 2^{k-1} becomes unstable. It is well known that these critical parameter values follow a universal law [15,16].

$$|r_k - r_\infty| \sim A \, \delta^{-k} \qquad (k \to \infty), \qquad (3.2)$$

where $\delta = 4.66920...$, is a universal constant and A a nonuniversal prefactor. The chaotic regime sets in at the accumulation point r_∞ denoted r_c in Fig. 4. Depending on the parameter value, the x values are confined to $1,2,4,8,...$ intervals ("bands"). There are band-merging parameters $\tilde{r}_k$ where 2^k bands merge into 2^{k-1} bands. These parameters $\tilde{r}_k$ also follow Eq. (3.2) and accumulate at r_c [17].

In contrast to what Fig. 4 suggests, the region above r_c is not characterized by chaotically scattered points within the bands. Instead there is an infinity of parameter intervals (windows) with stable periodic attractors. On the other hand, the set of parameter values with chaotic behavior contains no intervals (although it has positive Lebesgue measure). In Fig. 4, some of the periodic windows with small periods are indicated. The widest of these windows is the period-3 window. There is a universal sequence in which these periods occur when the parameter r is changed [18]. The onset of periodic windows is described by an equation similar to Eq. (3.2) [19].

$$|r_{k,q} - \tilde{r}_k| \sim C_k \gamma_k^{-q} \qquad (q \to \infty), \qquad (3.3)$$

where $r_{k,q}$ denotes the onset of a period $q \cdot 2^k$ in a 2^k-band regime and C_k are nonuniversal prefactors. The rate of bifurcations is determined by the constants γ_k, which converge to a universal constant $\gamma = 2.94805...$. Although an infinity of periodic windows exists, only a few are visible in Fig. 4. Because the periods can be very large, the number of different windows of a certain period, p, increases strongly with p. Fig. 4 thus cannot sufficiently resolve the different periodic windows along the r-axis and this is why the region above r_c appears more chaotic than it really is. For more details on the chaotic region see e.g., Ref. 20.

Let us now add noise to the system, i.e., let us consider maps of the type

$$x_{t+1} = r(x_t - x_t^2) + \sigma \xi_t. \qquad (3.4)$$

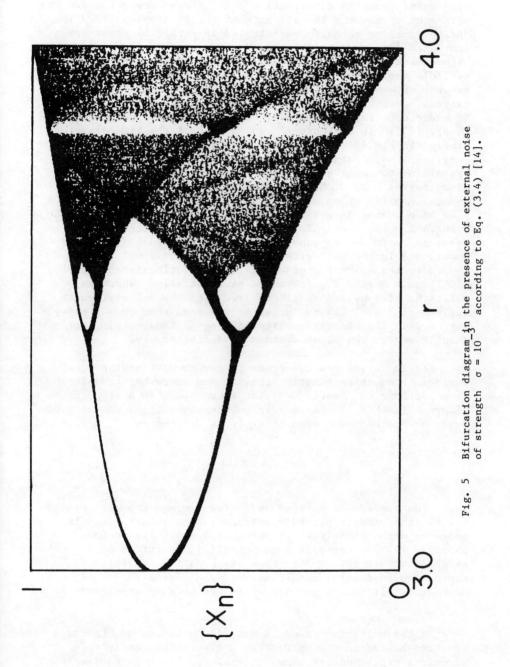

Fig. 5   Bifurcation diagram in the presence of external noise of strength $\sigma = 10^{-3}$ according to Eq. (3.4) [14].

Here, ξ_t is a δ-correlated random variable (white noise) with zero mean, Gaussian distribution and standard deviation 1. The parameter σ measures the strength of the external noise. The period-doubling scenario was studied in detail by Mayer-Kress and Haken [21] and Crutchfield, Huberman and Farmer [14]. The bifurcation diagram for $\sigma = 10^{-3}$ is shown in Fig. 5. In the periodic regime, one recognizes noisy periods 2 and 4. It is apparent that the orbits are more or less susceptible to the external perturbation, depending on the parameter r. This can be understood in terms of a linear and nonlinear response theory [22]. In the chaotic region most of the fine structure present in Fig. 4 has disappeared. Only a noisy period 3 has survived, the others are covered up by noise. Also, the transition to chaos now appears to proceed directly from a noisy period 4 to a chaotic 4-band situation, i.e., noise induces a bifurcation gap such that attractors with period p > 4 or band number larger than 4 are no longer observable in the bifurcation sequence. This was observed by Crutchfield and Huberman for a driven anharmonic oscillator [23]. It can be understood in the following way: In a period doubling sequence, the splitting of orbits which distinguishes a period 2^{k+1} from a period 2^k, decreases exponentially. When the splitting becomes too small compared with the noise strength σ, a period 2^{k+1} can no longer be distinguished from a noisy period 2^k. The unobservability of long periods within the chaotic region can be understood in a similar way.

Fig. 5 raises the questions of how to distinguish noisy periodic from noisy chaotic behavior and where the transition from periodic to chaotic behavior takes place in a bifurcation diagram like Fig. 5. An answer to these questions can be given based on the Lyapunov exponent. It is defined as

$$\lambda = \langle \ln \left| \frac{dx_{t+1}}{dx_t} \right| \rangle, \qquad (3.5)$$

where the average is defined as a time average along an orbit x_t. In the case of periodic attractors, neighboring orbits converge exponentially to it and thus $\lambda < 0$. If, on the average, $\lambda > 0$, there is an exponential separation of neighboring orbits. $\lambda > 0$ thus measures the sensitive dependence on initial conditions and the degree of unpredictability of a chaotic attractor as discussed in Sect. 2.

In the noise-free case, λ was first studied by Shaw as a function of the control parameter r [24]. The result of a numerical experiment is shown in Fig. 6a. As r increases in a sequence of period-doubling bifurcations, λ becomes zero at

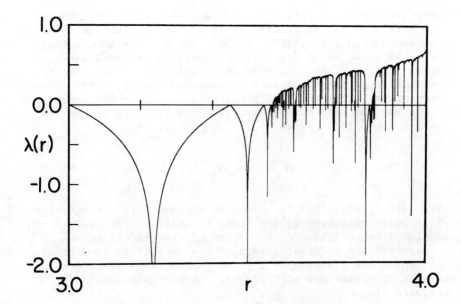

Fig. 6 a) Lyapunov exponent λ (Eq. (3.5)) versus control parameter r calculated for 30 000 iterations of Eq. (3.1) [14].

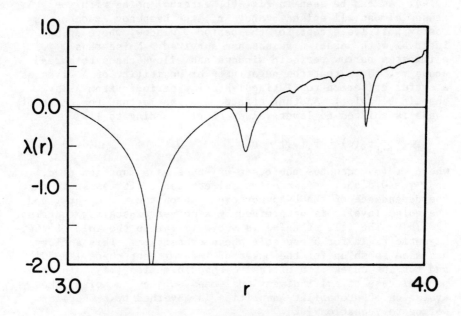

b) Lyapunov exponent in the presence of external noise ($\sigma = 10^{-3}$). The average is carried out along a perturbed orbit according to Eq. (3.4) [14].

every bifurcation point, μ_k, and has logarithmic divergences to $-\infty$ in between (at superstability). These divergences are also present in the periodic windows above r_c, although they cannot be resolved in Fig. 6a. However, there are a large number of spikes with negative values of λ indicative of the presence of periodic and narrow windows. Near r_c, the Lyapunov exponent exhibits universal properties [25,26]. For example, at the band merging parameters r_k, where it is positive, it follows a power law [25]

$$\lambda(\tilde{r}_k) \sim (\tilde{r}_k - r_c)^\beta, \qquad (3.6)$$

with $\beta = \ln 2/\ln \delta = 0.4498...$ This is analogous to the behavior of an order parameter near a phase transition. In fact, λ can be viewed as the disorder parameter of the chaotic region, as it measures the increasing sensitivity to initial conditions and unpredictability above r_c. Of course, this is true only on the set of parameters with chaotic behavior and not in the periodic windows.

The Lyapunov exponent was also studied in the presence of noise [21,14]. In numerical experiments the average in Eq. (3.5) was taken along a perturbed orbit generated by Eq. (3.4). As can be seen in Fig. 6b, external noise with $\sigma = 10^{-3}$ erases almost all spikes. Above r_c, the Lyapunov exponent is mostly positive except for the period 3 window, where an interval with negative values has survived. Noise thus eliminates narrow periodic windows and allows chaos to spread above r_c. Note that the negativity or positivity of λ gives us a useful criterion to distinguish noisy periodic from noisy chaotic behavior. Another effect of noise is that the onset of chaos is shifted to lower values of r according to [14]

$$\left| r_c(\sigma) - r_c(0) \right| \sim \sigma^\gamma \qquad (3.7)$$

where $r_c(\sigma)$ denotes the onset of chaos as a function of σ, and $\gamma = 0.8153...$ is a universal constant. It was shown that the dependence of the Lyapunov exponent on the parameter r and the noise level σ is determined by a universal scaling function [27,28]. The role of noise is very similar to the role of the magnetic field for a magnetic phase transition. This analogy has also shown up for the onset of intermittent chaos [29-30] and for the onset of a diffusive chaotic motion [31]. In the general case of multi-dimensional maps subject to noise, the evolution of probability densities is governed by a Chapman-Kolmogorov equation [32].

4. Intermittent Chaos

The observer of an irregular signal will often notice a particular statistical behavior. As discussed in Sect. 2, statistical properties are sufficiently determined by the return map associated with a dynamical system or a chaotic signal. Depending on the map, the statistical behavior may consist in a band structure as in the period-doubling case, or it may appear as a random walk [31]. It may also consist in an alternating sequence of seemingly periodic motion and short chaotic bursts. The latter phenomenon is called intermittency and was clarified by Pomeau and Manneville [33,34]. Similar signals are observed in astrophysics e.g. from X-ray bursters. It is still an open question, however, whether the Pomeau-Manneville intermittency can account for them since other explanations are possible. The following section may help to elucidate this question.

Intermittent signals may look very different from each other. The signal shown in Fig. 7 is an optical signal stemming from a hydrodynamic experiment by Dubois et al. [35]. The duration of different episodes appears to be random. Here as in Sect. 2, one may ask whether the shape of the signal is due to chaos or noise. An answer was obtained in the following way. Successive maxima were recorded and denoted by I_n. Then I_{n+2} was plotted versus I_n as shown in Fig. 8 [35]. The data points clearly define part of a return map and thereby demonstrate the deterministic origin of the signal. The map has a fixed point $I_{n+2} = I_n$, where its slope is very close to 1. At some distance from the fixed point, the map has parts acting as a reinjection mechanism to the vicinity of the fixed point. These are not included in the figure.

Maps of this type were previously studied by Manneville and shown to generate 1/f-noise [34]. (It is not possible to avoid the term noise here; note that it is not meant in the same sense as in Sects. 2 and 3). Let us assume for simplicity that the slope in the fixed point is exactly 1 and consider only the right-hand part of the map (even n). Its expansion around the fixed point then reads

$$x_{t+1} = x_t + ax_t^2 + \ldots \quad (x \geqslant 0) \quad (4.1)$$

where $x = I - I_F$ denotes the distance from the fixed point I_F. The time variable n is now denoted by t as before, and $a > 0$. The closer an orbit starts to the origin, the longer it remains in its vicinity. For the long-time behavior, therefore, only the motion near x=0 is relevant. There the displacements per time step $x_{t+1} - x_t$ become arbitrarily small. In a continuum approximation for the long time limit they can be replaced by

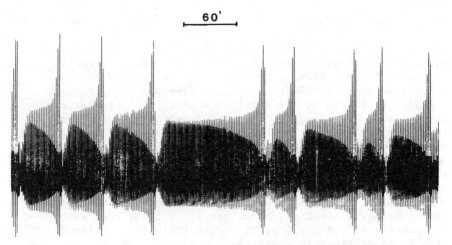

Fig. 7 Optical signal obtained in a Rayleigh-Bénard experiment displaying type-III intermittency [35].

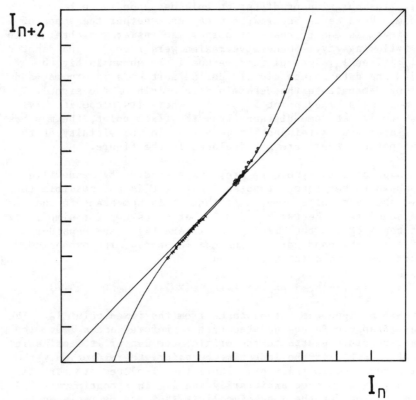

Fig. 8 Return map derived from the maxima I_n of Fig. 7. Squares stand for even n, crosses for odd n [35].

the derivative dx/dt, whereby Eq. (4.1) turns into a differential equation

$$\frac{dx}{dt} = ax^2. \qquad (4.2)$$

Its solution as a function of time t and starting point x_o is

$$x_t = [\frac{1}{x_o} - at]^{-1}, \qquad (4.3)$$

which is a monotonously increasing function. We may ask how long x_t stays near $x=0$, or more precisely what is the waiting time T until x_t reaches a given boundary x_c. From $x_T = x_c$ and Eq. (4.3) one obtains

$$T = \frac{1}{a}(\frac{1}{x_o} - \frac{1}{x_c}). \qquad (4.4)$$

We can take x_c to be the boundary beyond which the map reinjects the orbit to a new value x_o near the origin and a new episode according to Eq. (4.3) starts. The details of the reinjection mechanism are not crucial. In fact, it is sufficient to assume that the injection includes a vicinity of the origin and that the distribution of injection points is smooth enough to be expanded in a power series [36]. The distribution of waiting times $\Psi(T)$ then follows as $|dx_o/dT|$ from Eq. (4.4)

$$\Psi(T) = 2a(2 + aT)^{-2}. \qquad (4.5)$$

Such an expression was first given by Manneville [34]. It represents the distribution of durations of individual episodes like those in Fig. 7. This distribution has a long-time tail, i.e. it decays like T^{-2}, less than exponentially. This was also observed in the experiment by Dubois et al. [35]. The long-time tail leads to a 1/f-type power spectrum, as was noted by Manneville

$$S(\omega) \sim \frac{1}{\omega \ln^2 \omega} \qquad (\omega \to 0). \qquad (4.6)$$

A generalization can be made to maps, in which the term ax_t^2 of Eq. (4.1) is replaced by ax_t^z with $z > 3/2$ [37].

Above, I have outlined what is called type-III intermittency. Pomeau and Manneville have also distinguished other types, in particular type-I intermittency, which is found more frequently [33,38]. It arises before the onset of periodic windows of period-doubling systems (Fig. 4). In contrast to the above case it does not exhibit 1/f-spectra. The influence of external noise was studied in refs. 38,29,30.

In the present paper I have tried to illustrate techniques that can be applied to an irregular signal in order to gain insight into its chaotic or noisy origins. I have shown how return maps can be extracted and how they determine statistical properties. At present it is still a question to what extent chaos plays a role in astrophysical observations. However, very detailed records of observations are available similar to those by Makino et al. on X-ray bursters [39]. Their evaluation with the techniques described here is possible and hopefully will soon elucidate the question.

Acknowledgements

The author acknowledges a Heisenberg Fellowship and the support of the Deutsche Forschungsgemeinschaft.

REFERENCES

1. e.g., B. A. Huberman and J. P. Crutchfield, Phys. Rev. Lett. 43, 1743 (1979).
2. For an introductory article see D. Ruelle, La Recherche 11(2), 132 (1980), english translation in Math. Intelligencer 2, 126 (1980).
3. D. Ruelle and F. Takens, Comm. Math. Phys. 20, 167 (1971); 23, 343 (1971).
4. J. P. Eckmann, Rev. Mod. Phys. 53, 643 (1981).
5. For a survey see T. Geisel, Nature 298, 322 (1982).
6. N. H. Packard, J. P. Crutchfield, J. D. Farmer, and R. S. Shaw, Phys. Rev. Lett. 45, 712 (1980).
7. H. Froehling, J. P. Crutchfield, D. Farmer, N. H. Packard, and R. Shaw, Physica 3D, 605 (1981).
8. F. Takens, in Dynamical Systems and Turbulence, eds. D. A. Rand, L. S. Young, Lect. Notes in Math. Vol. 898 (Springer, Berlin 1981), p. 366.
9. For a review see J.-C. Roux, J. S. Turner, W. D. McCormick, and H. L. Swinney, in Nonlinear Problems: Present and Future, eds. A. R. Bishop, D. K. Campbell, and B. Nicolaenko (North Holland, 1982), p. 409; J.-C. Roux, R. H. Simoyi, and H. L. Swinney, Physics 8D, 257 (1983).
10. P. Grassberger, this volume.
11. J. Guckenheimer, this volume.
12. D. Farmer, J. Hart, and P. Weidman, Phys. Lett. 91A, 22 (1982).
13. A. Brandstäter, J. Swift, H. L. Swinney, A. Wolf, J. D. Farmer, E. Jen, and J. P. Crutchfield, Phys. Rev. Lett. 51, 1442 (1983).

14. J. P. Crutchfield, J. D. Farmer, and B. A. Huberman, Phys. Rep. 92, 45 (1982).
15. M. J. Feigenbaum, J. Stat. Phys. 19, 25 (1978); 21, 669 (1979).
16. P. Coullet and C. Tresser, C. R. Acad. Sci. 287, 577 (1978); J. Phys. (Paris), Colloq. 39, C5-25 (1978).
17. S. Grossmann and S. Thomae, Z. Naturforsch. 32a, 1353 (1977).
18. N. Metropolis, M. L. Stein, and P. R. Stein, J. Comb. Theory A 15, 25 (1973).
19. T. Geisel and J. Nierwetberg, Phys. Rev. Lett. 47, 975 (1981).
20. T. Geisel and J. Nierwetberg, in Dynamical Systems and Chaos, ed. L. Garrido, Lecture Notes in Phys. Vol. 179 (Springer, Berlin 1983). p. 93.
21. G. Mayer-Kress and H. Haken, J. Stat. Phys. 26, 149 (1981).
22. J. Heldstab, H. Thomas, T. Geisel, and G. Radons, Z. Phys. B 50, 141 (1983);
    T. Geisel, J. Heldstab, and H. Thomas, Z. Phys. B 55, 165 (1984).
23. J. P. Crutchfield and B. A. Huberman, Phys. Lett. 77A, 407 (1980).
24. R. Shaw, Z. Naturforsch. 36a, 80 (1981).
25. B. A. Huberman and J. Rudnick, Phys. Rev. Lett. 45, 154 (1980).
26. T. Geisel, J. Nierwetberg, and J. Keller, Phys. Lett. 86A, 75 (1981).
27. J. P. Crutchfield, M. Nauenberg, and J. Rudnick, Phys. Rev. Lett. 46, 933 (1981).
28. B. Shraiman, C. E. Wayne, and P. C. Martin, Phys. Rev. Lett. 46, 935 (1981).
29. J. E. Hirsch, B. A. Huberman, and D. J. Scalapino, Phys. Rev. A 25, 519 (1982).
30. J. P. Eckmann, L. Thomas, and P. Wittwer, J. Phys. A14, 3153 (1982).
31. T. Geisel and J. Nierwetberg, Phys. Rev. Lett. 48, 7 (1982).
32. H. Haken and G. Mayer-Kress, Phys. Lett. 84A, 159 (1981); Z. Phys. B 43, 185 (1981).
33. P. Manneville and Y. Pomeau, Phys. Lett. 75A, 1 (1979); Physica 1D, 219 (1980); Comm. Math. Phys. 74, 189 (1980).
34. P. Manneville, J. Phys. (Paris) 41, 1235 (1980).
35. M. Dubois, M. A. Rubio, and P. Bergé, Phys. Rev. Lett. 51, 1446 (1983).
36. T. Geisel and S. Thomae, Phys. Rev. Lett. 52, 1936 (1984).
37. I. Procaccia and H. Schuster, Phys. Rev. A 28, 1210 (1983).
38. G. Mayer-Kress and H. Haken, Phys. Lett. 82A, 151 (1981).
39. F. Makino et al., seminar given at this meeting.

CLUES TO STRANGE ATTRACTORS

John Guckenheimer

ABSTRACT

This lecture is a brief review of attempts to characterize the types of data produced by strange attractors. These efforts have produced crude but workable criteria for the analysis of time series data in settings typified by laboratory fluid experiments. I discuss here my personal perceptions of how useful these techniques might be in situations where the data are less complete or accurate than data obtained from fluid experiments and numerical computations.

Underlying the concept of an attractor are mathematical models and a set of assumptions about the physical system being modeled. The basic models are evolution equations defined on some state (or phase) space. Information about the past and future of the physical system as it is observed is represented by a curve in state space parameterized by time. To the extent that the model accurately describes the physical system, this curve is approximated by a solution trajectory of the evolution equation. There are two implicit assumptions which are usually invoked for analysis of time series data in these terms. The first assumption is that of stationarity. One assumes that observations can be continued forever without significant change of statistical averages. The system will return at some future time to approximate its current state, and statistical averages on a longer time scale than this recurrence time will not change appreciably in time. The second assumption is one of "genericity" or reproducibility. Different observed evolutions from near the same initial conditions should have the same long term statistical averages. These two assumptions are not always satisfied, but this is not the place for an extended discussion of when they are valid. Indeed, there are serious problems in verifying when they hold even in the idealized world of systems of differential equations

(Guckenheimer and Holmes 1983, Chapter 5, Grebogi et al., 1983) but observational experience with numerical computations for chaotic systems has been satisfactory in this regard.

Starting from this viewpoint, one accepts time series of experimental observations as representing typical trajectories in an attractor (once transients have had an opportunity to decay). The question facing one is then a determination of the properties of an attractor which give an efficient description of the data. Power spectra have been the principal statistics which have been used to study such data. They have the advantages that (1) fluctuations coming from observational errors have a minor influence on the statistics and (2) they are economical to compute. There has been a strong tendency to equate chaos with the property that a time series has a continuous part to its power spectrum. However, there are drawbacks to reliance upon power spectra as a diagnostic tool. The most severe of these is that the power spectrum is sensitive to aperiodicity but does not distinguish its causes. One need only peruse the lore of "1/f noise" to appreciate that substantially different systems can yield similar spectra.

The following example is a simple illustration of how power spectra fail to detect important aspects of strange attractors. The discrete dynamical system given by the transformation $f(x) = 2x \pmod 1$ on the unit interval has the properties (1) that it is deterministic and (2) that the power spectrum of a typical trajectory is a δ-function at frequency 0. A typical sequence of independent trials of a random variable uniformly distributed in the unit interval has the same spectrum. Thus the spectra fail completely to make a distinction between these two very different models for a physical process. Other tools are required if one is to make a judgment as to whether a "strange attractor" or "noise" is lurking below the surface of a system whose behavior is aperiodic.

Strange attractors are creatures which arise in the study of ordinary differential equations. The characteristics of ordinary differential equations in evolutionary processes are determinacy, finite dimensionality, and differentiability (Arnold 1973 p. 1). Therefore, the issue of deciding whether an ODE model is appropriate for a set of data will revolve around the questions of determinacy and dimension. In terms of dealing with time series data, the question of dimension comes first because one needs to know how much information serves to distinguish the different observed states of the physical system. Technically, the question of dimension is a difficult one to deal with because attractors can have a complicated local geometric structure (i.e., they may be <u>fractals</u> (Mandelbrot)) and the distribution of a typical trajectory in an attractor may be highly concentrated. The development of data analysis techniques has proceeded with these technical difficulties in clear focus, but it is unclear whether there is yet, or ever will be, a definitive set of optimal procedures. The use of statistics here as elsewhere calls for subjective judgment.

The concept of dimension has a lengthy mathematical history and may be formulated from several points of view (Farmer et al. 1983). The viewpoint which has emerged as a practical basis for analyzing experimental time series data is based upon geometrical scaling properties. The essential fact which provides the basis for the method is that the volume V of a ball of radius r in a d-dimensional space is proportional to r^d as d varies. Thus, log V is a linear function of log r which has slope d. We want to extend this idea into a setting involving strange attractors that will allow one to estimate a dimension for an attractor efficiently and in a manner suitable for use with experimental time series data.

Trajectories within a strange attractor Λ wind around chaotically, spending varying proportions of time in different regions of the attractor. In accord with the modeling assumptions described above, one assumes that almost all trajectories have the same long term distribution of where they spend their time. Mathematically, this distribution is an <u>asymptotic measure</u> μ which describes the probability that a trajectory be found in a given subset of the attractor Λ at a specific time. Using the assumption of stationarity, μ also describes the proportion of time a trajectory spends in the subset. The measure μ can be described in terms of physical observables. As ϕ varies over integrable functions on the state space which represents observed quantities, then the identity

$$\lim_{T \to \infty} \frac{1}{T}\int_0^T \phi(x(t))dr = \int \phi d\mu$$

(for almost all initial conditions x(0) in a neighborhood of Λ) characterizes the measure μ by the property that time averages along typical trajectories are given by spatial averages of μ over Λ.

The existence of such an asymptotic measure μ is a nontrivial question, and much effort has been devoted to determining when an asymptotic measure does exist. The Sinai-Ruelle-Bowen theory (Bowen 1975) establishes the existence for uniformly hyperbolic systems, and the theory of Jakobson-Collet-Eckmann (Collet and Eckmann 1983) gives a similar result for a large class of one dimensional mappings which are not uniformly hyperbolic. Otherwise, little is known. In numerical and experimental studies one assumes that an asymptotic measure μ exists and that an observed trajectory has time averages over large finite times that are good approximations to μ measure. Statistical assumptions about the nature of this approximation play an important role in the process of estimating dimensions.

For an attractor Λ with an invariant measure μ, one can define the <u>pointwise dimension</u> of μ at $x \in \Lambda$ as

$$\lim_{r \to 0} \frac{\log \mu(B_x(r))}{\log r} = d_x(\mu) ,$$

where $B_x(r)$ is the ball of radius r centered at x in state space with respect to a specified metric (distance function). If Λ has an asymptotic measure μ and if the limit above does exist and is independent of $x \in \Lambda$ for μ, then we call this limit the pointwise dimension of Λ. To apply this definition to a trajectory observed for N time increments, one approximates μ by μ_N, the average of N delta functions at the observed positions. Then, the quantity $B_r(x)$ is simply the number of observations within distance r of x divided by N. By computing the distances from each observation to x and sorting these, it is easy to obtain the step function $\mu_N(B_x(r))$, thereby obtaining an estimate for $d_x(\mu)$.

The statistical robustness of these procedures is open to question and experimentation. There are several sources of fluctuations which are discussed in Guckenheimer (1984). From a practical standpoint, there is an insurmountable recurrence problem when one is studying an attractor whose dimension is moderate or large. To estimate the limit of $r \to 0$ in the definition of the pointwise dimension, one needs a good estimate for the volume of balls of small radius. However, for a time series of N observations, the expected ratio between the largest and smallest observed distances is of the order of $N^{1/d}$. If d is 10, for example, then time series of 10^{10} observations will be needed to achieve neighbors to x whose distance is of order one tenth the diameter of the attractor. Achieving estimates of the μ-measure of small balls requires volumes of data which are unobtainable.

This recurrence problem presents an interesting philosophical issue that deserves more discussion. Experimental systems can seldom be prepared with precise initial conditions lying on a strange attractor. The separation of trajectories within an attractor makes it appear that states on an attractor can only be reproduced by observing the system long enough for them to recur. How long is long enough depends upon the precision one wants and the dimension of the attractor – in a way that grows exponentially with dimension. In many situations, one will have the stamina or even the lifetime to observe recurrences only in attractors of small dimension. Determinism in a system is the property that only one evolution is possible from a given state. If recurrences of a given state are unobservable because they occur too infrequently, then one simply cannot tell whether the system is behaving deterministically or not.

From the point of view of data analysis, the boundary between small and large dimensions seems disappointingly small because the type of detailed analysis of chaotic dynamics which has been carried out for many low-dimensional attractors does not appear feasible for attractors of large dimension even if they are deterministic. Thus, studies of transitions and bifurcations, the measurement of Lyapunov exponents and entropies, and other studies which involve the local behavior of trajectories in state space all carry an implicit restriction that the dimensions of the attractors involved are very small. What then are the telltale signs of a low-dimensional strange attractor lurking behind a time series and what can be learned from the recognition that this is the case? These are the final questions addressed here.

Return to the examples which I introduced at the beginning: (1) the iteration of the function $f(x) = 2x(\bmod 1)$ and (2) a sequence of independent random variables uniformly distributed $[0,1]$. Presented with time series of moderate length from these two systems, there is an easy way to tell them apart. The time series $x(t)$ has an approximate recurrence when there are times t_0 and t_1 for which $x(t_0)$ and $x(t_1)$ are close to one another. Here, if $|(x(t_0 + 1) - x(t_1 + 1))|$ is larger than $2|x(t_0) - x(t_1)|$ (and $1/2$ is not between $x(t_0)$ and $x(t_1)$ in this example), then the time series is random. <u>In a strange attractor, approximate recurrences in state space give rise to recurrent patterns in the time series</u>. With a low dimensional attractor, these recurrent patterns appear far more frequently than would be expected in a random model. Visual recognition of recurrent patterns is a clue to the possibility of low-dimensional strange attractor models.

The occurrence of repetitive patterns can be analyzed more systematically. For a time series $x(t)$, plots of k-dimensional vectors $(x(t), x(t + 1), \ldots, x(t + k - 1))$ may show whether there is a tendency for clustering onto sets of smaller dimension than k. The estimates of pointwise dimension described above give one statistical test to see aspects of

this clustering. Experience with various procedures of these
kinds is growing, and they do form an adequate base for
considering the question of whether a low-dimensional model
appears to be appropriate for a given time series. However,
the questions of how much data and what relative precision of
measurement is necessary for employing these methods still
remain. Here the answers are unknown. Comparison of recent
studies involving the transition to chaos in fluid systems
(Guckenheimer and Buzyna 1983, Brandstadter et al. 1983) can
give the reader some impression of the state of this art. One
thing apparent from these efforts is that substantial
measurement errors (say larger than one percent) are a
significant obstacle in visualization attempts. The
sensitivity of the pointwise dimension estimates to appreciable
measurement errors has not been thoroughly tested. In
situations where a random perturbation of a low-dimensional
attractor is observed, the techniques described in
(Guckenheimer 1982, Guckenheimer 1983) give one approach to the
estimation of the amplitude of the random perturbations.

I do feel that there is a significant opportunity for the
development of the statistical methods I have described here
for the analysis of chaotic data in a diverse set of applications. However, the prevalence of systems whose evolution is
governed by low-dimensional attractors in their state space is
much more open to question. To make this determination
requires that one be able to observe a system for a long time
relative to its natural time scales with measurements of
moderately high precision. Whether there are chaotic astronomical phenomena describable by low-dimensional strange
attractors remains a fascinating question.

Acknowledgement

This research has been partially sponsored by the Air
Force Office of Scientific Research, the National Science
Foundation and the Guggenheim Foundation. The U. S. Government
is authorized to reproduce and distribute reprints for
Governmental purposes not withstanding any copyright notation
thereon.

REFERENCES

1. Arnold, Ordinary Differential Equations, MIT Press (1973).
2. Bowen, R., Equilibrium States and Ergodic Theory of Anosov
   Diffeomorphisms, Springer Lecture Notes in Mathematics,
   470 (1975).
3. Brandstadter, A., Swift, J., Swinney, H., Wolf, A.,

Farmer, J., Jen, E. and Crutchfield, J., Low-dimensional Chaos in a System with Avogadro's Number of Degrees of Freedom, Phys. Rev. Lett. 51, 1442-1445 (1983).
4. Collet, P., and Eckmann, J. P., Positive Liapunov Exponents and the Absolute Continuity for Maps of the Interval, Ergod. Th. and Dynam. Sys. 3, 13-46 (1983).
5. Farmer, J. D., Ott, E., and Yorke, J. A., The Dimension of Chaotic Attractors, Physica 7D, 153-180 (1983).
6. Grebogi, C., Ott, E. and Yorke, J. A., Fractal Basin Boundaries, Long-Lived Transients, and Unstable-Unstable Pair Bifurcations, Phys. Rev. Lett. 50, 935-938 (1983).
7. Guckenheimer, J., Noise in Chaotic Systems, Nature 298, 358-361 (1982).
8. Guckenheimer, J., Strange Attractors in Fluid Dynamics, Lecture Notes in Physics 179, 149-157 (1983).
9. Guckenheimer, J., Dimension Estimates for Attractors, Contemporary Mathematics 28 (1984).
10. Guckenheimer, J. and Buzyna, G., Dimension Measurements for Geostrophic Turbulence, Phys. Rev. Lett. 51, 1438-1441 (1983).
11. Guckenheimer, J. and Holmes, P., Nonlinear Oscillations, Dynamical Systems and Bifurcation Theory, Springer Verlag (1983).
12. Mandelbrot, B., The Fractal Geometry of Nature, Freeman, San Francisco (1982).

INFORMATION ASPECTS OF STRANGE ATTRACTORS

P. Grassberger

ABSTRACT

The unpredictability of chaotic motion is related to an information flow associated with it. Topics discussed in the present talk are: (i) the analogy with the second law of thermodynamics, (ii) the relation between the information flow rate (= Kolmogorov entropy), Lyapunov exponents, and dimensions of the attractor, (iii) the discussion of generalized dimensions and entropies related to the flow of Renyi informations, (iv) the measurements of these in real and computer experiments, and (v) the flow on repellers and along unstable orbits, often leading to long-lived metastable transients.

1. Introduction

A scientist is typically faced with the problem that he wants to describe a very complex system (consisting of $\gtrsim 10^{20}$ particles, say) by only a few variables. In the typical textbook examples of classical mechanics, this approach works perfectly and leads to deterministic equations with computable solutions.

In other examples, like Brownian motion, the effect of those degrees of freedom which are not treated explicitly is seen as random noise. Thus, one has non-deterministic equations of motion, but the origin of the non-determinacy is well understood and seemingly not fundamental; <u>in principle,</u> one could avoid the "coarse-graining", and solve the deterministic system of $\gtrsim 10^{20}$ equations of motion.

This is the point of view stressed already by Laplace (1). His point of view had to be modified by the advent of quantum mechanics, but apart from that, it was the canonical point of view until some 20 years ago. According to his theory, phenomena like hydrodynamic turbulence are manifestations of a large number of collectively excited modes, whose behavior is unpredictable only from a practical but not from a fundamental point of view (2). Conversely, if seemingly unpredictable behavior occurred in systems with few degrees of freedom, this was attributed to some external or internal noise.

But during the last years, it was realized more and more that an essentially different type of coarse-graining plays a crucial role in many systems as well. It results from the fact that we not only cannot deal with too many variables, but that we cannot deal with infinitely precise numbers. If we accept that space-time is continuous, this means that we <u>must</u> cut off the digital expansions for all coordinates <u>somewhere</u>.

In the textbook examples mentioned above, the cut-off has no effect. But there are formally deterministic systems - and even very simple ones - where trajectories emerging from nearby initial conditions diverge exponentially. Due to this "sensitive dependence on initial conditions", the ignorance about the insignificant cut-off digits spreads with time towards the significant digits, leading to a "chaotic", and essentially unpredictable, behavior.

In this review I shall concentrate on <u>dissipative</u> chaotic systems called strange attractors. Usually they are not only very sensitive to the initial conditions but also to small changes in control parameters and to externally added noise.

Let me stress again that chaotic behavior is a fundamental problem. The unpredictability cannot be avoided by just making more precise measurements of the initial conditions. Assume we want to measure the initial conditions very precisely, say with the same error $\pm \varepsilon$ in each variable of state space. We first need to build a suitable measuring device. But this device can later be used to measure the state again with the same precision, while the equations of motion are unable to predict it with the same accuracy.

If the distance between nearby trajectories increases exponentially - and this is expected by scale invariance, as long as this distance is much smaller than all typical length scales in the problem, - the lack of information is independent of ε and proportional to the elapsed time.

INFORMATION ASPECTS OF STRANGE ATTRACTORS

Consider now a piece of trajectory between times t_1 and t_2. Assume that t_1 is sufficiently large so that transients have died out, and that the system is on its attractor with some time-invariant probability distribution. The information needed to specify this trajectory to an error $\pm\varepsilon$ consists then of two parts: (a) the information, S_ε, needed to specify it at time t_1, and (b) the information needed to fix up the ignorance leaking in with constant rate due to the ignorance about originally insignificant digits. Thus,

$$S_\varepsilon[t_1, t_2] \approx S_\varepsilon + (t_2-t_1)\cdot K \text{ for } \varepsilon \to 0 \text{ and } t_2-t_1 \to \infty. \quad (1.1)$$

The constant K is called the Kolmogorov-Sinai (3) or "metric" entropy. It is essentially the information flow rate in the limit of nearly error-free measurements (4).

For predictable systems, one would have K=0. For Brownian motion[1], at the other extreme, one has $K = \infty$; even perfect knowledge of the state at some instant would not be sufficient to predict it, even in the near future.

Another system with finite information flow is a coin flipped again and again at a constant rate. Assume that the information flow rate in a deterministic chaotic system is K bits per second. Then, it is possible to build a purely deterministic machine which uses the actual state of the system as input, and turns a coin K times per second in such a way that the turnings are indistinguishable from random flips. Thus, chaotic systems are indeed "as random as a coin flip" (5), but it is useful to keep in mind that systems with continuous variables and white noise are infinitely more random than that. Thus, chaotic systems form a genuine third class of systems, located between deterministic and stochastic ones.

Let us now look at the dependence of S_ε on the error ε. If we want to specify a point x on a fixed interval with accuracy $\pm\varepsilon$, the needed information (i.e., the number of significant digits) behaves for $\varepsilon \to 0$ like $\log 1/\varepsilon$ (if we measure information in bits, i.e., if we use a binary representation of x, the logarithm has to be to the base 2; this will be understood throughout the following). For a point in D-dimensional space, the information is D times as big. Thus, we expect that

$$S_\varepsilon \sim D \cdot \log \frac{1}{\varepsilon} \text{ for } \varepsilon \to 0, \quad (1.2)$$

where D is called the information dimension of the attractor.

We have to be somewhat more precise. The estimate $\log 1/\varepsilon$ for the information stored in a point on an interval is

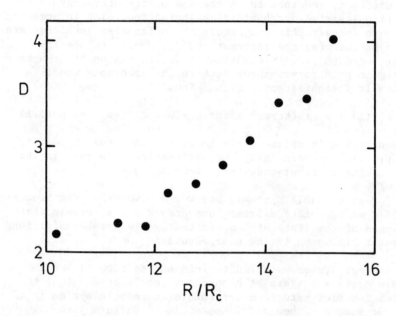

Fig. 1: Information dimension for Taylor-Couette flow, as a function of Reynolds number (adapted from Ref. (8)).

true only if we know a priori that the point is indeed on the interval, and nothing more. Analogously, Eq. (1.2) assumes a priori that the state is on the known attractor (any transient has died out), and distributed according to some invariant "measure" (= distribution), which also is assumed to be known from previous observations.

The typical case, assumed throughout the following, is that the system is ergodic and mixing. In that case, nearly all initial conditions within some suitable basin of attraction lead to the same invariant distribution, called the "natural measure" in accordance with Ref. (6). But in general, there are also other invariant distributions, reached from initial conditions of measure zero. Quantities like D and K actually refer to one particular distribution. If nothing else is said, this will be the natural measure.

The most striking feature of strange attractors is that D is in general non-integer. Since D is closely related to the Hausdorff dimension (see Sec. 3), the attractor is a "fractal" in the sense of Mandelbrot (7). The observation of non-integer dimensions in Couette-Taylor flow (8) (see Fig. 1), and in other hydrodynamic systems (9), represents indeed the most clear-cut proof that strange attractors are relevant for astrophysics.

In addition to S_ε and $S_\varepsilon[t_1, t_2]$, one can also consider the information $S_\varepsilon(t_2/t_1)$ needed to specify $\vec{x}(t_2)$, provided $\vec{x}(t_1)$ was already known. (I shall denote by $\vec{x}$ collectively all variables of the system, be they space or momentum coordinates, temperatures, densities,...). If the system is mixing, the distributions of $\vec{x}(t_1)$ and of $\vec{x}(t_2)$ become independent for $t_2 - t_1 \to \infty$. Thus,

$$S_\varepsilon(t_2/t_1) = \begin{cases} S_\varepsilon & \text{for } |t_2 - t_1| \to \infty \\ 0 & \text{for } t_x = t_1 \\ \sim K \cdot (t_2 - t_1) & \text{for } 0 \ll t_2 - t_1 \ll S_\varepsilon/K. \end{cases} \quad (1.3)$$

The ratio S_ε/K is a measure of the time for which any meaningful prediction is possible (4). It increases logarithmically with $1/\varepsilon$.

Furthermore, since information about $\vec{x}(t_1)$ can only become less and less useful in fixing $\vec{x}(t_2)$, we have

$$\frac{\delta}{\delta t_2} S_\varepsilon (t_2/t_1) \geqslant 0 \quad . \quad (1.4)$$

This is essentially the second law of thermodynamics, adapted to the present case. Indeed, the natural measure is completely analogous to the macrostate in statistical mechanics (whence it is called the "Gibbs measure" in the Russian literature), while the actual state corresponds to the microstate.

The fact that we have obtained a second law even for non-closed systems driven from outside results from the fact that we have considered only the stationary case, after transients have died out. During the initial period of attraction, $S_\varepsilon(t_2/t_1)$ and $S_\varepsilon[t_1,t_2]$ both will decrease with t_2, and Eqs. (1.1)-(1.4) will not be true.

As we have already stressed, the motion on a strange attractor is unpredictable due to the divergence of nearby trajectories. This divergence is measured by the so-called Lyapunov exponents. In order to define them, consider an infinitesimally small sphere around some point $\vec{x}$ with radius ε. After a time, $t \gg 0$, this sphere is transformed into an ellipsoid with semi-axes

$$\varepsilon_i(t) \approx \varepsilon \cdot e^{\lambda_i t} \quad (i = 1,2,\ldots f) \quad , \quad (1.5)$$

for nearly all $\vec{x}$. In the following, we shall always assume that the λ_i are ordered by magnitude $\lambda_1 \geq \lambda_2 \geq \ldots$. Since the system is dissipative, their sum is < 0. But since it is chaotic by assumption, at least one of the λ_i is positive, $\lambda_1 > 0$. (The distance between two nearby points will in general behave like

$$|\Delta\vec{x}(t)| \propto e^{\lambda_1 t} \quad).$$

At nearly all points in the basin of attraction, the semi-axes ε_i define a Vielbein[2]. Those directions with $\lambda_i > 0$ are called unstable directions and those with $\lambda_i < 0$ stable. Notice that these directions are tangent to the stable and unstable manifolds passing through $\vec{x}$. For continuous time systems, the velocity vector $\dot{\vec{x}}$ defines a further direction with $\lambda = 0$.

In the next section, we shall provide more precise definitions of Kolmogorov entropy and information dimension. In particular, we shall see that we can attribute a dimension D_i to each of the stable and unstable directions, such that $D = \Sigma_i D_i$. The Kolmogorov entropy will then be related to the D_i and λ_i, and a relation between D and the λ_i's proposed by Kaplan and Yorke, will be seen to follow from an "information conservation law".

INFORMATION ASPECTS OF STRANGE ATTRACTORS 199

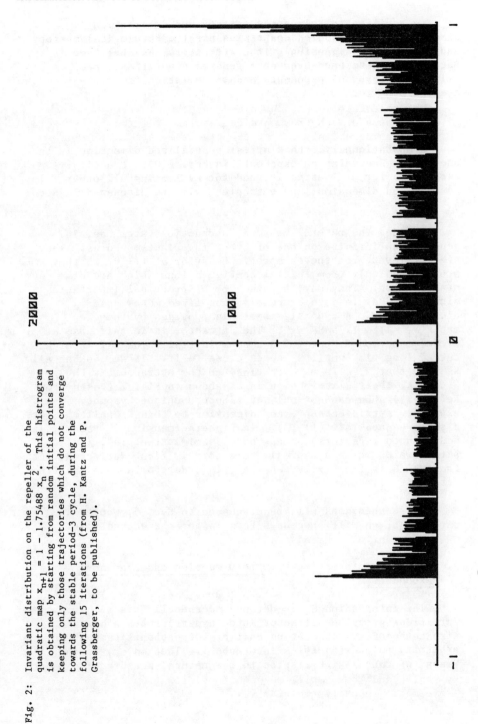

Fig. 2: Invariant distribution on the repeller of the quadratic map $x_{n+1} = 1 - 1.75488\, x_n^2$. This histogram is obtained by starting from random initial points and keeping only those trajectories which do not converge towards the stable period-3 cycle, during the following 15 iterations (from H. Kantz and P. Grassberger, to be published).

In addition to Kolmogorov entropy and information dimension, often-studied quantities are the topological entropy and the fractal dimension of the attractor. We shall see in Sec. 3 how these (and even more general quantities), are related to Lyapunov exponents and to deviations from the average behavior

$$\varepsilon_i(t) \approx \varepsilon\, e^{\lambda_i t}.$$

Computationally, the simplest generalized dimension is the one called "correlation exponent" in Ref. (10). Practical algorithms for evaluating it, and for evaluating all other generalized dimensions and entropies, will be discussed in Sec. 4.

Besides the natural measure, a chaotic system has, in general, an infinite number of other invariant measures. They are irrelevant for the asymptotic behavior at $t \to \infty$, as they are reached only from initial configurations which are of measure zero. They may, however, be of practical importance since they may lead to extremely long-lived metastable distributions. A typical example is provided by the map $x_{n+1} = 1 - 1.75488\, x_n^2$. The attractor is in this case a cycle of period 3, but for a random initial value x_0, it might take a long time until x_n comes close to it. If we look at all x_n such that x_{n+k} is not yet close to the attractor, with $0 \ll n,k$, their distribution is as shown in Fig. 2 (taken from Ref. (11)). We can not only attribute a unique Lyapunov exponent, metric entropy, and dimension to this "repeller", but also an escape rate (11,12). In higher dimensional systems, in addition to repellers, we can have "semi-attractors", i.e., sets towards which a typical trajectory is first attracted, but from which it finally is repelled, in order to land on the true attractor.

These phenomena will be discussed in Sec. 5, where, in particular, we will present a formula for the escape rate.

2. Relating Information Flow to Dimension and Lyapunov Exponents

The information S_ε is defined technically via a partitioning of the attractor into (hyper-) cubes of size ε (see Fig. 3). Let us call p_i the probability that an arbitrary point $\vec{x}(t)$ falls into cube i. That is, p_i is the "mass" of cube i with respect to the natural measure μ:

$$p_i = \int_{\text{cube } i} d\mu(\vec{x}) \,. \qquad (2.1)$$

INFORMATION ASPECTS OF STRANGE ATTRACTORS

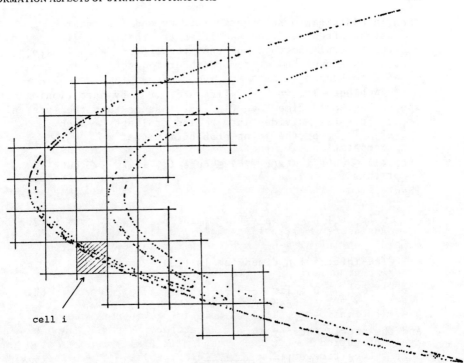

Fig. 3: Partitioning of space (Hénon map).

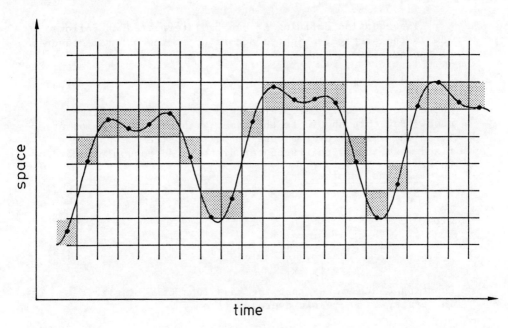

Fig. 4: Partitioning of space-time.

Then, S_ε is defined as

$$S_\varepsilon = - \sum_i p_i \log p_i . \qquad (2.2)$$

Analogously, we can define $S_\varepsilon[t_1,t_2]$ by partitioning not space, but space-time (see Fig. 4). Assume that the interval $[t_1,t_2]$ has been divided into n bins of length τ each. Let $p_{\{i_1 \ldots i_n\}}$ be the joint probability that $\vec{x}(t_1+\tau) \in$ cube i_1, $\vec{x}(t_1+2\tau) \in$ cube $i_2, \ldots \vec{x}(t_2) \in$ cube i_n.
Then

$$S_\varepsilon[t_1,t_2] = - \sum_{\{i\}} p_{\{i_1 \ldots i_n\}} \log p_{\{i_1 \ldots i_n\}} . \qquad (2.3)$$

The information dimension is (13)

$$D = \lim_{\varepsilon \to 0} \frac{S_\varepsilon}{\log 1/\varepsilon} \qquad (2.4)$$

and the metric entropy is

$$K = \lim_{\substack{\varepsilon \to 0 \\ \tau \to 0}} \lim_{t_2 \to \infty} \frac{1}{t_2 - t_1} S_\varepsilon[t_1,t_2] . \qquad (2.5)$$

The equation defining S_ε can be interpreted as follows: S_ε is the average, weighted according to the natural measure, of the quantity

$$- \log p_i = - \log \int d\mu(\vec{x}) , \qquad (2.6)$$

where the integral goes over a cube of size ε in a fixed mesh. But this should be the same, within a constant of order 1, as the average over a cube or ball of size ε, not in a fixed mesh, but with its center distributed randomly according to μ:

$$S_\varepsilon \approx - \int d\mu(\vec{x}) \cdot \log M_\varepsilon(\vec{x}) \equiv - \langle \log M_\varepsilon \rangle \qquad (2.7)$$

with

$$M_\varepsilon(\vec{x}) = \int_{|\vec{y}-\vec{x}| < \varepsilon} d\mu(\vec{y}) . \qquad (2.8)$$

Equation (2.4) states that the mass in a ball of size ε increases, in geometric average, like ε^D.

INFORMATION ASPECTS OF STRANGE ATTRACTORS

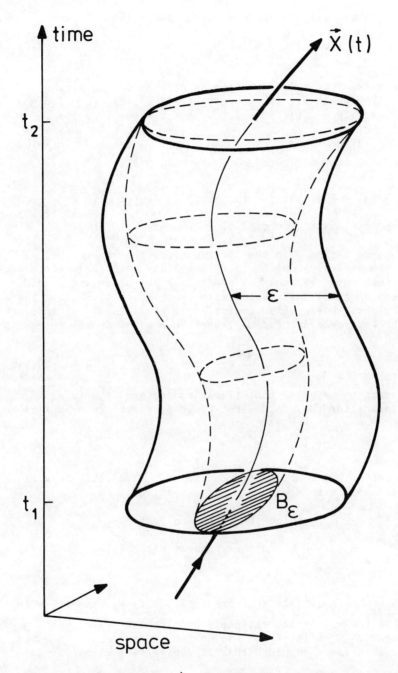

Fig. 5: The domain $B_\varepsilon(\vec{x})$ (shaded region) consists of those points, from which the trajectories emerge, staying within an ε-sausage around $\vec{x}(t)$.

A completely analogous argument shows that

$$S_\varepsilon[t_1,t_2] \approx - \langle \log M_\varepsilon[t_1,t_2] \rangle \tag{2.9}$$

where $M_\varepsilon(\vec{x}, [t_1, t_2])$ is the mass of that domain around $\vec{x}(t_1)$ from which all the emerging trajectories stay within a distance ε from $\vec{x}(t)$ for all $t_1 < t < t_2$ (see Fig. 5):

$$M_\varepsilon(\vec{x}, [t_1,t_2]) = \mu(B_\varepsilon(\vec{x})) \tag{2.10}$$

with

$$B_\varepsilon(\vec{x}) = \{ \vec{y}(t_1): |\vec{y}(t) - \vec{x}(t)| < \varepsilon \text{ for all } t \in [t_1,t_2] \}. \tag{2.11}$$

As we had said in the introduction, the stable and unstable manifolds define in (nearly) any point a complete set of directions. We can thus generalize the above and consider, instead of balls, ellipsoids with axes $\varepsilon_1, \varepsilon_2, \ldots$ along these directions. During time evolution, each ellipsoid will rotate (which is irrelevant for us), and the semi-axes will change as

$$\varepsilon_i \to \varepsilon_i \cdot e^{\Lambda_i(\vec{x},t)}. \tag{2.12}$$

The exact behavior of the dilatation factors Λ_i is also irrelevant for the moment, but their averages are by definition the Lyapunov exponents:

$$\lambda_i = \frac{1}{t} \int d\mu(\vec{x}) \Lambda_i(\vec{x},t). \tag{2.13}$$

Let us now consider the domain $B_{\varepsilon_1 \varepsilon_2 \ldots}(\vec{x})$. It will be an ellipsoid with semi-axes

$$\approx \begin{cases} \varepsilon_i & \text{along the stable directions,} \\ \varepsilon_i \, e^{-\Lambda_i(\vec{x},t)} & \text{along the unstable directions.} \end{cases} \tag{2.12a}$$

If $S_{\varepsilon_1 \varepsilon_2 \ldots}[t_1,t_2]$ is to increase linearly with t_2-t_1, with rate K, the mass of an ellipsoid should then scale like a power in each ε_i, at least for the unstable directions. However, scale invariance suggests that it scales with each ε_i,

$$\langle \log M_{\varepsilon_1 \varepsilon_2 \ldots} \rangle \sim \sum_i D_i \log \varepsilon_i \text{ as } \varepsilon_i \to 0, \tag{2.14}$$

from which one obtains, using Eqs. (2.5), (2.9), and (2.13)-(2.14)

$$K = \sum_i{}' D_i \lambda_i. \qquad (2.15)$$

Here, the sum extends over all directions with positive λ_i only.

The constants D_i can obviously be considered (information) dimensions along the i-th (un-) stable manifold. The dimensions clearly satisfy

$$0 \leqslant D_i \leqslant 1 \qquad (2.16)$$

and

$$D = \sum_i D_i. \qquad (2.17)$$

Equation (2.14) shows that the attractor factorizes essentially in a direct product of continua (corresponding to $D_i = 1$), discrete points ($D_i = 0$), and Cantor sets ($D_i \neq 0 \neq 1$), with orientation according to the (un-) stable directions. As an example, we show in Fig. 6 the Hénon (15) attractor with b = .3 and a = 1.4, in which case $D_1 = 1$ and $D_2 \approx 0.25$.

From Eq. (2.15) we see that $K \leqslant \sum_i' \lambda_i$. This was first derived (rigorously) by Ruelle (14), who also showed that the inequality is saturated if the natural measure is absolutely continuous along the unstable directions. He made the conjecture that the latter is always true for the natural measure. (It is in general not true for other invariant measures, for which Eq. (2.15) still should hold.)

For a system whose time reversed evolution is again deterministic, we can apply exactly the same argument to the reversed system, and obtain in analogy to Eq. (2.15)

$$K = - \sum_i{}'' D_i \lambda_i. \qquad (2.18)$$

Here, the sum extends over all directions with negative λ_i.

The physical meaning of Eqs. (2.15) and (2.18) should now be obvious. Eq. (2.15) represents the rate of uncertainty flowing into the significant digits due to the divergence of trajectories, while Eq. (2.18) represents the outward flow due to the convergence in the stable directions. The equality between both shows that information is conserved - it is just flowing through the system.

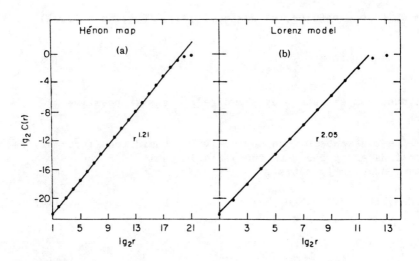

Fig. 6: Attractor of the Hénon map $(x,y) \to (1+y-ax^2, bx)$ with $a = 1.4$, $b = .3$. The arrows indicate the stable resp. unstable directions.

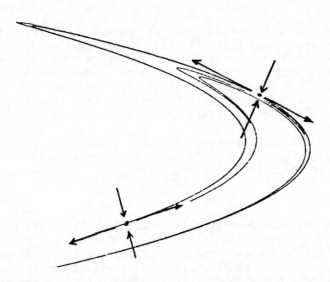

Fig. 7: Correlation integral, estimated from series of 15.000 iterations of the Hénon map and the Lorenz model (from Ref. (10)).

Combining Eqs. (2.15) and (2.18), we obtain

$$\sum_i D_i \lambda_i = 0. \quad (2.19)$$

From this and from the constraints $0 \leq D_i \leq 1$ and $D = \Sigma D_i$, it is easy to obtain an upper bound on D. Let j be that index for which

$$\sum_{i=1}^{j} \lambda_i \geq 0 > \sum_{i=1}^{j+1} \lambda_i . \quad (2.20)$$

Then one finds (16)

$$D \leq D_{Lyap} \equiv j + \frac{\sum_{i=1}^{j} \lambda_j}{|\lambda_{j+1}|}, \quad (2.21)$$

and the inequality is saturated if, and only if,

$$D_i = \begin{cases} 1 & \text{for } i \leq j \\ 0 & \text{for } i > j+1 \end{cases} \quad (2.22)$$

such that the attractor is Cantor-like only along the (j+1)-st direction.

In two dimensions (i = 1,2), this is certainly fulfilled, and accordingly (17) $D = D_{Lyap}$. In higher dimensions, one knows a number of examples where $D \neq D_{Lyap}$, but all are untypical (i.e., they correspond to a set of measure zero in some control parameter). Indeed, Frederickson, Kaplan and Yorke (18) have conjectured that $D = D_{Lyap}$ in all typical cases. If true, this represents by far the easiest way of computing D in analytically defined models. In experimental situations, it is much less useful.

Equations (2.15) and (2.19) were given first in Ref. (19), but with a different derivation. The present derivation is not only simpler, it also shows that they should hold for invariant distributions other than the natural one. Also, with obvious modifications, they should hold in situations where the evolution is either non-invertible (as for maps of the interval onto itself), or non-deterministic.

An example of a fractal invariant measure which is not an attractor, and on which the evolution is not invertible, was shown in Fig. 2. It is centered on the repeller of the map $x_{n+1} = 1 - 1.75488\, x_n^2$, and one finds (11) K = 0.4557 ± .0003 (= .657 bits/iteration), λ = 0.4886 ± .0004, and D = .9335 ± .0003. Thus Eq. (2.15) is indeed fulfilled numerically.

3. Generalized Entropies and Dimensions

Shannon's definition of information is not the only one possible. It fulfills a number of important conditions (the Khinchin axioms (20,21,22), but if one relaxes these conditions somewhat, a number of other information-like quantities can be defined.

The most important of these generalized entropies are the order-α Renyi informations, defined by (21)

$$S_\varepsilon^{(\alpha)} = \frac{1}{1-\alpha} \log \sum_i p_i^\alpha. \qquad (3.1)$$

Here, α is any positive real number $\neq 1$, and the p_i's are the same probabilities as in the preceding section. By de l'Hôpital's rule, one finds

$$S_\varepsilon = \lim_{\alpha \to 1} S_\varepsilon^{(\alpha)}, \qquad (3.2)$$

while the derivative with respect to α can be written as

$$\frac{\partial S_\varepsilon^{(\alpha)}}{\partial \alpha} = -(1-\alpha)^{-2} \sum_i z_i \log \frac{z_i}{p_i} \qquad (3.3)$$

with $z_i = p_i^\alpha / \Sigma p_j^\alpha$. Since $\sum_i z_i = 1$, we can interpret the right-hand side of Eq. (3.3) as a Kullback information gain (21,22), which is well known to be non-negative. Thus, the $S_\varepsilon^{(\alpha)}$ are indeed generalizations of the Shannon information, and they provide upper bounds (for $\alpha < 1$) and lower bounds (for $\alpha > 1$) for it.

The same construction can be made for the spatio-temporal informations $S_\varepsilon[t_1,t_2]$, and order-α dimensions $D^{(\alpha)}$. Information flow rates $K^{(\alpha)}$ can be defined (23) again by Eqs. (2.4) and (2.5). Due to the monotonicity of $S_\varepsilon^{(\alpha)}$, both $D^{(\alpha)}$ and $K^{(\alpha)}$ are also montonically decreasing with α, and $D = \lim D^{(\alpha)}$ and $K = \lim K^{(\alpha)}$ as $\alpha \to 1$.

Of particular interest is the limit $\alpha \to 0$. In this limit, $S_\varepsilon^{(\alpha)}$ becomes the logarithm of the number of non-empty cubes, and thus $D^{(0)}$ is just the fractal dimension (7) of the attractor, while $K^{(0)}$ is the topological entropy (Billingsley, Ref. (3)) of the flow on the attractor. (In most discussions of topological entropy, the flow in the whole basin of attraction is discussed. If this basin has fractal boundaries, the topological entropy receives a finite contribution from the flow near the boundaries.)

Finally, all $D^{(\alpha)}$ and $K^{(\alpha)}$ are invariant under a smooth coordinate change (23), i.e., they do not depend on the particular choice of coordinates. This is very important, as the choice of suitable coordinates is, in general, not unique (see Sec. 4).

Like the Shannon information, the Renyi information can be approximately related to the mass of ε-balls. The difference is that now one has to use a different kind of averaging,

$$S_\varepsilon^{(\alpha)} \approx \frac{1}{1-\alpha} \log \langle M_\varepsilon^{\alpha-1} \rangle \qquad (3.4)$$

instead of the geometrical average, and consequently

$$\langle M_\varepsilon^{\alpha-1} \rangle \sim \varepsilon^{(\alpha-1)D^{(\alpha)}} \quad \text{as } \varepsilon \to 0 \qquad (3.5)$$

and, similarly,

$$\langle M_\varepsilon[t_1,t_2]^{\alpha-1} \rangle \sim \varepsilon^{(\alpha-1)D^{(\alpha)}} e^{(\alpha-1)(t_2-t_1)K^{(\alpha)}} \qquad (3.6)$$

as $t_2 - t_1 \to \infty$ and $\varepsilon \to 0$.

Equations (3.5) and (3.6) have two important applications.

First, consider the case $\alpha = 2$. In this case, Eq. (3.5) simplifies to

$$\iint d\mu(\vec{x}) \, d\mu(\vec{y}) \, \Theta(\varepsilon - |\vec{x} - \vec{y}|) \sim \varepsilon^{D^{(2)}}. \qquad (3.7)$$

The left-hand side is simply the integral over the two-point correlation function. The equation measures the probability that two random points $\vec{x}$ and $\vec{y}$ — random, but distributed according to the measure μ — are within a distance ε. Because of that, $D^{(2)}$ was called "correlation exponent" in Ref. (10). Its measurement proceeds by taking a long time series

$\{\vec{x}(t), \vec{x}(t+\tau), \vec{x}(t+2\tau),\ldots\}$ and counting the number of pairs $[\vec{x}(t+n\tau), \vec{x}(t+k\tau)]$ with distance less than ε. This number should scale like $\varepsilon^{D(2)}$. For the Hénon map and the Lorenz model, this scaling is shown in Fig. 7, and it holds equally well for other systems (10).

The order-2 entropy $K^{(2)}$ can be obtained in exactly the same way by starting from Eq. (3.6) (19,24). We need only count the number of pairs $[\vec{x}(t+n\tau), \vec{x}(t+k\tau)]$ with the requirement that

$$\left| \vec{x}(t+n\tau) - \vec{x}(t+k\tau) \right| < \varepsilon$$

$$\left| \vec{x}(t+(n+1)\tau) - \vec{x}(t+(k+1)\tau) \right| < \varepsilon$$

$$\vdots$$

$$\left| \vec{x}(t+(n+d)\tau) - \vec{x}(t+(k+d)\tau) \right| < \varepsilon. \quad (3.8)$$

This number should decrease $\sim e^{-d\tau K^{(2)}}$, tests of which can be found in Ref. (19).

Modifications of these algorithms, (needed if the state space has very high dimensions,) and similar algorithms for estimating D and K directly, will be given in the next section.

For the second application of Eqs. (3.5) and (3.6), consider first their generalizations from ε-balls to ellipsoids. In complete analogy to Eq. (2.14), and as suggested by scale invariance, we assume that

$$\left\langle M_{\varepsilon_1 \varepsilon_2 \ldots}^{\alpha-1} \right\rangle \underset{\varepsilon_i \to 0}{\sim} \prod_i \varepsilon_i^{(\alpha-1)D_i^{(\alpha)}}. \quad (3.9)$$

Here again

$$0 < D_i^{(\alpha)} < 1,$$

$$D^{(\alpha)} = \sum_i D_i^{(\alpha)}, \quad (3.10)$$

and $D_i^{(\alpha)}$ can be considered as the order-α dimension of the measure along the i-th direction. Direct numerical checks of Eq. (3.9) are not easy. The only system for which it has been checked numerically (25) is the Mackey-Glass delay equation (26).

Now we want to derive the analoga of the information flow equation (2.15), and the information conservation law (2.19). But there is a subtle problem. As we have said, the Renyi information does not satisfy one of the Khinchin axioms, namely postulate IV on p. 548 of Ref. (21). This postulate states that the information in a joint distribution is the sum of the information corresponding to the single distributions. The Renyi information satisfies it if, and only if, the single distributions are independent, i.e., if the joint distribution factorizes. In our case, this means that we can expect something like conservation of Renyi entropies only if the flow is mixing (which should be the case for natural measures).

For a mixing flow, the expansion coefficients $\Lambda_i(\vec{x},t)$ should become for $t \to \infty$ like the sum of t independent random variables:

$$\langle \Lambda_i(\vec{x},t) \rangle \sim t \cdot \lambda_i , \quad (3.11a)$$

$$\langle (\Lambda_i - \lambda_i)(\Lambda_k - \lambda_k) \rangle \sim t \cdot Q_{ik} , \quad (3.11b)$$

etc.

In terms of the generating function, this means

$$\langle e^{\Sigma z_i \Lambda_i} \rangle \sim e^{t\, g(z_1, z_2, \ldots)} \quad \text{as } t \to \infty \quad (3.12)$$

with

$$g(\vec{z}) = \sum_i z_i \lambda_i + \frac{1}{2} \sum_{i,k} z_i z_k Q_{ik} + \ldots . \quad (3.13)$$

While Eq. (3.11a) is just the definition of Lyapunov exponents, Eq. (3.11b) and the analogous equations for the higher moments are non-trivial. They have been tested in Ref. (19), where very precise values of the covariance matrix Q were obtained for several attractors. For some models, even third order correlation matrices could be computed, with Eq. (3.12) always satisfied.

The time invariance of the measure, the Ansatz (3.9) for the mass of an ellipsoid, and Eq. (3.12) can now be taken together to obtain the very simple relation

$$g((1-\alpha)\vec{D}^{(\alpha)}) = 0 \quad (3.14)$$

or, equivalently,

$$\sum_i D_i^{(\alpha)} \lambda_i = \frac{\alpha - 1}{2} \sum_{i,k} D_i^{(\alpha)} D_k^{(\alpha)} Q_{ik} \pm \ldots \quad . \tag{3.15}$$

This is obviously the generalization of Eq. (2.19). In order to obtain the analogon of (2.15),

$$K = \sum_i' D_i^{(\alpha)} \lambda_i - \frac{\alpha - 1}{2} \sum_{i,k}' D_i^{(\alpha)} D_k^{(\alpha)} Q_{ik} \pm \ldots , \tag{3.16}$$

one has to start from Eq. (3.6), generalized first to ellipsoids.[3]

Both (Eq. 3.15) and Eq. (3.16) have been tested numerically in several examples, Refs. (19) and (29), and (11). While they were always fulfilled for the flow on the natural measure, they were not obeyed for flows on other, non-mixing, meaures.

4. Measuring Information Flow And Dimension

The most straightforward way to measure K and D would be to use the definitions based on box counting. For systems with low-dimensional state space (i.e., with few variables), this is indeed feasible, but for $\gtrsim 4$ variables the storage requirements become prohibitive, independent of the dimension of the attractor.

In that case, all practical methods are based on estimating the mass $M_\varepsilon(\vec{x})$ of typical ε-balls, (for estimating D), or of typical domains $B_\varepsilon(\vec{x})$ (see Eq. (2.11) and Fig. 5), for estimating K.

In order to estimate $M_\varepsilon(\vec{x})$, one counts the number of points $\vec{y}_n$ in a time series $\{\vec{y}_1 = \vec{y}(t), \vec{y}_2 = \vec{y}(t + \tau), \ldots \vec{y}_N\}$ which satisfy $|\vec{y}_n - \vec{x}| < \varepsilon$. For the distance, one can use either the Euclidean norm or the norm given by the largest difference in any coordinate. In the latter case, one counts these $\vec{y}_n$ for which all $|y_{n,k} - x_k| < \varepsilon$ for all k. The point $\vec{x}$ should be arbitrarily placed on the attractor (in general, it will be one of the $\vec{y}_n$'s). In order to obtain the information dimension, taking one single point $\vec{x}$ is enough in principle (13,28), but then the time series has to be excessively long. Thus, in practice, one will average over several points $\vec{x}_1, \ldots \vec{x}_M$.

INFORMATION ASPECTS OF STRANGE ATTRACTORS   213

When determining D (and any $D^{(\alpha)}$ with $\alpha \neq 2$), one should take first the limit $N \to \infty$ (estimating thus $M_\varepsilon(\vec{x}_k)$ for fixed $\vec{x}_k$), and afterwards $M \to \infty$. The errors in $M_\varepsilon(\vec{x}_k)$ due to the finiteness of N, cause systematic errors (due to the non-linear averaging) for small ε, where the scaling laws (2.14) and (3.5) should be most precise. Thus, these scaling laws can only be tested down to that ε for which every ball around $\vec{x}_k$ contains $\gg 1$ point.

This problem is absent only for $\alpha = 2$, in which case one averages the $M_\varepsilon(\vec{x}_k)$ linearly, and in which case the role of $\vec{y}_n$ and $\vec{x}_k$ is completely symmetric. In that case, scaling can be tested further down to values of ε such that the <u>sum</u> of all balls around all $\vec{x}_k$'s contains $\gg 1$ point. For that reason, the correlation exponent $D^{(2)}$ is the easiest generalized dimension to estimate, even if it is not the most interesting. Attractors with dimension $D^{(2)} \approx 7$ have been successfully analyzed in this way, from time series of $\sim 10^4$ points (10). Efficient algorithms (Ref.10) are essential in keeping computer time low.

The same remarks apply to generalized entropies, in which case one has to count the number of sequences $\{\vec{y}_n, \vec{y}_{n+1}, \ldots \vec{y}_{n+d}\}$ in the time series for which

$$|\vec{y}_{nm} - \vec{x}_{km}| < \varepsilon$$
$$|\vec{y}_{n+1} - \vec{x}_{k+1}| < \varepsilon \qquad\qquad (4.1)$$
$$\vdots$$
$$|\vec{y}_{n+d} - \vec{x}_{n+d}| < \varepsilon \ .$$

Here, we have assumed that the $\vec{x}_k$ are also obtained from a time series (eventually $\vec{x}_k = \vec{y}_k$) with the same delay between successive measurements. The mean number of such sequences should decrease (for small ε) like $e^{-d\tau K^{(\alpha)}}$, where α depends on the method of averaging.

A variant of the above methods of measuring $D^{(\alpha)}$ consists of the following. One measures the radii $R_j(\vec{x}_k)$ of the smallest balls around $\vec{x}_k$ containing exactly j points $\vec{y}_{n_1} \ldots \vec{y}_{n_j}$.
Their logarithmic average (averaged over all $\vec{x}_k$) should behave like (8,28,29)

$$\langle \log R_j \rangle \sim \frac{1}{D} \log \frac{j}{N} \ . \qquad\qquad (4.2)$$

The averages of powers of R_j should scale like (29,39)

$$\langle R_j^q \rangle \sim \left(\frac{j}{N}\right)^{q/D^{(\alpha)}}, \qquad (4.3)$$

with α given by

$$q = (1 - \alpha) D^{(\alpha)}. \qquad (4.4)$$

For this method, one can test scaling between the two ranges discussed above. For small distances (corresponding to $j \approx 1$), one has systematic deviations from Eqs. (4.2) and (4.3), but this time the deviations can be computed exactly (30). The main drawback of the method is that one has to order the points $\vec{y}_n$ according to their distances from $\vec{x}_k$. This has to be done for each k, which enhances computation time and storage requirements considerably. Nevertheless, it seems that this method is quite efficient for computing Kolmogorov entropy (31) and information dimension (8).

In the above discussion, we have assumed that we have determined all dynamical variables (i.e., all components of $\vec{x}$) in the time series. In all practical cases, this is rather unrealistic. The way out of this dilemma was pointed out by Takens (32) and Packard et al. (33).

First, one notices that a D-dimensional attractor should be faithfully representable in any R^d with dimension d > D. Actually this is not quite true, since information dimension is a purely measured theoretical concept, and the small R^d in which a D-dimensional attractor is embeddable might have an arbitrarily high dimension. But, in practice, such pathological cases do not seem to occur.

Secondly, a R^d useful for that purpose is spanned by d sucessive measurements of a (generic) time sequence of any single observable. Since $D_i^{(q)}$ and $K^{(q)}$ are invariants, they will be independent of the choice of observable and of the time sequence, except in singular cases.

Successful estimates of different $D^{(\alpha)}$'s for various systems, both real ones and simulated, have been performed in this way in Refs. (8),(9),(10),(19),(23),(34). In general, these references found that the embedding dimension d had to be considerably bigger than $D^{(\alpha)}$, otherwise the latter was systematically underestimated.

As a last remark, let us consider the influence of low-level noise on an otherwise deterministic attractor. Assume that the noise is white, with amplitude of order δ. Unless the

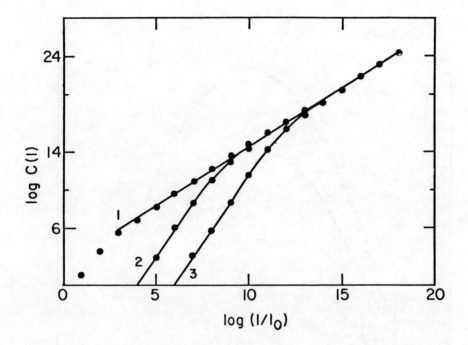

Fig. 8: Correlation integral of the Hénon map with added random noise. Noise levels are $|\delta x_n| = 0$ (curve 1), $|\delta x_n| \lesssim .5 \times 10^{-3}$ (curve 2), and $|\delta x_n| \lesssim 5 \times 10^{-2}$ (curve 3) (from Ref. (36)).

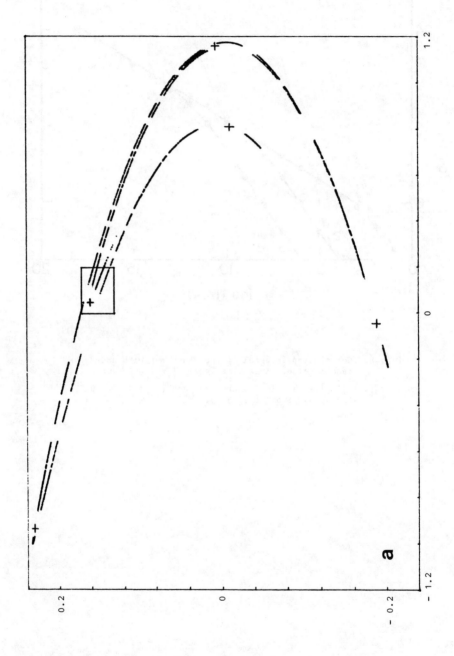

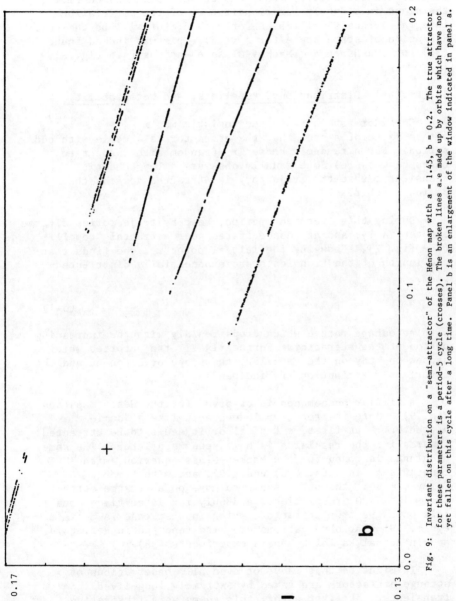

Fig. 9: Invariant distribution on a "semi-attractor" of the Hénon map with a = 1.45, b = 0.2. The true attractor for these parameters is a period-5 cycle (crosses). The broken lines a,e made up by orbits which have not yet fallen on this cycle after a long time. Panel b is an enlargement of the window indicated in panel a.

system amplifies this noise excessively (35), the noise will not be felt on a length scale $\gg \delta$. On the other hand, on length scales $\ll \delta$, , any deterministic structure is washed out completely, and the system is not confined to the D-dimensional attractor. Thus, the effective dimension (i.e., the slope in a doubly-logarithmic plot of the correlation integral versus ε) will be D for $\varepsilon \gg \delta$, but will be the full dimension of state space for $\varepsilon \ll \delta$. This was indeed found (36) for the Hénon and Mackey-Glass equations (see Fig. 8).

5. Unstable Distributions, Repellers, And Metastability

Consider for instance the logistic map $x_{n+1} = 1 - ax_n^2$, with a value of $\underline{a}$ such that the attractor is a cycle with odd period. For a randomly chosen x_0 (random with respect to Lebesgue measure), one usually observes a long chaotic transient before the trajectory is attracted towards the periodic orbit.

During this transient period, near-by trajectories diverge exponentially, and are concentrated near a fractal "repeller" (see Fig. 2). Changing the initial point x_0, one finds that the length of the transients are exponentially distributed (12),

$$P(T) \sim e^{-\alpha T} , \qquad (5.1)$$

with an escape rate α which drops rapidly with the increasing period of the attractor. For nearly all trajectories which happen to stay on the repeller, the Lyapunov exponent and the invariant distribution are independent of x_0.

A similar phenomenon is observed for the Hénon map (see Fig. 9), where the repeller indeed acts like a "semi-attractor": at first, a typical orbit seems to be attracted towards it; the repulsion is only seen much later. The same occurence is found for the Mackey-Glass equation, with constants as in Refs. (26) and (19), and with time delay $\tau = 90$: the orbit seems to move on a strange attractor for $\sim 10^3 - 10^4$ revolutions, and only for later times does it fall onto the true attractor, which in that case is a limit cycle. Similar observations are also reported in Ref. (37), and in actual Rayleigh-Bénard experiments (38).

It might be that many, or even all, observations of strange attractors are faked by extremely long-lived transients. At first sight, this seems very frustrating. But I claim that it is not. The reason is that in all of the above cases there seems to be a unique invariant distribution on the

repeller, and all previous considerations apply to that
distribution as well. For the distribution shown in Fig. 2,
this was checked in detail; for Fig. 9, it agrees with
preliminary results.

In addition, one can relate the escape rate α to
dimensions and Lyapunov exponents. Along those directions for
which $\lambda_i < 0$, the invariant set acts as an attractor. Along
those with $\lambda_i > 0$ but with $D_i = 1$ (i.e., with an essentially
continuous measure), it is not attracting, but there is no
possible escape. Escape occurs only along these directions for
which $\lambda_i > 0$ and $D_i \neq 1$. In Ref. (11), we show that the escape
rate is

$$\alpha = \sum_i{}' \lambda_i (1 - D_i) + \frac{1}{2} \sum_{i,k}{}' Q_{ik}(1 - D_i)(1 - D_k) + \ldots \quad (5.2)$$

where the sum goes over all directions with $\lambda_i > 0$, in
general. For the escape from the distribution shown in Fig. 2,
we found $\alpha = .0327 \pm .0002$ and $Q_{11} = .09 \pm .01$, in perfect
agreement with this conjecture.

We conclude that observation of long-lived transients,
with well-defined Lyapunov exponents and dimensions, and with
Equations (2.15), (2.19), (3.11b), and (5.2) satisfied, is
indicative of chaotic behavior. This is chaos in the sense of
Li and Yorke (39). Indeed, Fig. 9 shows just the repeller of a
Smale's horseshoe (40).

6. Conclusion

We have discussed the information flow leading to the
unpredictability of chaotic motion. We have seen that it is a
fundamental problem, rendering obsolete Laplace's "superior
intelligence" who can predict the fate of the universe from its
initial conditions. Indeed, Lorenz's famous paper (41) arose
from our inability to predict even the weather of our earth for
more than a few days.

We have seen that this information flow is closely related
to the second law of thermodynamics, and to the sensitive
dependence on initial conditions. The violation of the second
law for non-equilibrium systems, in its naive version, is due
to the fact that a dissipative system orders itself on a lower-
dimensional set in state space - the attractor. The
information flow rate was then found to be essentially the
product of the divergence rates for nearby trajectories (the
Lyapunov exponents) times the dimensions of the attractor along
these expanding directions, summed over all directions. The

information balance equation leads then to a connection between
dimension and Lyapunov exponents (a weakened form of the
Kaplan-Yorke conjecture).

In this discussion, we have used the premise that motion
is stationary in the mean, i.e., that the distribution is
invariant under time translations. It need not be
attractive. Accordingly, we obtain the same results for
repellers or other invariant distributions, occasionally
leading to metastability.

A remark about Hamiltonian systems is in order. All above
arguments apply to Hamiltonian systems as well, but according
to conventional folklore, most of the above arguments are
trivial in that context. This folklore claims that single
chaotic orbits fill regions of finite volume on the energy
shell. (In this case, all $D_i^{(\alpha)}$ are equal to 1 for a chaotic
orbit, and Eqs. (2.19) and (3.15) are equivalent to Liouville's
theorem.) For systems with a 3-dimensional energy shell, this
is indeed true. For higher-dimensional systems, the sum of all
chaotic orbits fill a finite volume, and one assumes that all
chaotic regions are connected by Arnold diffusion. But the
latter need not be true (42), and numerical simulations (43)
suggest indeed that typical chaotic trajectories fill fractal
regions, with fractal dimension much less than the dimension of
the energy shell. If this proves to be correct, the above
considerations should also apply non-trivially to Hamiltonian
systems.

Throughout the present paper, I have assumed that some
coarse-graining is done, and have looked at the behavior when
this coarse-graining is made finer and finer. Without coarse-
graining, it seems impossible to define information or
entropy. Without considering the limit of infinitely fine
coarse-graining, it seems hard to distinguish between
deterministic chaos and stochastic processes. There are ,of
course, also other indications of deterministic chaos, like
Feigenbaum sequences, or flows which resemble Lorenz
attractors. But the most clear-cut sign of determinism is that
effective Kolmogorov entropy flow rates and effective
dimensions stay finite in the limit of fine coarse-graining.
Whether existing observations are precise enough to recognize
this behavior is an open question which definitely deserves
further study.

FOOTNOTES

[1] Or, more precisely, for its mathematical description as a Wiener process.

[2] Up to the signs of the basis vectors.

[3] The sum here should extend over that set of indices which maximizes the r.h.s. This need not be the set of unstable directions, but in general it will be.

REFERENCES

1. Pierre Simon Marquis de Laplace, Théorie analytique des probabilités, , 1795 (oeuvres completes tome 7, Gauthier-Villars, Paris 1886).
2. L. D. Landau and E. M. Lifshitz, Fluid Mechanics (Addison-Wesley, Reading 1968).
3. A. N. Kolmogorov, Dokl. Akad. Nauk SSSR 124, 768 (1959);
   Ya. G. Sinai, Dokl. Akad. Nauk SSSR 124, 768 (1959);
   P. Billingsley, Ergodic Theory and Information (Wiley, New York, 1965).
4. R. Shaw, Z. Naturforsch. 36a, 80 (1981);
   J. D. Farmer, Thesis (Univ. of California at Santa Cruz, 1981).
5. J. Ford, "How random is a coin toss?", Physics Today, April 1983, p. 1.
6. R. Bowen and D. Ruelle, Inventiones Math. 29, 181 (1975).
7. B. B. Mandelbrot, The Fractal Geometry of Nature (Freeman, San Francisco 1982).
8. A. Brandstäter et al, Phys. Rev. Lett. 51, 1442 (1983).
9. B. Malraison et al., J. Physique-LETTRES 44, L897 (1983).
10. P. Grassberger and I. Procaccia, Physics 9D, 189 (1983);
    Phys. Rev. Lett. 50, 346 (1983).
11. H. Kantz and P. Grassberger, to be published.
12. G. Pianigiani and J. A. Yorke, Trans. Amer. Math. Soc. 252, 351 (1979).
13. J. Balatoni and A. Renyi, Publ. Math. Inst. Hung. Acad. Sci. 1, 9 (2956);
    D. Farmer, Z. Naturforsch. 37a, 1304 (1982);
    D. Farmer, E. Ott, and J. A. Yorke, Physica 7D, 153 (1983).
14. D. Ruelle, in Bifurcation Theory and its Application in Scientific Disciplines, New York Acad. Sci. 316 (1979).
15. M. Hénon, Commun. Math. Phys. 50, 69 (1976).
16. F. Ledrappier, Commun. Math. Phys. 81, 229 (1981).
17. L. -S. Young, "Dimension, entropy, and Lyapunov exponents", preprint.

18. P. Frederickson, J. L. Kaplan, and J. A. Yorke, J. Diff. Egn., to be published; J. L. Kaplan and J. A. Yorke, in Lecture Notes in Math. 730, p. 288 (Springer, Berlin 1978).
19. P. Grassberger and I. Procaccia, Physica D, to be published.
20. A. J. Khinchin, Mathematical Foundations of Information Theory (Dover, N.Y., 1957).
21. A. Renyi, Probability Theory (North-Holland, 1970).
22. S. Kullback, Ann. of Math. Statistics 22, 79 (1951); F. Schlögl, Physics Reports 62, 267 (1980).
23. P. Grassberger, Phys. Lett. 97A, 227 (1983).
24. F. Takens, "Invariants related to dimension and entropy", in Atas. do 13° Coloquis Brasileiro de Matematics.
25. D. Farmer and P. Grassberger, to be published.
26. M. C. Mackey and L. Glass, Science 197, 287 (1977); D. Farmer, Physics 4D, 366 (1982).
27. P. Grassberger, preprint WU B 84-5 (1984).
28. J. Guckenheimer and G. Buzyna, Phys. Rev. Lett. 51, 1438 (1983).
29. R. Badii and A. Politi, Firenze preprint (1983).
30. P. Grassberger, submitted to Phys. Rev. Lett.
31. Y. Termonia, Phys. Rev. A29, 1612 (1984).
32. F. Takens, in Proc. Warwick Symp. 1980; D. Rand and B. S. Young, eds.; Lecture Notes in Math. 898 (Springer, Berlin 1981).
33. N. H. Packard et al., Phys. Rev. Lett. 45, 712 (1980).
34. Y. Termonia and Z. Alexandrowicz, Phys. Rev. Lett. 51, 1265 (1983).
35. D. Farmer, Los Alamos preprint (1983).
36. A. Ben-Mizrachi et al., Phys. Rev. A29, 975 (1984).
37. J. A. Yorke and E. D. Yorke, J. Stat. Phys. 21, 263 (1979); C. Grebogi, E. Ott and J. A. Yorke, Phys. Rev. Lett. 50, 935 (1983) (erratum 51, 942 (1983)); Physica 7D, 181 (1983); S. Takesue and K. Kaneko, Progr. Theor. Phys. 71, 35 (1984).
38. P. Berge and M. Dubois, Phys. Lett. 93A, 365 (1983).
39. T. Li and J. A. Yorke, Am. Math. Month. 82, 985 (1975).
40. S. Smale, Bull. Amer. Math. Soc. 73, 747 (1967).
41. E. N. Lorenz, J. Atmos. Sci. 20, 130 (1963).
42. L. Galgani, private discussion.
43. M. Pettini et al., preprint.

ON THE RAPID GENERATION OF MAGNETIC FIELDS

S. Childress and A. M. Soward

ABSTRACT

We will consider the problem of computing the mean emf generated by the steady velocity field $\mathbf{u}^* = (\sin(y)+\cos(z), \sin(z)+\cos(x), \sin(x)+\cos(y))$ in the presence of a mean magnetic field, limited by a of large magnetic Reynolds number R. An asymptotic procedure relying on a presumed geometry of flux ropes and sheets is used. Our results suggest that alpha has a finite nonzero limit as $R \to \infty$. We express this limit in terms of the geometry of certain two-dimensional manifolds emerging from stagnation points of $\mathbf{u}^*$.

1. Introduction

Although it is now generally agreed that many naturally occurring magnetic fields are sustained by fluid dynamos [1,2] there are as yet few satisfactory models of the process. This is especially true if the increase of magnetic energy is <u>rapid</u>. In that case the material responds as if its electrical conductivity were very large. The precise conditions, expressed in terms of length and time scales for the motion and magnetic diffusivity η, is that the magnetic Reynolds number $L^2/T\eta$ is large compared to unity. Alfvén [3] argued that at large R, the magnetic lines of force are almost "frozen" into the fluid, in the manner of the vortex lines of a perfect fluid flow.

This property makes the dynamo process difficult to understand at large R. On the one hand, tubes of flux can be freely stretched and distorted so as to increase the magnetic energy density; on the other, complicated fields of flow can form regions of high shear where the nominally small diffusive effects become greatly enhanced. Also, finite diffusivity will not only introduce dissipation, but will also allow perhaps crucial changes in the topology of flux tubes. It is not even clear, for example, whether or not some suitable measure of the intensity of the dynamo process will increase or decrease with R when this parameter is large.

The simplest setting in which to address some of these issues is the <u>kinematic</u> dynamo theory, where the velocity field $\mathbf{u}(\mathbf{x},t)$ is a prescribed function. Induction on the time scale of most dynamos produces insignificant electromagnetic radiation, and if material properties are taken to be constant the equations for the magnetic field $\mathbf{b}(\mathbf{x},t)$ are

$$\frac{\partial \mathbf{b}}{\partial t} = \nabla \times (\mathbf{u} \times \mathbf{b}) + R^{-1}\nabla^2 \mathbf{b} \equiv L_u \mathbf{b}, \quad \nabla \cdot \mathbf{b} = 0, \qquad (1.1)$$

where we have given the dimensionless form taking L/T as a typical fluid speed. Whenever (1.1) admits solutions which satisfy associated linear conditions, and which grow like $e^{\lambda t}$ with $\mathrm{Re}(\lambda) > 0$ we say that $\mathbf{u}$ is a kinematic dynamo. Mathematically, $\mathbf{u}$ is a kinematic dynamo whenever the associated linear operator L_u in (1.1) has spectrum in the right half-plane. In a natural system, of course, the exponential growth of the field is presumably interrupted by the dynamic feedback of the field onto the flow through the Lorentz stresses. Nevertheless, kinematic theory is useful for studying the early phase of growth as well as for trying to understand what classes of $\mathbf{u}$ are particularly well-suited to dynamo action.

At large R, the distinction between a dynamo effect in the above sense, involving uninterrupted exponential growth, and the simple production of large energy for some period of time, is an essential one. There are simple two-dimensional motions (e.g., "pastry-chef" movements which stretch and fold flux tubes) which can produce large energy densities for a finite time at the expense of a proliferation of surfaces of high shear. Abruptly, diffusion breaks down the shear field and the magnetic energy ultimately decays to zero. At the same time, there are reasons to believe that sufficiently complicated three-dimensional motions may exhibit dynamo action in the kinematic sense for arbitrarily large R. Alfvén [4] suggested the process shown in Figure 1. In one cycle, the volume of the tube remains fixed but a rapid movement will essentially double the field intensity. If, at the outset, the tube is slender, the "X-type" kink should have little influence on energy and, during periods of "stasis" where the magnetic field decays freely, we should ultimately find an axial field with radial profile

$$b \sim \frac{b_0}{\tilde{t}} e^{-r^2/4\tilde{t}}, \quad \tilde{t} = Rt \gg 1. \qquad (1.2)$$

ON THE RAPID GENERATION OF MAGNETIC FIELDS 225

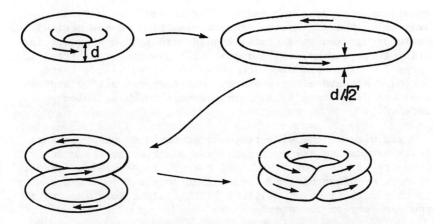

Fig. 1 A hypothetical dynamo operating on a flux ring at large R. After one cycle the magnetic energy is increased by a factor of approximately four.

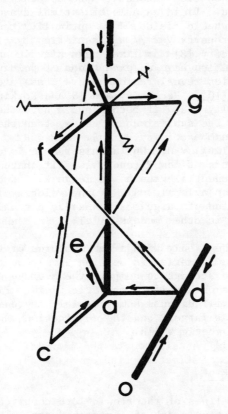

Fig. 2 Geometry of the stagnation points and straight heteroclinic streamlines of **u***, see Section 2.

Providing the power necessary to drive the movement is
available, the exponential growth resulting from a periodic
cycle of movement and stasis should more than compensate for
dissipative losses, and a dynamo will result for arbitrarily
large R. Other examples are known [5], involving a fixed
periodic structure for **u** and **b**, where dynamo action appears
and then disappears as R is increased through a finite
interval.

In view of the wide range of possible behavior at large R,
it is natural to try to relate dynamo action to the underlying
geometry of the flow field, and thereby to determine "generic"
classes of flows having a similar effect on **b** [6]. As a step
in this direction, we consider in this paper some aspects of
the geometry of the steady flow

$$\mathbf{u}^* = (\sin y + \cos z, \sin z + \cos x, \sin x + \cos y), \tag{1.3}$$

which may bear on the dynamo process at large R. The
motion **u*** emerges from kinematic dynamo theory at <u>low</u> R as the
simplest periodic field capable of optimal dynamo action on
large-scale magnetic fields [7]. Optimality results here from
the Beltrami property $\nabla \times \mathbf{u}$ // $\mathbf{u}$, satisfied by (1.3), which here
maximizes helicity [8] for fixed mean kinetic energy. The flow
lines of (1.3) also have an exceptional topology [9], and
numerical studies reveal regions where these are apparently
space-filling [10]. Recently, Arnold and Korkina [11] have
shown numerically that, if both **b** and **u** have the same
period (2π) in the spatial coordinates, then there is a finite
"window" of dynamo activity $R_1 \simeq 8.9 < R < R_2 \simeq 17.5$ where
$\text{Re}(\lambda) > 0$ and $\text{Im}(\lambda) \neq 0$. While this result is consistent with
smoothing methods at low R (and, also, if there were no other
windows of exponential growth, at large R), it does show that
dynamo action on a local scale of the velocity field can mask
average induction of large-scale fields, a problem that has
already arisen in other models [12,13].

We shall deal only with a few features of the geometry of
(1.3), which we summarize in Section 2. These are used in
Section 3 in an attempt to compute the average emf generated
when (1.3) acts on a uniform magnetic field. Finally, in
Section 4, we will discuss some possible ingredients of a
dynamo theory at large R and the application of such a theory
to the solar magnetic field.

2. <u>GEOMETRY OF u*</u>

The streamlines of the flow **u*** are complicated and do not
conform to any simple set of streamsurfaces. The general

Fig. 3 A model of the mirror image flow **u***(-x,-y,-z). The node A may be compared to the point a of Figure 2. Extending upward and to the left from A is a structure containing six nodes; the latter are equivalent to points a-h of Figure 2. Note that the reflection has reversed the sense of twist of this structure, as can be seen from the secondary streamlines such as df.

nature of the periodic structure in which these lines are
distributed can be understood rather easily, however, by
looking at the stagnation points, where $\mathbf{u}^* = 0$, and certain
heteroclinic streamlines which connect them. If we set
$(x,y,z) = (-\pi/r, -\pi/4, -\pi/4) + (x^*,y^*,z^*)(\pi/2)$, then (1.3) has
the form

a	b	c	d	e	f
(0,0,0)	(2,2,2)	(0,-1,-1)	(-1,0,1)	(1,-1,0)	(2,3,1)

g	h
(1,2,3)	(3,1,2)

Table 1. Values of (x^*,y^*,z^*) for the points of Figure 2

It can also be shown by reference to (2.1) and Table 1,
that the straight lines connecting these points, as shown in
Figure 2, are streamlines of the flow. For example, it is seen
that when $\mathbf{u}^*$ is evaluated at $ad + k\, df$ we get a multiple of
df. The direction of flow along these lines is indicated in
the Figure. There are three kinds of lines, of lengths
$\sqrt{2}$, $\sqrt{12}$, and $3\sqrt{2}$ (in starred coordinates), respectively. The
shortest links are represented by ac, ad, ae, and the parallel
triad emerging from b. Other such links are not shown, but
emerge, for example, from d in the plane determined by ad and
df. The intermediate links are represented in the Figure by
ab; we also show a piece of another intermediate link passing
through d orthogonal to the plane of ad and df. The longest
links are, for example, ch, df, and eg. The helicity of the
flow is evident from these connections. Three such links also
emerge from a and b as continuations of the triads of short
links, and similarly at other points. In fact it can be seen
that all stagnation points are equivalent under a translation
and rotation, which bring into coincidence, with proper
orientation, the three kinds of links connected to the points.

From the direction of the flow along these heteroclinic
streamlines we may isolate two kinds of manifolds. The short
and long links are associated with two-dimensional manifolds
(stable at a, unstable at b), while the intermediate links
belong to one-dimensional manifolds, unstable at a and stable
at b.

We will next discuss the nature of the flow in the
vicinity of these straight heteroclinic streamlines.
Let (ψ,η,ζ) be a right-handed Cartesian system with
the ζ-axis on the streamline, and let (ζ,ρ,θ) be corresponding
cylindrical polar coordinates. We then have locally

$$(u^*_\zeta, u^*_\rho, u^*_\theta) = [f(\zeta), \rho G(\zeta,\theta), \rho H(\zeta,\theta)] + o(1,\rho,\rho). \quad (2.2)$$

The incompressibility of the flow and the ζ-component of the equation $\nabla \times \mathbf{u}^* + \mathbf{u}^* = 0$ then imply that

$$f' + 2G + \frac{\partial H}{\partial \theta} = 0, \quad 2H + \frac{\partial G}{\partial \theta} = -f. \quad (2.3)$$

It thus follows that

$$(u^*_\zeta, u^*_\rho, u^*_\theta) = [f(\zeta), -\frac{\rho}{2}f' + \rho(g(\zeta)\cos 2\theta + h(\zeta)\sin 2\theta),$$
$$-\frac{\rho f}{2} + \rho(h(\zeta)\cos 2\theta - g(\zeta)\sin 2\theta)] + o(1,\rho,\rho). \quad (2.4)$$

Now, applying (2.4) to specific lines, consider first ab in Figure 2. The invariance under 120° rotation implies that in this case $g = h = 0$ in (2.4) and we obtain, if ζ has values 0 and $\sqrt{3}\pi$ at a and b,

$$f = \sqrt{6}\sin(\frac{\zeta}{\sqrt{3}}), \quad g = h = 0. \quad (2.5)$$

The other lines of interest are typified by ac and its extension ac', say, in the opposite direction. From Table 1 we take the directions of increasing ξ, η, and ζ to be $(1,0,0)$, $(0,-1,-1)$, and $(0,1,-1)$ respectively. It is then straightforward to establish (2.4) with

$$f = 1 - \cos(\zeta/\sqrt{2}) - \sin(\zeta/\sqrt{2}), \quad (2.6a)$$

$$g = (1/2\sqrt{2})(\sin(\zeta/\sqrt{2}) - \cos(\zeta/\sqrt{2})), \quad (2.6b)$$

$$h = -(1/2)(1 + \cos(\zeta/\sqrt{2}) + \sin(\zeta/\sqrt{2})). \quad (2.6c)$$

We thus see that there is a further classification of the straight heteroclinic streamlines based upon local flow structure. The primary streamlines, such as ab, have locally a converging-diverging axisymmetric swirling flow, while the <u>secondary</u> streamlines, such as ac and df, are distinguished by the presence of non-axisymmetric components in the transverse flow.

Near the point a, the flow along ab, near the point a, reduces to a simple axisymmetric stagnation point flow:

$$(u^*_\zeta, u^*_\rho, u^*_\theta) \sim \sqrt{2}\,(\zeta, \rho/2, 0). \quad (2.7)$$

This flow determines how the streamlines near a primary streamline fan out into the vicinity of an unstable two-dimensional manifold (i.e., at d). The global structure of such nearby orbits is very involved [10], and there are currently studies under way [14] of the Lagrangean flow for the generalized Beltrami field $\mathbf{u}^*(A,B,C) = (B\sin)(y) + C\cos(z),..,..)$. We confine attention here to one feature of the geometry of importance in steady dynamo theory, namely the structure of an unstable two-dimensional manifold M. We shall refer to a point such as, a, as a point of convergence, and to a point such as, d, as a point of divergence. The unstable manifolds emerge from points of divergence, and to study their geometry we can imagine the fate of a small circular ring of marked particles, say $\zeta = 0$, $\rho = \varepsilon$ in suitable coordinates. For $\mathbf{u}^*$ there is no simple characterization of the motion of the particles in terms of streamsurfaces, and the information which can be gained from the local flow (2.4) is extremely limited. In Figure 4, we show a numerical simulation of a 60° segment of the particle ring originating at d in the plane of da and df. The control of the surface by the secondary streamlines df and da and the primary streamlines ab and ff' is evident. As the surface emerges from the convergence at a, the tangent plane is essentially in the plane of ad and ab. Along ab, the tangent element rotates at a rate $(1/2)|\nabla \times \mathbf{u}^*| = (1/2)u^*$ and so, as b is approached it has rotated $\sqrt{3}\pi/2$ radians or 155.9°. Thus, it emerges from the divergence at b somewhere between bf and the extension of gb. Thereafter, the fate of particles initially close to ab is no longer determined by primary or secondary streamlines, although the former continue to provide a network of exceptional streamlines. Note that at both f and f' in Figure 4 there are apparently portions of the manifold going in both directions from a stagnation point. Because of the lack of any obvious restrictions on the domain of the manifolds, we conjecture that their extensions fall within the region of space-filling trajectories noted by Hénon [10].

3. Mean emf Generated By $\mathbf{u}^*$.

Our object in the present section is to indicate how the symmetric pseudo-tensor $\boldsymbol{\alpha}$ defined by

$$\langle \mathbf{u}^* \times \mathbf{b} \rangle = \boldsymbol{\alpha} \cdot \langle \mathbf{b} \rangle, \quad \langle \cdot \rangle = \text{volume average}, \tag{3.1}$$

can be evaluated in the limit of large R in terms of the geometry of the manifolds M. The tensor $\boldsymbol{\alpha}$ arises in mean-field electrodynamics [1], and $\boldsymbol{\alpha} \cdot \langle \mathbf{b} \rangle$ can be viewed as the average emf created by magnetic induction (here, under steady conditions), when there is a uniform mean magnetic field $\langle \mathbf{b} \rangle$. Note that $\boldsymbol{\alpha}$ has the dimensions of a velocity.

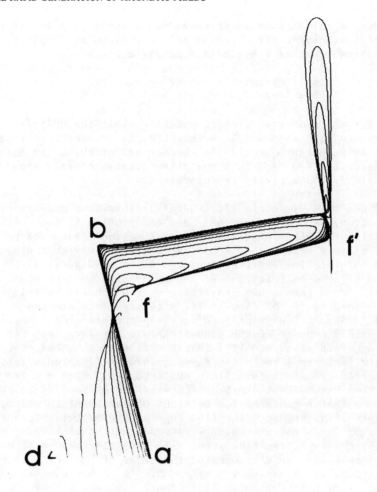

Fig. 4 Views of a marked line of particles carried by the flow
u*. Initially, the line is a 60° sector of the circle
of radius .01 centered at d and lying in the plane
adf. A polygonal line consisting of n(t) points is
advanced using a classical Runge-Rutta procedure
applied to the vertices. n(0) = 21, and a link whose
length exceeds twice the initial length is divided,
creating a new vertex. The time step is .0707, and
views are shown every 10 steps. the run was stopped at
n = 5000. The resulting surface, if continued
indefinitely, approximates 1/6 of the manifold M.

Such a calculation involves a double limit in R and in the size of the domain. If V is a domain of volume v (a sphere for example), we then seek a constant matrix α_0 such that

$$\lim_{R,v\to\infty} \left[\frac{1}{v} \int_V (\mathbf{u}^* \times \mathbf{b} - \alpha_0 \mathbf{b}) dV\right] = 0, \qquad (3.2)$$

where for the moment we are not specific about the order of the limits. We have supposed here that $\alpha(R)$ is bounded for large R, and an important question is whether or not the α_0 in (3.2) vanishes. If it does, the alpha effect vanishes with large R and rapid dynamo activity is suppressed.

Since $\boldsymbol{\alpha}$ is symmetric [7], and $\mathbf{u}^*$ is invariant under cyclic permutation, we have $\alpha_{ij} = \alpha \delta_{ij}$, and the problem is that of estimating the behavior of the scalar function $\alpha(R)$ at large R. If a limit exists and is nonzero, it is likely that dynamo action persists for arbitrarily large R, as we discuss in Section 4. The calculations we are describing suggest that, for $\mathbf{u}^*$, α does have a nonzero limit, α_0, which can be related to the geometry of the flow field. The arguments are based upon a fundamental assumption concerning the presumed asymptotic structure of the magnetic field for large R. We have suggested elsewhere [15] that for flows such as $\mathbf{u}^*$, the magnetic field is largely confined to intense flux ropes in the vicinity of certain heteroclinic trajectories, which may be identified here with the primary streamlines of Section 2. We shall see that because of the Beltrami character of the flow, at points of divergence, the flux rope undergoes twisting and pulling out onto the unstable two-dimensional manifold M. For large R, the resulting flux sheet will have thickness of order $R^{-1/2}$ and will, over distances of the order of the periodicity of the flow, lie close to a manifold M. As is the case for a related axisymmetric problem [15], these flux sheets are the sites of intense induction, which leads to $O(1)$ contributions to emf and a nonvanishing α in the limit of large R.

The resulting basic magnetic structure, near a divergence such as b, is shown in Figure 5a. We associate with b the manifold M(b) and the segments ba and ba' of length $\sqrt{3}\pi$ extending in each direction from b. The piece of flux tube aa' carries a flux F_b parallel to aa'. The magnetic structure consisting of the portion of tube and the flux sheet emanating from it is defined as $\mathbf{B}(\mathbf{x};\mathbf{x}_b,F_b)$. We indicate the meaning of the structure $\mathbf{B}$ in Figure 5b.

If a uniform field in a certain direction initially fills space and the flow $\mathbf{u}^*$ is switched on, we ultimately expect to find structures of the form $\mathbf{B}$ distributed over space, at points of divergence $\mathbf{x}_i$ and carrying flux vectors F_i. The latter are

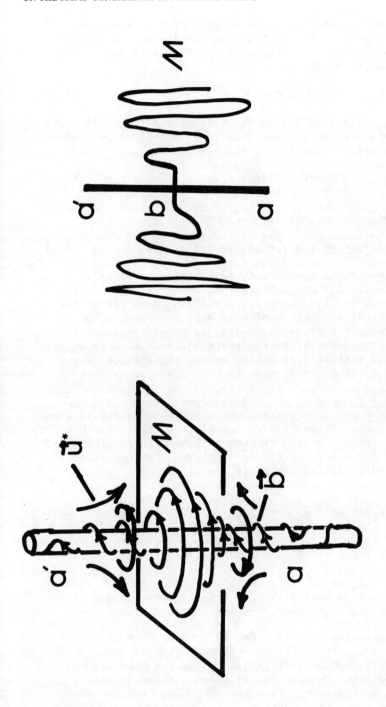

Fig. 5 (a) Local geometry of the magnetic and velocity fields near a point of divergence b (see Figure 2) on a primary streamline. (b) The manifold M issuing from b.

presumably determined by the orientation of the associated primary streamlines relative to the direction of the initial field. If we restrict attention to x_i which fall within the control volume V, we may write

$$\mathbf{b} \sim \sum_i \mathbf{B}(\mathbf{x}; \mathbf{x}_i, \mathbf{F}_i). \tag{3.3}$$

Since, as we shall show below, the manifolds arise as in Figure 5 by deformation of a flux tube, there is no effect on average flux and we have

$$\frac{1}{V} \int_V \mathbf{b} dV \sim \sum_i 2\sqrt{3}\pi \, \mathbf{F}_i. \tag{3.4}$$

To find α_0 we must, similarly, evaluate

$$\frac{1}{V} \int_V \mathbf{u}^* \times \mathbf{B}(\mathbf{x}; \mathbf{x}_i, \mathbf{F}_i) dV \equiv \mathbf{E} \tag{3.5}$$

for large v and R. A convenient way to treat (3.5) is to interchange summation and integration, but this leads to a possible divergence for large V since there is no reason to expect the infinite integral of $\mathbf{u}^* \times \mathbf{B}$ to converge. Thus, in order to formulate a means of finding α_0 we must examine the structure of the $\mathbf{B}$ for large R. Since all such structures are equivalent under rigid body motion and multiplication by a constant, we treat a single case in the polar coordinates of Section 2.

The flux rope. We consider the segment ab of Figure 2 and the local representation (2.5). Locally, the flux rope has the structure determined by the related axisymmetric problem[15] and the magnetic field for $0 < \zeta < \sqrt{3}\pi$ follows from the axisymmetric results [16]. Taking $|\mathbf{F}_b| = 1$, we have

$$(b_\zeta, b_\rho, b_\theta) = (\rho^{-1} \frac{\partial A}{\partial \rho}, -\rho^{-1} \frac{\partial A}{\partial \zeta}, \rho B), \tag{3.6a}$$

$$A = (2\pi)^{-1}(1-e^{-\eta}), \quad \eta = R\rho^2 f(\zeta)/4\zeta, \tag{3.6b}$$

$$B = -Rh(\zeta)e^{-\eta}, \quad h(\zeta) = (8\pi\zeta^2)^{-1} \int_0^\zeta \zeta f'(\zeta) d\zeta. \tag{3.6c}$$

Note the progressive twisting that is evident from (3.6c) as ζ increases from 0. To obtain the corresponding expressions for A and B for the rope ba', $\sqrt{3}\pi < \zeta < 2\sqrt{3}\pi$, we make these functions even with respect to the plane $\zeta = \sqrt{3}\pi$. This symmetry can be understood from the fact that flux persists down the tube, and the induction of b_θ is proportional to the derivative of u_θ with respect to ζ.

Near $\zeta = \sqrt{3}\pi = \zeta_0$ we have from (3.6c)

$$B \sim R(24\pi^3)^{-1} \int_0^{\zeta_0} f(\zeta) d\zeta \, \exp[-R\rho^2(\zeta_0-\zeta)/2\sqrt{2}], \quad \zeta < \zeta_0. \qquad (3.7)$$

Within a domain of size $R^{-1/3}$, the values of A and B are conserved through the divergence on the streamlines of the local stagnation point flow [15], the local streamfunction being

$$\psi = (\zeta_0-\zeta)\rho^2/\sqrt{2}. \qquad (3.8)$$

From (3.6b) and (3.7), we obtain the functions of ψ which characterize A and B on streamlines near the origin of the manifold M, and which therefore determine the structure of the flux sheet associated with M.

<u>The flux sheet</u>. We now consider the structure of the flux sheet as it emerges from the $R^{-1/3}$ region. A boundary layer approximation is then possible, involving a reduced coordinate $(\zeta_0-\zeta)R^{1/2}$. The result is that radial diffusion may be neglected, and from (1.1) we obtain the local equation for B:

$$\sqrt{2}(\zeta_0-\zeta)\frac{\partial B}{\partial \zeta} + \frac{\rho}{\sqrt{2}}\frac{\partial B}{\partial \rho} - \frac{1}{R}\frac{\partial^2 B}{\partial \zeta^2} \sim 0. \qquad (3.9)$$

Writing $B = B(\phi,\psi)$ where $\phi = \rho^4/4\sqrt{2}$ and ψ is given by (3.8), (3.9) takes the form

$$\frac{\partial B}{\partial \phi} - \frac{1}{R}\frac{\partial^2 B}{\partial \psi^2} = 0. \qquad (3.10)$$

On $\phi = 0$, (3.7) supplies the initial condition

$$B(0,\psi) = \frac{R}{2\sqrt{2}\pi^3} e^{-\frac{R}{2}|\psi|}, \qquad (3.11)$$

where the absolute value sign accounts for the contribution from the field in the region $\zeta_0 < \zeta < 2\zeta_0$. Since (3.11) implies that ψ is $O(R^{-1/2})$ when ϕ is $O(1)$, when R is large we may replace the right-hand side of (3.11) by $\sqrt{2}\pi^{-3}\delta(\psi)$. Consequently, once ϕ is finite we have

$$B \sim (R/2\pi^7\phi)^{1/2} \exp(-(R\psi^2/4\phi)). \tag{3.12}$$

Note that ρ cancels out of the exponential so that the flux sheet emerges from the $R^{-1/3}$ region with a fixed thickness of order $R^{-1/2}$. Continuing along the manifold, the thickness increases approximately as the square root of the distance traveled by a fluid particle.

The subsequent evolution of the sheet cannot, of course, be determined explicitly, since we cannot exhibit the geometry of the manifold M. For any fixed R, eventually the flux sheets from different points of divergence may overlap, and it is the gradual decay of flux as a function of position and R which must be understood in order to evaluate the double limit in (3.2).

We can show, that for the purpose of finding α_0, the detailed structure of the flux sheet can be replaced by a geometrical property of $\mathbf{u}^*$. In the sheet, we can replace $\nabla^2 \mathbf{b}$ in (1.1) by the normal derivatives, and if homogeneous conditions on the dominant field are imposed at $\pm \infty$ along the normal, diffusion integrates to zero and one obtains an integral normal to the sheet which is conserved along the sheet. Alternatively, the flux sheet boundary layer equations can be utilized in local orthogonal curvilinear coordinates [17]. If $\mathbf{b}_{//}$ and $\mathbf{u}^*_{//}$ denote the projections of $\mathbf{b}$ and $\mathbf{u}^*$ onto the local tangent plane, and n measures distance along the normal, we obtain

$$(\mathbf{u}^*_{//} \times \int_{-\infty}^{+\infty} \mathbf{b}_{//} \, d\bar{n}) \cdot \mathbf{n} = \text{constant} = K, \quad \bar{n} = nR^{1/2} \tag{3.13}$$

provided that the improper integral exists. To evaluate the constant in (3.13), we orient the manifold so that the unit normal at b points is in the direction ab (that is, in the direction of $\mathbf{F}_b$). Near b, the curvilinear system can be defined to agree with the local cylindrical polars. The dominant contribution to the integral in (3.13) comes locally from the terms $u^*_\rho b_\theta$, since u^*_θ vanishes as the manifold is crossed. Thus, we find, using (3.12),

$$K = (R/2\pi^7\phi)^{1/2} \int_{-\infty}^{+\infty} \exp(-(R\psi^2/4\phi)) d\psi = \sqrt{2}/\pi^3. \tag{3.14}$$

It follows from the estimates of the axisymmetric problem [15], that the dominant contributions to the integral of $\mathbf{u}^* \times \mathbf{B}$ are the $O(1)$ contributions coming from the flux sheet at M. From (3.13) and (3.14) we then have, if M_V is that part of M contained within a volume V,

$$\int_V \mathbf{u}^* \times \mathbf{B} \, dV = K \int_{M_v} \mathbf{n} \, dM. \tag{3.15}$$

In this way the calculation of α_0 is directly related to the projected area of the manifolds M.

<u>Evaluation of α_0 in terms of M.</u> It should be emphasized that we expect the right-hand of (3.15) to diverge with increasing V. In Figure 6, we show the arithmetical projection of area of the growing manifold of Figure 4, onto a plane orthogonal to the primary streamline from which it emerges. This and other numerical simulations suggest that the projected area develops a growing oscillation, as elements of M more and more distant from its origin are included. It is therefore inappropriate to try to calculate projected area of any one manifold; rather, we must sum up the contributions from all manifolds in a given volume element.

To study this question, we first consider the values of $\mathbf{F}_i$. Let $\mathbf{e}(\mathbf{x}_i)$ be a unit vector parallel to $\mathbf{F}_i$. The mapping $\langle \mathbf{b} \rangle \to \mathbf{F}_i$ is onto a fixed direction, and clearly $\mathbf{F}_i = 0$ when $\langle \mathbf{b} \rangle$ and $\mathbf{e}(\mathbf{\bar{x}}_i)$ are orthogonal. Thus,

$$\mathbf{F}_i = \beta \mathbf{e}(\mathbf{x}_i) \mathbf{e}(\mathbf{x}_i) \cdot \langle \mathbf{b} \rangle, \tag{3.16}$$

where β is a constant. To evaluate β, we recall from Section 3:

$$\langle \mathbf{b} \rangle \sim \frac{1}{v} \int_V \sum_i \mathbf{B}(\mathbf{x}; \mathbf{x}_i, \mathbf{F}_i) dV = 2\sqrt{3}\pi \, \rho(\frac{1}{N} \sum_i \mathbf{F}_i) \tag{3.17}$$

where $\rho = N/v$ is the number density of the points of divergence. For $\mathbf{u}^*$, the matrices $(e_i e_j)$ are equally distributed among the four choices $\sqrt{3}\mathbf{e} = (1,1,1)$, $(-1,1,1)$, $(1,-1,1)$, $(1,1,-1)$ and therefore:

$$\lim_{N \to \infty} \frac{1}{N} \sum_{i=1}^{N} \mathbf{F}_i = \beta \langle \mathbf{b} \rangle \tag{3.18}$$

and hence, $\beta = (2\sqrt{3}\pi\rho)^{-1}$.

We now want to consider $\mathbf{E}$ as defined by (3.5). As $R \to \infty$, we may take $\mathbf{u}^* \times \mathbf{B}$ as essentially concentrated on M_i with vector density $K \, \mathbf{n} |\mathbf{F}_i|$ (cf. (3.15)). If M_i is oriented so that $\mathbf{n} = \mathbf{e}(\mathbf{x}_i)$ at $\mathbf{x}_i$, we may represent this distribution in the form $\beta \, K \, T(\mathbf{x}; \mathbf{e}(\mathbf{x}_i), \mathbf{\bar{x}}_i) \, \mathbf{e}(\mathbf{x}_i) \cdot \langle \mathbf{b} \rangle$. Thus, (3.5) will be rewritten

$$E_j - (\beta K/v) \int_V \sum_i T_j(\mathbf{x}; \mathbf{e}(\mathbf{x}_i), \mathbf{x}_i) e_k(\mathbf{x}_i) dV \, \langle b_k \rangle. \tag{3.19}$$

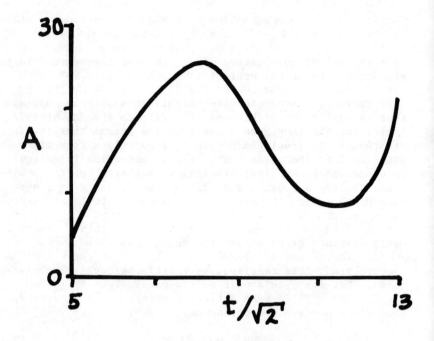

Fig. 6 Projected area versus time for the numerical run shown in Figure 4. Here A is six times the (signed) projected area of the closed polygonal line d v_1 v_2 ... v_n f, where the v_i are vertices.

Consider now the contribution from all the **T**'s to some small volume element ΔV. It is likely that in certain regions of space, there are few, if any, intersections with any of the M_i. In the region of space-filling streamlines, however, it is likely that the intersections are in some sense dense. Nevertheless, we expect the orientation of the M_i to oscillate when they are dense, since contributions which are "distant" from the points of origin of the manifolds should not reflect a preferred direction. We therefore will assume that the contributions of all the T's to a fixed volume element can actually be summed, and that there is a spatially-periodic second-order tensor function $\mathbf{m}(\mathbf{x})$ such that

$$\int_{\Delta V} \sum_i T_j e_k \, dV = \rho \int_{\Delta V} \mathbf{m} \, dV. \qquad (3.20)$$

The appearance of ρ is a convenient normalization. We cannot justify the assumption (3.20) rigorously, nor can we define precisely the integral involved. It is perhaps appropriate to assume that it be in the sense of Lebesque.

With this assumption, we have from (3.2) and (3.19),

$$\alpha_0 \delta_{jk} = \frac{1}{\sqrt{6}\pi^4} \lim_{v \to \infty} \frac{1}{v} \int_V m_{jk}(\mathbf{x}) dV. \qquad (3.21)$$

The existence of **m**, and its spatial periodicity, would seem to imply that α_0 as defined by (3.21) must then exist, but again we have no proof of this. Physically, it is reasonable to associate with each point of divergence a localized region of excitation, which we might approximate by a corresponding contribution

$$\mathbf{m}_i = \frac{A}{\rho} \delta(\mathbf{x}-\mathbf{x}_i) \mathbf{e}(\mathbf{x}_i) \cdot \mathbf{e}(\mathbf{x}_i) \qquad (3.22)$$

to **m**. In this case (3.21) yields a relation between α_0 and the constant A:

$$\alpha_0 = \frac{A}{\sqrt{6}\pi^4}. \qquad (3.23)$$

We can identify A with a projected area of the active manifold onto a plane orthogonal to $\mathbf{e}(\mathbf{x}_i)$. From Figure 6, we take A = 15 for $\mathbf{u}^*$, yielding the estimate $\alpha_0 \sim .02$.

<u>Estimation of **m**</u>. The most essential and interesting mathematical question raised by our study is that of constructing $\mathbf{m}(\mathbf{x})$. For spatially-periodic fields such as $\mathbf{u}^*$, it is tempting to look for some ergodic property associated with the region of space-filling streamlines, allowing a deduction from a Poincare map (e.g., intersection with the

plane $z = 0$ mod 2π). Since we must follow a manifold, however, the map must track line segments, representing intersections by a narrow ribbon. It is not clear, however, what, if any, connection there is between return maps onto a plane and the volume density **m**.

Another approach is to change the problem from a fixed spatially-periodic field to a random field. The difficulty is then to connect a realization of **u** to a realization of $\mathbf{x}_i$ and M_i. Figure 4 suggests that M_i may have some properties of a random walk (of a surface), which can be exploited in the construction of **m**. One possibility is to attempt a mean-field theory, in which a selected $\mathbf{x}_i$ introduces its manifold into a background turbulent flow **u**.

A related scheme, applicable to **u***, adopts the exact form of the manifold but randomizes over the $\mathbf{x}_i$. Consider a sea of elements **B** having random orientation and position, with uniform statistics. Select the region V where **m** is to be constructed, and enclose V in a much larger region V_0. We shall distribute the $\mathbf{x}_i$ uniformly over V_0 and take all such points into account in calculating **m** over V. In (3.19), we first take the integral over V, which yields intersection areas in place of the T's. Next, we average over V_0; since the points are independently distributed, this yields N_0 equivalent integrals, where N_0 = number of $\mathbf{x}_i$ in V_0. Finally, we average over orientation. As v_0 = volume of $V_0 \to \infty$ with $N_0/V_0 = \rho$ we obtain

$$\frac{1}{v} \int_V \mathbf{m}\, dV = \frac{1}{3v} \int A(\mathbf{x}) d\mathbf{x} \qquad (3.24)$$

where $A(\mathbf{x})$ is a projected area of the intersection of an M issuing from **x** (inside or outside V) with V:

$$A(\mathbf{x}) = \mathbf{e}(\mathbf{x}) \cdot \int_V \mathbf{T}(\mathbf{y}; \mathbf{e}(\mathbf{x}), \mathbf{x}) dV_y. \qquad (3.25)$$

Thus, we expect the integrand in (3.18) to fall off rapidly as **x** leaves V and the probability of M hitting a distant region becomes small. When **x** is within V, however, (3.18) effectively averages over oscillations. Again, from Figure 4 we might estimate the right-hand side of (3.18) to be about 5, leading to the previous estimate of α_0.

4. Discussion

This paper has been concerned with a possible physical mechanism for developing $O(1)$ mean emf at large magnetic Reynolds number R. This mechanism is summarized in Figure 5 and depends upon both the concentrating of the magnetic field, leading to rope-like structures, and the subsequent twisting and pulling out of azimuthal field near a point of divergence of the flow. Our analysis has been highly idealized in that we have restricted attention to a special velocity field and have treated only a steady-state problem. We wish now to comment on these results in relation to models which might be closer to naturally occurring dynamos. We have in mind principally, the application of large-R theory to the solar magnetic field.

The principal feature of the observable magnetic field of the solar photosphere-intense local fields of the order of thousands of times the mean field- is consistent with the present flux-rope assumption if an R of 10^3 of 10^4 is assumed, although an incompressible model provides a very incomplete description of the photosphere. Other estimates suggest $R \sim 10^4$ as a reasonable figure [1,18], based on typical speeds of the order 1 km/s on the length scale of the smallest granulation, $L \sim 10^3$ km. The resulting eddy turnover time, $T \sim 10^3$ s, is not an unreasonable time scale for the magnetic fine-structure. If 1 km/s is taken as an rms velocity fluctuation, then, since (1.3) gives an rms velocity of $\sqrt{3}$, the estimate of α_0 given above yields $.02/\sqrt{3} \sim 10^{-2}$ km/s as the value of alpha for the Sun. Stix [18] has noted that consistency with α-ω solar dynamo models leads to values of alpha in the range 10^{-5}-10^{-4} km/s. It is therefore likely that a theoretical value of α_0 for $\mathbf{u}^*$ will be larger than this range by a factor of 100 or so.

We suggest, however, that an overestimation of this size is not unexpected for a steady state theory. We might expect flux tubes to form in the manner assumed here, by flow convergence and divergence near a primary streamline, and it is perhaps possible for these to be relatively long lived in certain flow geometries, but it is surely not reasonable to expect time dependent flows to develop extensive deformation of the tubes as in Figure 5. In this connection, the construction of $\mathbf{m}$ suggested in Section 3 is something of an academic exercise.

A better model would be time-dependent; this modification might be made by switching on $\mathbf{u}^*$ and following the formation of tubes and their stretching onto the manifolds simultaneously, in order to determine the emf developed within a time of the order of eddy time scale T. It is possible to estimate a rough time scale for the formation of flux ropes and sheets from numerical simulations where such structures are found. Galloway and Proctor [19] find that hexagonal convection cells produce flux ropes in a few eddy turnover times at magnetic Reynolds numbers in the range 200-400, but the essential completion of the steady-state structure takes somewhat longer, the time scale being consistent with the estimated $R^{1/3}$ x eddy turnover time which was found by Weiss [20]. These results suggest that, while the lifetime of an eddy may permit the substantial field concentrations observed in the solar field, the drawing out of a flux sheet will be very incomplete by the time the structure breaks up, so only a small fraction of the emf otherwise available will be realized. At the same time, this limited development of sites of induced emf would tend to isolate them spatially.

This picture of weakly interacting sites of induction, both spatially and temporally limited, harks back to the early postulate of random "cyclonic events" introduced by Parker [21]. There, however, the action of the site was to lift and twist a mean field penetrating the eddy in a direction orthogonal to the axis of the twist. In the present scenario, the creation of a flux tube and its interaction with a stagnation pont flow are the crucial ingredients. We feel that it would be very useful to carry out a parallel analysis of an "event" based upon the present mechanism.

It is interesting to speculate on the possible relation between the present calculation and finite-R dynamo action found numerically for (1.1) by Arnold and Korkina [11]. In that calculation, the exponentially growing magnetic mode has the periodicity of **u***. To produce exponential growth in the present context, we must cancel the average emf generated quasi-steadily by the process we have described, by a slow time dependence of the fluxes $\mathbf{F}_i$ and their slow modulation in space. In this way we can realize a mean field $\langle \mathbf{b} \rangle$ satisfying

$$\frac{\partial \langle \mathbf{b} \rangle}{\partial t} = \nabla \times \alpha_0 \langle \mathbf{b} \rangle + R^{-1} \nabla^2 \langle \mathbf{b} \rangle. \qquad (4.1)$$

If we look for spatially-periodic solutions of (4.1) with wavenumber $n \ll 1$, these will grow at a rate $\sim \alpha_0 n$. We may think of n as an independent parameter in the specification of the magnetic mode of excitation, extending from 2π down to arbitrarily small values. Suppose that we try to achieve dynamo action at large R with $n = 2\pi$. In the present theory, we could again deal with fluxes $\mathbf{F}_i$ which are identical when the associated ropes lie on a common line, but which yield a zero mean field over a cell of period 2π. But the very fact that the structures **B** interact weakly through their associated flux sheets suggests that there should be little, if any, dynamo action and perhaps even a (slow) decay of the magnetic field. This last possibility would be consistent with the finite "window" of exponential growth which was found as R was increased [11].

We conclude by asking to what extent **u*** as defined by
(1.3), gives us "generic" properties of dynamo action at large
R. If the magnetic structures envisaged in the present paper
have time to be more or less realized, then the question really
reverts to the genericity of excitation at a single site, by
way of a converging-diverging flow which twists and distorts a
flux tube in the way we have described. We therefore suggest
that for large R this matter of genericity is equivalent
mathematically to the universality of points such as b of
Figure 2, at least for times of the order of an eddy lifetime. In this sense, **u*** which supplies a regular array of such
points, could be generic. **u*** is special in the regular
ordering of helicity at these sites, but it is not clear to
what extent helicity statistics beyond the mean value affect
the value of alpha which is developed as $R \to \infty$.
J. Guckenheimer has remarked to us that the heteroclinic
connections between adjacent stagnation points on the same
primary streamline are <u>not</u> generic, in the sense that they are
structurally unstable. There is thus some reason for selecting
the magnetic structures **B** of Section 3 as defining a site of
induction. It would be interesting to make analogous
deductions from genericity properties of time-dependent three-
dimensional flows.

Acknowledgements

A portion of this work was done at the University of
Newcastle upon Tyne during tenure (by S.C.) of an SERC Visiting
Fellowship Grant. One of us (S.C.) was supported for part of
the work by the National Science Foundation under Grant MCS-
8301809 at New York University. The authors are indebted to U.
Frisch and D. Galloway for helpful discussions.

REFERENCES AND NOTES

1. Moffatt, H. K., <u>Magnetic Field Generation in Electrically
   Conducting Fluids</u>, Cambridge University Press (1978).
2. Parker, E. N., <u>Cosmical Magnetic Fields</u>, Clarendon Press
   (1979).
3. Alfven, H., On the existence of electromagnetic-
   hydromagnetic waves, Arkiv. f. Mat. Astron. Fysik. <u>29B</u>, no.
   2 (1942).
4. Alfven, H., Origin of solar magnetic fields, Tellus <u>2</u>, 74
   (1950).
5. Roberts, G. O., Dynamo action by fluid motions with two-
   dimensional periodicity, Phil. Trans. Roy. Soc. A <u>271</u>, 411-
   454 (1972).

6. Arnold, V. I., Zel'dovich, Ya. B., Rusmaikin, A. A., and Sokolov, D. D., Sov. Phys. JETP 54, (6), 1083-1086 (1981).
7. Childress, S., A Class of solutions to the magnetohydrodynamic dynamo problem, In The Application of Modern Physics to the Earth and Planetary Interiors, S. K. Runcorn, ed., Wiley-Interscience, 629-48 (1969).
8. Moffatt, H. K., The degree of knottedness of tangled vortex lines, J. Fluid Mech. 35, 117-29 (1969).
9. Arnold, V. Sur la geometrie differentielle des groupes de Lie de dimension infinie et ses applications a l'hydrodynamique des fluides parfaits, Annales l'Institute Fourier, vol. XVI, No. 1, 319-361 (1966).
10. Hénon, M. Sur la topologie des lignes de courant dans un cas particulier, C. R. Acad. Sc. Paris 262, 312-314 (1966).
11. Arnold, V. I. and Korkina, E. I., Amplification of a magnetic field in three-dimensional stationary flow of an incompressible fluid, Bect. Mock. yh-ta. Mat. Mech. Ser. 1, No. 3, 43-46 (1983).
12. See ref. 1, section 7.11.
13. Childress, S., Stationary induction by intermittent velocity fields. In Stellar and Planetary Magnetism, The fluid Dynamics of Astrophysics and Geophysics, Vol. 2, A. M. Soward, ed. Gordon and Breach, 81-90 (1983).
14. U. Frisch, M. Hénon, A. M. Soward and others are involved with this project.
15. Childress, S., Alpha-effect in flux ropes and sheets, Phys. Earth Planet. Int. 20, 172-180 (1979).
16. We take this opportunity to correct two misprints in ref. 15. The factor R^{-1} should be omitted from the left of equation (2.23), and the factor $f'(z)/2$ should be inserted on the right. The derived solution (2.24) is correct.
17. Rosenhead, L. (ed.), Laminar Boundary Layers, Oxford University Press (1963).
18. Stix, M., Theory of the solar cycle, Solar Phys. 74, 79-101 (1981).
19. Galloway, D.J. and Proctor, M. R. E., Magnetic flux expulsion in hexagons, loc. cit. ref. 13, 99-111.
20. Weiss, N. O., The expulsion of magnetic flux by eddies, Proc. Roy. Soc A 293, 310-328 (1966).
21. Parker, E. N., Hydromagnetic dynamo models, Astrophys. J. 122, 293-314 (1955).

ORDERED AND CHAOTIC MOTIONS IN HAMILTONIAN SYSTEMS AND THE
PROBLEM OF ENERGY PARTITION

Luigi Galgani

ABSTRACT

In the present paper a review is given of the mathematical
and the numerical research on the existence of ordered motions
in nearly integrable Hamiltonian systems. Some considerations
on a possible interpretation of such motions are also included.

I. Introduction

As most people doing research in the field of qualitative
properties of Hamiltonian systems, I am tempted to say that we
are still living in a revolutionary period which began in the
year 1954 with two works of a completely different spirit and
attitude: the theoretical work of Kolmogorov on invariant tori
in nearly integrable Hamiltonian systems [1] and the numerical
work of Fermi, Pasta and Ulam (FPU) on the partition of energy
in systems of weakly coupled oscillators [2]. In the early
sixties, the proof of Kolmogorov's theorem was made complete by
Arnold and Moser [3],[4], while the numerical work of
Contopoulos [5], originating in problems of astronomical
interest, led in 1964 to the discovery by Hénon and Heiles [6]
of the existence of an abrupt transition to stochasticity in
Hamiltonian systems of two degrees of freedom. This occurred
one year after the analogous discovery of chaotic motions in
dissipative systems by Lorenz. The theoretical understanding
of the origin of stochasticity through the break up of
separatrices for perturbations of integrable systems was made
clear analytically by Melnikov, and in its geometrical
counterpart by Smale's horseshoe. For chaotic motions, one
also has the works of Anosov on hyperbolic systems, which in a
sense are the most chaotic ones among natural Hamiltonian
systems. Anosov's works became popular among physicists after
the announcement of the proof of ergodicity for a hard sphere

gas given by Sinai, although the complete proof had been
given in the very similar problem of ergodicity for billiards.

All these things are rather well known and most of them
have been recently described in a review paper by Hénon [7],
one of the pioneers in this field, to whom I may refer the
reader. In my opinion, however, not enough emphasis has been
put on a very important work by Nekhoroshev [8] (a pupil of
Arnold), which is much more recent (1976) and can be considered
in a sense the culmination of classical perturbation theory.
Moreover, as will be illustrated below, such a theorem allows
one to interpret the numerical works on the partition of
energy, initiated by FPU.

Obviously, no revolution is actually fully a revolution.
Indeed, in a sense almost everything can already be found
in Poincaré [9]; the homoclinic phenomenon leading to chaos is
explicitly described; the possibility of the existence of
invariant tori with good frequencies satisfying suitable
diophantine conditions is explicitly considered and it is shown
that one cannot exclude it. And, even before the explicit use
of diophantine conditions in a strongly related problem by
Siegel in 1942, (where one has the essence of Kolmogorov's
theorem) one should look at the whole work of Birkhoff and at
the works of Hedlung. As for the problem of the partition of
energy, the results of Fermi, Pasta and Ulam when interpreted
in the light of Nekhoroshev's theorem, were in a sense
anticipated by Boltzmann (1895) [10] and Nernst (1916) [11].

Nevertheless, I believe that one is authorized to speak of
a revolution, at least in a sociological sense. Indeed it is a
fact that the scientific community is not really aware of those
results, as is witnessed by textbooks which are, as Kuhn
correctly says, the actual reference point of scientific
communities at any time. In this connection, it is of interest
to remark that possibly the present state of awareness can be
understood if one takes into account the role played by the
availability of high speed numerical computers. Indeed the
speculations of Poincaré on homoclinic points were probably too
advanced, even for the community of pure mathematicians, and
only after the exhibition of their effects by Hénon did people
become acquainted with them. In an analogous way, si parva
licet componere magnis, only after Columbus came back from his
trip and exhibited some indios, did people become convinced
that the existence of the antipodes was not only a problem of
intellectual ingenuity, but a factual reality. From this point
of view, I believe that the coincidence of the year 1954 for
the works of Fermi and of Kolmogorov should be interpreted as a
sign of a nonfortuitous correspondence.

2. Kolmogorov's and Nekhoroshev's Theorems Versus Poincaré's and Fermi's Theorems

The theorems of Kolmogorov and of Nekhoroshev are concerned, in a complementary way, with the problem of the persistence of order for perturbations of integrable systems, in the sense that the first one insures a kind of stability for all times but only for very special initial data (on the invariant tori), while the latter one insures another kind of stability (for finite, very long times) in open regions.

First of all, one can agree to qualify as integrable a Hamiltonian system which admits action-angle variables I, ϕ such that its Hamiltonian H^0 depends only on the actions, $H^0 = H^0(I)$. More precisely, the dynamical variables might be defined in a set of the form $B \times T^n$, where $B \subset R^n$ is open (space of actions I) and T^n is the n-dimensional torus (space of the angles ϕ). Alternatively, one can think of the space $B \times R^n$ instead of $B \times T^n$, all functions being 2π-periodic in each of the variables $\phi_1, \ldots, \phi_n$. As H^0 does not depend on ϕ, $\dot{I} = 0$ and $\dot{\phi} = \omega(I)$, where $\omega = \partial H^0 / \partial I$ so that one has the general solution $I(t) = I(0)$, $\phi(t) = \omega(I(0))t + \phi(0)$. The phase space is correspondingly foliated by tori (I = const), each of which is invariant under the flow, and is run with a constant frequency $\omega(I)$ (translation of the torus). One considers now a perturbed Hamiltonian of the type $H(I,\phi) = H^0(I) + \varepsilon H^1(I,\phi)$, $|\varepsilon| \ll 1$, which is analytic in $B \times T^n$. The problem is then how much of the integrable structure persists for such a perturbation.

The first negative result was given by Poincaré at the end of the last century when he showed that generically, in some appropriate sense, the perturbed system should possess no integral of motion apart from the Hamiltonian itself. A sharpening of this result was provided in the year 1923 by Fermi [12] who showed that, under the same conditions, there could not even exist a single $(2n-1)$ dimensional invariant surface distinct from one of the "energy surfaces" H – const. (Recall that the existence of an integral of motion would imply the existence of a continuous foliation into invariant surfaces.) For a sharpening of Fermi's result, see ref. [13].

Essentially, the negative results of Poincaré and of Fermi are due to the generic density of resonances, as I will try to explain in few words. Let us assume that the system is, as one usually says, nondegenerate, i.e., that det $\partial \omega / \partial I \neq 0$, or in other words, that the frequencies can be taken as coordinates in place of the actions. Moreover, assume that the perturbation $H^1(I,\phi)$ contains harmonics of any class, where, having written the Fourier series $H^1(I,\phi) = \Sigma_k h_k(I) \exp(ik \cdot \phi)$,

$k, k' \in Z^n$ are in the same class if k' is a multiple of k (infinitely many harmonics have to be present); and this is the genericity condition. The problem is then, that in trying to build up an integral of motion $F(I,\phi) = F^0(I) + \varepsilon F^1(I,\phi) + O(\varepsilon^2)$, with $F^1(I,\phi) = \Sigma_k f_k e^{ik\cdot\phi}$, $k \in Z^n$, one has the equation $(H,F) = 0$, (where $(,)$ is the Poisson bracket), namely $(H^0, F^0) + \varepsilon[(H^0, F^1) + (H^1, F^0)] + O(\varepsilon^2) = 0$, which leads to the equation $\omega \cdot \partial F^0/\partial \phi = 0$, and then immediately, for the unknown f_k, to the equation $f_k = $ (known term)$/\omega \cdot k$. The expression on the right hand side does not make sense when the denominator vanishes, which for the hypotheses made occurs in a dense subset of B. This is the classical problem the small denominators. One also says that the relation $\omega \cdot k = 0$ represents a resonance. Actually, as $\omega(I)$ represents the frequency with which the torus with action I is run in the unperturbed motion, such a relation characterizes also those unperturbed tori for which the unperturbed solutions are not dense on the whole torus, but only on a torus of lower dimension.

Because of such theorems, people became convinced that the integrable structure should be completely destroyed after the imposition of any small perturbation. This would be a striking and incomprehensible situation. Actually, people were so convinced of this that the ergodic hypothesis came to be taken for granted, even if in fact it had never been proven. We recall that according to the ergodic hypothesis there should be no invariant set of nontrivial measure or any energy surface for the perturbed system. We note, however, that there is no contradiction concerning the ergodicity of the hard sphere problem since the latter cannot be considered a small perturbation of an integrable system.

The way out from such a paradoxical situation was provided by Kolmogorov, who was able to understand in which sense continuity could be restored; and the fact that this sense is just measure theory should not appear strange if one thinks of Kolmogorov's work on the foundations of probability in terms of measure theory. This great insight was to take seriously what Poincaré himself had already conceived, namely that one could think in terms of invariant surfaces which might not a priori fill up the whole of phase space. There should then be analytic invariant surfaces which do not fill any open subset of phase space, and this for an analytic Hamiltonian. More precisely, only the "highly nonresonant" tori should persist (apart from being slightly distorted) as invariant surfaces, while the resonant tori should in general give rise to chaotic motions by the breaking up of separatrices, in a way analogous to the one for systems of two degrees of freedom as conceived by Poincaré and Birkhoff. But even forgetting for now the possible arising of chaotic motions in the complement of the set of invariant tori, Kolmogorov understood that the set of invariant tori should have full measure when the perturbation ε tends to zero. And this is in fact the sense in which continuity is restored. If $\mu(\varepsilon)$ denotes a suitably normalized Lebesgue measure of the set of invariant tori for the Hamiltonian with perturbation ε, has $\mu(\varepsilon) = 0$ for ergodic systems and $\mu(\varepsilon) = 1$ for integrable systems, while, according to Kolmogorov, for nearly integrable systems one has $\mu(\varepsilon) \to 1$ as $\varepsilon \to 0$; moreover, according to the numerical computations of Hénon (for many systems) $\mu(\varepsilon)$ remains very near to the value 1 up to a macroscopic value of ε, and then tends to 0 as ε gets sufficiently large.

For two recent proofs of Kolmogorov's theorem I refer to refs. [14] and [15], while for a proof according to the original Kolmogorov's line, to ref. [16]. In fact, the original line of proof of Kolmogorov, which concentrates the attention on each single torus, allows one to see that there are indeed invariant tori. A correct appreciation of their measure is better obtained by following the line of Arnold, which is in a sense more global, because it looks at the whole of phase space by taking out only the noninvariant regions, and thus constructs a bridge towards Nekhoroshev's theorem.

Thus, Kolmogorov's theorem shows that the integrable structure (invariance for all times) is preserved in measure under perturbation, but it says nothing about the complement of the invariant tori. A different point of view is taken in Nekhoroshev's theorem, where the attention is given to the whole of phase space, but with results valid only for finite times. The framework is the same as for Kolmogorov's theorem; but, instead of the nondegeneracy condition, the so-called steepness condition is required. Without entering into too many details, I will only recall that steepness is independent of nondegeneracy, and is a generic property. It is satisfied if, for example, H^0 as a function of I, is convex. Then, Nekhoroshev's theorem reads as follows:

Let $H(I,\phi)_o = H^0(I) + \varepsilon H^1(I,\phi)$ be analytic in $B \times T^n$, $B \subset R^n$ open, with H steep. Then there exists a positive ε_0 and positive constants a, b, such that, for any $\varepsilon \in [0, \varepsilon_0]$ one has

$$|I(t) - I(0)| < I\varepsilon^b$$

for all $t \in [0,T]$, where

$$T(\varepsilon) = \tau \frac{1}{\varepsilon} \exp\left(\left(\frac{1}{\varepsilon}\right)\right)^a$$

I and τ are dimensional constants characteristic of the system, and $|\ldots|$ a norm in the space of actions.

So one sees that while one has $I(t) - I(0) = 0$, for $\varepsilon = 0$. For small enough perturbations, the actions change very little with time (or, one could also say, are frozen) with a uniform bound in phase space, up to a time which increases exponentially with $1/\varepsilon$.

The proof of the theorem is based on two main tools: an analytical one and geometrical one. The main idea underlying the analytic part is that if one considers only a finite number of steps of perturbation theory, then only a finite number of resonances are relevant (at least if H° has a finite number of harmonics, otherwise the so-called Arnold's ultraviolet cutoff should be employed). Consequently, the space B of actions can be divided into a finite number of regions. In each such region a suitable normalization can be performed (elimination of angles, up to some suitable combinations which are adapted to each region). In such a way, up to a "noise" due to the remainder, which is small for finite times, the variation of the actions is completely controlled: they don't vary at all in the nonresonant regions, while they vary little and in a well defined way in the resonant regions (here the steepness condition is relevant). The geometric part is concerned with good "overlapping conditions" for the various regions. For a proof of the theorem, I refer to [17].

It is of interest to point out that neither Kolmogorov's nor Nekhoroshev's theorem is directly applicable to systems of weakly coupled oscillators, which have a Hamiltonian of the form $H = H^0 + H^1$, where H^0 is linear in the actions, $H^0 = \Sigma \omega_j I_j$ with suitable (usually positive) constants $\omega_1, \ldots, \omega_n$. Indeed, in such a case the degeneracy is obvious, while non-steepness can also be proven. The way out of this difficulty consists in making a different splitting of H into H^0 and H^1 through a sequence of so-called Birkhoff transformations [14]. Assume there are no low order resonances, i.e., there exists a positive integer l such that no relation of the form $k \cdot \omega = 0$ is satisfied with $k \in Z^n$ and $|k_1| + \ldots + |k_n| \leq l$. Then one can find a change of variables to new variables I', ϕ', where $I' = 1 + \ldots$ such that

$$H'(I', \phi') = H'^0(I') + H'^1(I', \phi') = \Sigma \omega_j I'_j + \ldots$$

is in general steep and H'^1 is of order $|I'|^{1-1/2}$. Notice in this connection that performing a Birkhoff transformation imposes the restriction of being sufficiently near the origin in the space of actions, so that one can only consider a region of the type $|I| \leq A$, where A is a suitable action. In such a way, both theorems can in general be applied to systems of weakly coupled oscillators. In particular, from Nekhoroshev's theorem one deduces again that the action I are frozen up to a time given by the formula reported above, with ε replaced by the quantity

$$\varepsilon_{eff} = \frac{|I(0)|}{A}$$

which depends on the initial datum. So one finds that the time T of freezing is given by

$$T = \tau \frac{A}{|I(0)|} e^{(A/|I(0)|)^a}$$

I will return to this point later.

3. Numerical computations for the Fermi Pasta Ulam problem and for models of continua

As recalled above, nearly integrable Hamiltonian systems were expected, before Kolmogorov, to be in general ergodic. In the case of weakly coupled harmonic oscillators, as is very well known, this would lead in particular to equipartition in time average. Precisely, if $H = \Sigma E_j + \ldots$, where $E_j = \omega_j I_j$, and one denotes $\overline{E_j}(t) = 1/t \int_0^t E_j(\tau)d\tau$, then in the approximation in which the higher order terms are neglected (although they are essential in possibly producing ergodicity) one should have that $\lim_{t \to \infty} \overline{E_j}(t)$ exists and is independent of j, namely is equal to kT, k being the Boltzmann constant and T the absolute temperature.

FPU considered a chain of n (=64) equal particles on a line with fixed boundary conditions, coupled by simple anharmonic forces (with a potential $V(r) = r^2/2 + ar^3 + \beta r^4$). As is well known, this systems equivalent to a system of weakly couled oscillators (or modes) with frequencies $\omega_j = \overline{\omega} \sin(j/(n+1) \pi/2)$, $j = 1, \ldots, n$, where $\overline{\omega}$ is a frequency independent of n. By numerical integration of the equations of motion with initial excitation of the lower frequencies, they found that energy did not flow the higher modes, at least within the times then accessible to computation. The most striking figure they gave, plots $\overline{E_j}(t)$ as a function of t for any j, and one sees there that such quantities apparently reach a limit which has nothing to do with equipartition; in fact, the "final values" decrease as the mode number j increases, in a way which is approximately exponential. In this connection, see also ref. [18].

One has then to wait until 1966 in order to find an interpretation of such result in terms of KAM theorem. This was provided by Izrailev and Chirikov [19] who, as subsequently Ford [20], made a great effort in making this subject popular among physicists. The idea beyond the paper of Izrailev and Chirikov is that KAM theorem should apply only for low enough initial energies, and that a threshold should exist for stochasticity, leading to equipartition. In fact, this is quite intuitive if one has in mind the previous result of Hénon and Heiles (which apparently was not known to those authors). But indeed one finds in that paper a much subtler and intriguing concept, namely the concept that each oscillator should have a proper threshold, which cannot be immediately inferred on the basis of KAM theorem.

The next important result was provided by Bocchieri, Scotti, Bearzi and Loinger [21] who, in a slightly modified FPU model (interaction through realistic molecular potentials, precisely the Lennard-Jones potential), found equipartition if the initial energy was high enough, say greater than a global critical energy E^c. Moreover, such critical energy was found to be extensive, i.e. proportional to n, up to the highest values of n considered (n = 100), at least if n was not too small (say n ⩾ 6). Precisely, E^c/n was found to be of the order of (1/100)-th of the height of the potential well, if the zero of energy was taken to coincide with the minimal energy.

After that, the extensivity of the critical energy was often confirmed in several technical way, [22] for example by methods related to the separation of nearby orbits in phase space, which led to an understanding of the role of Liapunov Characteristic Exponents [23] and sensitive dependence on initial conditions in such systems.

A quite different situation occurs instead if one considers models of continua, because apparently one never finds there equipartition in finite times. A very interesting model was studied by Bocchieri and Loinger [24]. This is an electromagnetic field inside a cavity with reflecting walls, the coupling among the modes being provided by a charged surface which can move vertically. Making the cavity infinitely extended in two directions, the problem turns out to be essentially one-dimensional and quite manageable (the modes having frequencies $\bar{\omega}_j, j = 1,3,5,...$). When a nonlinearity is added through a nonlinear spring attracting the charged surface to a fixed point, the problem could just be investigated numerically. The figure analogous to that of FPU mentioned above was given in ref. [25]. One clearly sees there that the partition of energy occurs in the following way; for very small times energy decreases exponentially with frequency; then, energy is equally shared among the modes of low frequency, and moreover every mode has a characteristic time (increasing exponentially with frequency) for entering the group of modes which share energy. One can also say that there is equipartition of energy among a drop of modes of low frequency, and that such drop extends towards the higher frequencies as time increases.

If this is the general mechanism, then one understands very well the difference between systems with frequencies in a bounded domain (such as solids, or the FPU model), and systems with unbounded frequencies. Indeed in the former case after a certain time one will have equipartition; in the latter case, instead, at any finite time this would be true only for low enough frequencies, while sufficiently high frequencies would not be able to take part in the sharing. In this sense one could say that at any time one would be in a kind of metaequilibrium state, with a dynamical cutoff of the high frequencies.

On the other hand, a general framework for understanding such mechanism based on simple properties of analytic functions has been proposed by Parisi et al. [27], on turbulence in dissipative systems. Consider for example the case of a finite string with periodic boundary conditions, and assume its configuration is described by a real analytic function, say $f(x)$, $x \in (0,1)$, which can be analytically continued in a strop of width δ in the complex plane around the real interval $(0,1)$. Then, as is well known, the Fourier coefficients c_k of f decay exponentially with $|k|$ m say $c_k \leq \text{const} \exp(-\delta|k|)$, and correspondingly the harmonic energy E_k of mode k decays exponentially with $|k|$. In such a way the approach to equipartition is seen to be possible only inasmuch as the "slope" δ tends to zero; but in any case the high frequencies should be necessarily cut off at any finite time. Very beautiful numerical computations well fitting such ideas were made in order to study the behavior of the slope as a function of time, and to compare it with analytical estimates [28]. On this matter I don't have time to enter here.

Another set of computations were addressed at estimating the thresholds for stochasticity which were first considered by Izrailev and Chirikov. At variance with those authors, it was suggested that such thresholds should increase with frequency, and actually even linearly, at least in a first approximation. Looking at the available numerical results in this connection is indeed quite an intriguing subject [30]; in my opinion, however, they support the conjecture of an increase with the frequency.

But the main support to such a conjecture comes now from the theorem of Nekhoroshev, which, as was recently remarked [29], appears to give a theoretical framework to understand the notion itself of a partial threshold. Indeed, as was recalled above, such theorem can be applied to coupled oscillators only if the initial conditions are such that the initial actions are below a critical action A, namely one has $[I(0)] < A$; and because of the relation $E_j = \omega_j I_j$ one has then that the theorem can be applied for initial conditions such that $E_j < A\omega_j$. Chaotic motions should occur if such conditions are violated, at least if one accepts the general philosophy that chaoticity should occur where stability is not guaranteed (recall that chaos was always assumed to occur, and that the stability theorems here considered can thus be regarded as just imposing restrictions to chaos).

Moreover, Nekhoroshev's theorem appears also the support the existence of characteristic times for relaxation to equipartition, and actually increasing exponentially with frequency, in agreement with the indications of the numerical computations recalled above. Indeed, taking initial conditions with only one oscillator excited, say the j - th one, the time of freezing reported above becomes

$$T = \tau \frac{A\omega_j}{E_j} e^{\left(\frac{A\omega_j}{E_j}\right)^a},$$

which is indeed increasing exponentially with frequency. So the general point of view recalled above turns out to be just a part of a general reasoning which also includes the fact that the critical energy increases exponentially with frequency.

As a matter of fact, this general reasoning, which is nothing but classical perturbation theory, i.e. essentially the adiabatic invariance of the action which goes back to Helmholtz and Boltzmann, was one of the guiding ideas beyond such conjecture. Indeed, if there is a critical energy which is a function of frequency. $E^c = E^c(\omega)$, because of the adiabatic invariance of the action the dependence on ω can just be such that $E^c/\omega = $ const, namely $E^c = A\omega$, with a constant action A characteristic of the system. This was particularly stressed by Cercignani, and in this way we passed from the conception of a global energy threshold [21], to the conception of partial energy thresholds [31]; another relevant point concerning such a conjecture is the analogy of the critical energy with the quantum zero point energy, which is also increasing linearly with frequency.

4. The problem of a physical interpretation of ordered motions

With the last remark reported above we are entering the quite delicate subject of a physical interpretation of the ordered motions. In fact, the first problem is whether such motions will subsist at the thermodynamics limit ($n \to \infty$) or not. Most people working in this field always stated the belief that the ordered motions should disappear, so that the standard conclusions for classical statistical mechanics should hold. Together with other people in Milan and Padova, I always aimed instead at showing that they subsist, and in my opinion the numerical indications [22] appear to favor the latter opinion.

I will now add just a few words on a possible physical interpretation of the ordered motions. As recalled above, already in the year 1971 it was suggested [31] that the critical energy $E^c(\omega)$ should be considered as the analog of quantum zero-point energy, the action A taking the place of Planck's $h/2\pi$; a point of view openly taken by Nernst [11] who also gave, along such line, a deduction of Planck's law. See also ref. [32]. Such deductions of Planck's law are inspired to the old idea of Boltzmann [10] (1895) that the stable motions should be considered as frozen, i.e. not taking part in the energy exchanges, so that their mechanical energy should not contribute to the thermodynamic energy. In the light of Nekhoroshev's theorem, this corresponds to making the approximation of considering as infinite all stability times. But this is the less true the more one goes towards the low frequencies, because the stability times increase exponentially with frequency; so it is quite conceivable that one should have deviations from Planck's law in that region (i.e. in the Rayleigh-Jeans region), in the sense that the actual radiation should there be higher than expected, for example showing a plateau followed by an exponential queu, as found in the numerical computations recalled above [25]. The experimental data on black bodies (which fit Planck's law only within 3% [33]) are not too much extended into the Rayleigh-Jeans region, as they just arrive at $x = 3$, where $x = h\nu/kT$; but for example the data for the Sun show a striking deviation just in the direction indicated above.

In conclusion, concerning the physical interpretation suggested above, I certainly have to admit that it presents a largely speculative character and that it might well be incorrect. I hope however people will concede at least that it plays a quite stimulating role, so that, in agreement with an old Italian adage, it might be appropriate to say of it: se non e vera e ben trovata.

REFERENCES

1. Kolmogorov A.N., Dokl. Akad. Nauk. SSSR, 98, N.4, 527-530 (1954).
2. Fermi E., Pasta J. and Ulam S., in Fermi E., Collected Works, Vol. II., 978-988 (Accademia Nazionale dei Lincei, Roma, 1965).
3. Arnold Vi., Russ. Math. Surveys 18, No. 5, 9-36 (1963), and 18, N. 6, 85-191 (1963).
4. Moser J., in Proc. Int. Conf. on Functional Analysis and Related Topics, 60-67 (University of Tokyo Press, Tokyo, 1969).
5. Contopoulos G., Astron. J. 68, 1-4 (1963).
6. Hénon M. and Heiles C., Astron. J. 69, 73-79 (1964).
7. Hénon M., Numerical Exploration of Hamiltonian Systems, in Ioos G., Helleman R.H.G. and Stora R. eds., Chaotic Behavior of Deterministic Systems (North Holland, Amsterdam, 1983).
8. Nekhoroshev N.N., Russ. Math. Surveys 32, 1-65 (1977).
9. Poincaré H., Les Méthodes Nouvelles de la Mécanique Céleste (Gauthier-Villars, Paris, 1892-1893-1899).
10. Boltzmann L., Nature 51, 413-415 (1895).
11. Nernst W., Verh. Dtsch. Phys. Ges. 18, 83 (1916).
12. Fermi E., Nuovo Cim. 26, 105 (1923).
13. Benettin G., Ferrari G., Galgani L., and Giorgilli A., Nuovo Cim. 72B, 137-148 (1982).
14. Poeschel J., Comm. Pure Apll. Math. 35, 653-695 (1982).
15. Gallavotti G., in Scaling and Self-Similarity in Physics (ed. Froelich J.), 359-426, (Birkhauser, Boston, 1983).
16. Benettin G., Galgani L., Giorgilli A., and Strelcyn J.-M., Nuovo Cim. 79B, 201-223 (1984).
17. Benettin G., Galgani L., and Giorgilli A., A Proof of Nekhoroshev's Theorem for Nearly Integrable Hamiltonian Systems, Preprint.
18. Galgani L. and Scotti, A., Phys. Rev. Lett. 28, 1173-1176 (1972).
19. Izrailev F.M. and Chirikov B.V., Sov. Phys. Dokl. 11, 30 (1966).
20. For J., The Statistical Mechanics of Classical Analytical Dynamics, in E.G.D. Cohen ed., Fundamental Problems in Statistical Mechanics, Vol. 3, 215-255 (North Holland, Amsterdam, 1975).
21. Bocchieri P., Scotti A., Bearzi B., and Loinger Phys. Rev. A2, 2013-2019 (1970).

22. Diana E., Galgani L., Casartelli M., Casati G., and Scotti A., Theor. Math. Phys. 29, 1022-1027 (1976); Casartelli M., Diana E., Galgani L., and Scotti A., Phys. Rev. A13, 1921-1925 (1976); Benettin G., Lo Vecchio G., and Tenenbaum A., Phys. Rev. A22, 1709-1719 (1980); Benettin G. and Tenenbaum A., Phys. Rev. A28, 3020-3029 (1983).
23. Benettin G., Galgani L., and Strelcyn J.-M., Phys Rev. A14, 2338-2345 (1976).
24. Bocchiere P., Crotti A., and Loinger A., Lett. Nuovo Cim. 4, 341 (1972). Bocchiere P. and Loinger A., Lett. Nuovo Cim. 2, 41-42 (1971).
25. Benettin G. and Galgani L., J. Stat. Phys. 27, 153-169 (1982).
26. Fucito F., Marchesoni F., Marinari E., Parisi G., Peliti C., Ruffo S., and Vulpiani A., J. Phys. (Paris) 43, 707-713 (1982). Bassetti B., Buttera P., Raciti M., and Sparpaglione M., "Complex Poles, Spatial Intermittency and Energy Transfer in a Classical Nonlinear String", Phys. Rev. A., in print.
27. Frisch U. and Morf R., Phys. Rev. A23, 2673 (1981). Frisch U., The Analytic Structure of Turbulent Flow, paper presented at the 6-th Kyoto Summer Institute on Chaos and Statistical Mechnics, September 1983.
28. Livi R., Pettini M., Ruffo S., Sparpaglione M., and Vulpiani A., Phys. Rev. A28, 3544-3552 (1983); Equipartition Thresholds in Nonlinear Large Hamiltonian Systems, Preprint.
29. Benettin G., Galgani L., and Giorgilli A., Boltzmann's Ultraviolet Cutoff and Nekhoroshev's Theorem on Arnold's Diffusion, Preprint.
30. Galgani L. and Lo Vecchio G., Nuovo Cim. 52B, 1-14 (1979). Callegari B., Carotta M.C., Gerrario C., Lo Vecchio G., and Galgani L., Nuovo Cim. 54B, 463 (1979). Carotta M.C., Ferrario C., Lo Vecchio G., and Galgani L., Phys. Rev. A17, 786-794 (1978).
31. Cercignani C., Galgani L., and Scotti A., Phys. Lett. A38, 403-404 (1972). Galgani L. and Scotti A., Rivista Nuovo Cim. 2, 189-209 (1972).
32. Galgani L. and Benettin G., Lett. Nuovo Cim. 35, 93-95 (1982).
33. Crovini L. and Galgani L., Lett. Nuovo Cim. 39, 210-214 (1984). Galgani L., Ann. Fond. L. de Broglie 8, 19-64 (1983).

THE TRANSITION TO CHAOS IN GALACTIC MODELS OF TWO AND THREE DEGREES OF FREEDOM

G. Contopoulos

ABSTRACT

We will discuss the onset of chaos in galactic models of two and three dimensions.

We will consider integrable systems on which various types of perturbations are added. If the perturbation is strong enough, most orbits are chaotic. In systems of two degrees of freedom, the onset of chaos is due to infinite bifurcations, or infinite gaps. In systems of three degrees of freedom, we have transitions of complex instability, or inverse bifurcations. An increase of stochasticity in the form of "Arnold diffusion" may be due to changes in the interconnections of the various families of periodic orbits at a "collision of bifurcations".

I. Introduction

Chaos is a greek word that denotes the primeval god from whom everything originated. The children of Chaos (Χάοσ) were "Ερεβοσ, Νύξ and Μοῖραι (Darkness, Night and Fate). The last word is, in fact, plural and corresponds better to our "Laws of Nature". Then came Οὐρανόσ and Γαῖα (Heavens and Earth), Κρόνοσ (Χρόνοσ=Time), Ζεύσ (Zeus) and the other gods of Olympus.

It is remarkable that the Greeks considered chaos as preceding order. The laws of nature were supposed to be generated from the primeval chaotic universe. However, at the moment we will try to establish the opposite sequence. Chaos from order, or "deterministic chaos", as we usually call it. Assuming certain well defined laws, we find that orderly motion is replaced by chaotic behavior as some parameters change beyond certain characteristic values.

This transition to chaos is seen, in particular, in galactic dynamics. The simplest galactic models are integrable. For example, in an axisymmetric flat galaxy we have two integrals of motion, the energy and the angular momentum. However, if we add a spiral perturbation we find small regions of chaotic motion and if the perturbation becomes large enough, the chaotic regions become predominant. In the present paper, we will consider the transition to chaos in galactic models of two and three dimensions. The mechanisms seem to be quite different in the two cases.

II. Two-Dimensional Galactic Models

We will deal first with infinitesimally flat galaxies, and assume that the motions are restricted on the galactic plane. The galaxies are assumed to be composed of an axisymmetric background and a spiral, or barred, perturbation. A model of a two-armed spiral Hamiltonian is

$$H \equiv \frac{1}{2}(\dot{r}^2 + J_o^2 r^{-2}) + V_o(r) + \varepsilon A(r) \cos[2\theta - \phi(r)] - \Omega_s J_o = h, \tag{1}$$

where r and $\dot{r}$ are the radial coordinate and velocity, J_o the angular momentum, θ the azimuthal angle, Ω_s the angular velocity of the system, h the value of the energy in the rotating system (Jacobi constant), $A(r)$ and $\phi(r)$ two given functions of r and ε a measure of the strength of the spiral perturbation. If $\phi(r)$=const. the galaxy is barred.

If $\varepsilon = 0$, we have an axisymmetric galaxy. In this case, we have only one family of stable orbits, the "central" family of circular periodic orbits x_1. In Figure 1a we give the characteristic of this family, i.e., the radius r as a function of the energy h. The maximum h along this characteristic appears at corotation (L_4) where the angular velocity orbit (Ω) is equal to the angular velocity of the system (Ω_s). Besides the circular orbits, there are also many resonant families of periodic orbits. The numbers in Figure 1 give the ratio of the frequency of the radial oscillations (λ) to the rotational frequency in the rotating frame ($\Omega - \Omega_s$). All of these families are unstable if $\varepsilon=0$. But, if $\varepsilon \neq 0$, the central family is split by gaps, and its various parts join the resonant families (Fig. 1b). Near these gaps we see the first appearance of chaotic motion.

As the perturbation ε increases, the regions of chaotic motion become larger and larger. The various families undergo period doubling bifurcations and the sequences of such bifurcations tend to a universal ratio $\delta=8.72$, appropriate for area preserving mappings (Benettin et al. 1980, Greene et al.

1981), and in particular for Hamiltonian systems (Contopoulos 1983a). An example is given in Figure 2. In this figure we see that the successive bifurcations -2/2, -4/4, etc, are period doubling.

The intervals Δh between successive bifurcations have a ratio that approaches the universal ratio $\delta = 8.72$. The various families become unstable beyond the corresponding bifurcations. Thus for h, a little beyond the bifurcation 4/4, we have an infinity of unstable families. Then the region of stochastic orbits (shaded in Fig. 2), becomes very large.

The ratio $\delta = 8.72$ is different from the bifurcation ratio found for dissipative systems by Feigenbaum (1978) and Coullet and Tresser (1978), which is equal to $\delta = 4.67$. On the other hand, we found a different bifurcation ratio between successive bifurcations and the same family of periodic orbits in a simple galactic model (Contopoulos and Zikides 1980). These bifurcations produce an infinity of successive stable and unstable intervals and their ratio tends to $\delta = 9.22$. However, this last ratio is non-universal and, as was proved by Heggie (1983), it is equal to $e^{\sqrt{\pi}/2}$.

All mechanisms that produce a large degree of instability lead to chaos. In fact, the non-periodic orbits are repelled by the unstable periodic orbits and if there is no stable orbit to attract them, the orbits wander in a chaotic way. Thus, all the mechanisms discussed above (infinite pitchfork bifurcations, infinite transitions to instability of the same family, and infinite gaps) lead to a large degree of chaos in Hamiltonian systems.

In the inner parts of a galaxy, chaotic motions cannot lead to escapes. The orbits of stars are restricted inside closed "curves of zero velocity" (equipotentials in the rotating frame of reference). The eccentricity of these curves is smaller than that of the bar density. Therefore, the role of chaotic motions is to limit the elongation of the barred galaxies. The same restriction applies to elliptical galaxies, which are believed to be mostly triaxial.

On the other hand, the chaotic motions nearby and beyond corotation lead to escapes of stars from the galaxies. Thus, large regions nearby and outside corotation are devoid of stars. This mechanism is believed to provide an explanation of the termination of bars near corotation in numerical experiments of the N-body type (Sellwood 1980, 1981, Schwarz 1981, Thielheim and Wolff 1981).

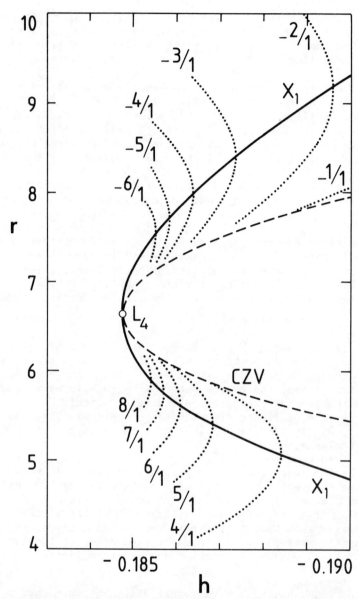

Fig. 1. Characteristics of families of periodic orbits in the axisymmetric case ($\varepsilon = 0$) and a weak bar case ($\varepsilon=0.00001$, or 1% extra density in the bar). (-----) stable and (....) are unstable families. The central family of circular orbits of the axisymmetric case (x_1) is split by gaps if $\varepsilon \neq 0$. Various resonant families are marked. L_4 is the Lagrangian point at corotation (when $\Omega = \Omega_s$). All motions are on the left of the curve of zero velocity (CZV). Chaotic motions appear for the first time near the gaps.

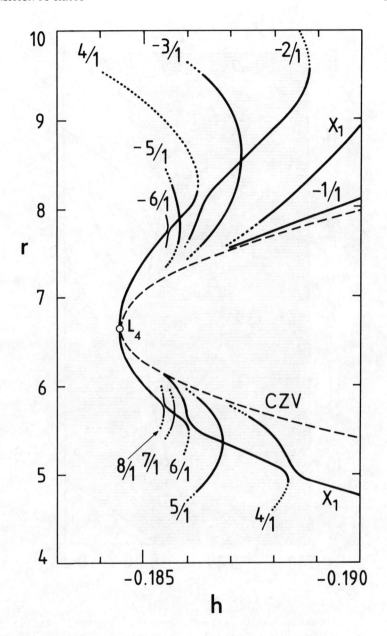

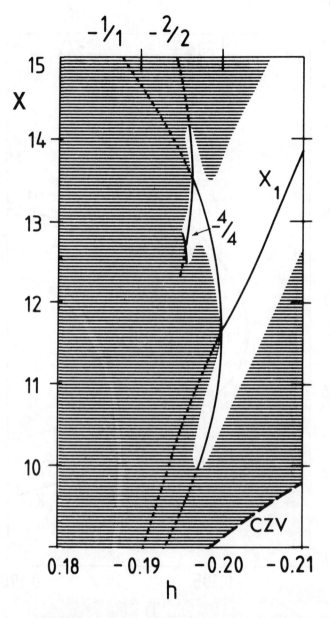

Fig. 2. Pitchfork bifurcations from the "central" family x_1 in a stronger bar model ($\varepsilon = 0.0001$, or bar density 10% of the background) outside corotation. The shaded region represents stochastic orbits. These orbits eventually escape from the galaxy.

Well outside corotation, there are still regions of ordered motions. In fact, the potential of a galaxy at large distances approaches that of a point mass. Therefore, the galactic model is approximately integrable and the motion is approximately quasiperiodic.

In our numerical experiments (Contopoulos 1983a), we found that in all rotating systems the sequences of infinite pitchfork bifurcations are followed by inverse sequences and thus they do not extend to infinity. In Figure 3 we see the continuation of the characteristics of the families of Figure 2. All of them join so that only one family of simple periodic orbits remains around the origin in large energies. This example shows that the chaotic regions in a galactic model are limited in space.

III. Three-Dimensional Galactic Models

In systems of three (or more) degrees of freedom, we encounter new phenomena that do not appear in 2-dimensional systems. Such phenomena are:
  (1) A new type of instability of periodic orbits (complex instability)
  (2) Collisions of bifurcations, and
  (3) Arnold diffusion (of the non-periodic orbits).

1) <u>Complex instability</u> appears when the eigenvalues of a periodic orbit are complex but not on the unit circle. Such a case cannot occur in systems of two degrees of freedom. In fact, in systems of two degrees of freedom we have either complex eigenvalues on the unit circle (stability) or real eigenvalues (instability). When a family of orbits becomes unstable as we vary a control parameter (e.g., the energy), we have bifurcation of another family that inherits its stability. However in systems of three degrees of freedom, two pairs of conjugate eigenvalues on the unit circle may collide and get out of the unit circle on the complex plane. This is a new type of transition to instability which is not followed by a bifurcation.

This instability also produces chaotic motions. However, we do not have the usual pattern of infinite oscillations of the asymptotic curves emanating from the unstable points of systems of two degrees of freedom.

A complex instability terminates a sequence of bifurcations. Namely, if along a sequence of pitchfork bifurcations we encounter a transition to complex instability, there is no bifurcation there to continue the sequence. This type of termination of bifurcation sequences is quite common

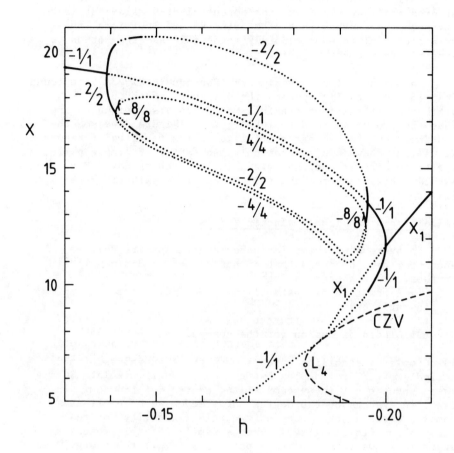

Fig. 3. Extension of the characteristics of Fig. 2 for large energies. The pitchfork bifurcations are followed by inverse bifurcations, forming an infinity of bubbles, one inside the other.

but not the only one. Another way of terminating a sequence of
bifurcations is through an inverse bifurcation (Fig.4). Most
bifurcations are direct and the bifurcating family is stable
(Fig.4a). However, as the control parameters vary, the
bifurcation may become inverse, and then the bifurcating family
is unstable (Fig. 4b). In such a case, the stability of the
original family is not transmitted continuously to a new
family. Thus a sequence of bifurcations stops when an inverse
bifurcation is encountered.

In a previous paper (Contopoulos 1983b), we presented
evidence that in genuine systems of three degrees of freedom
(i.e., systems not reducible to a system of two degrees of
freedom plus an independent oscillation), the sequences of
bifurcations stop after a finite number of steps, either by a
complex bifurcation or by an inverse bifurcation. It is still
an open question whether we ever have infinite pitchfork
bifurcations in such systems. If not, then the introduction to
stochasticity through infinite pitchfork bifurcations is
restricted to systems of two degrees of freedom only. This
behavior is different from that of dissipative systems where
the universal ratio δ = 4.67 does not depend on the number of
degrees of freedom (Feigenbaum 1980).

2) <u>Collisions of bifurcations.</u> In systems of three degrees of
freedom, a periodic orbit is characterized by two pairs of
eigenvalues, instead of one pair, as in two degrees of freedom.

The eigenvalues of a periodic orbit are given by a quartic
equation of the form

$$\lambda^4 \alpha_1 \lambda^3 + \beta_1 \lambda^2 + \alpha_1 \lambda + 1 = (\lambda^2 + b_1 \lambda + 1)(\lambda^2 + b_2 \lambda + 1) = 0. \qquad (2)$$

The values of b_1, b_2 depend on the control parameters of the
problem. An orbit is stable if both b_i are real and absolutely
smaller than 2, simply unstable if only one $|b_i| > 2$, doubly
unstable if both $|b_i| > 2$, and complex unstable if b_i are
complex. We have bifurcations whenever $|b_1| = 2$ or $|b_2| = 2$. A
<u>collision of bifurcations</u> appears when both b_1 and b_2 are
absolutely equal to 2 for the same set of control parameters.

We have studied in detail the bifurcations of the
Hamiltonian system

$$H \equiv \frac{1}{2}(\dot{x}^2+\dot{y}^2+\dot{z}^2+Ax^2+By^2+Cx^2) - \varepsilon xz^2 - \eta yz^2 = h \qquad (3)$$

that represents a distorted elliptic galaxy (Contopoulos
1984). Here A, B, C, and h are considered constant and we vary
the control parameters ε and η. As an example, we can consider

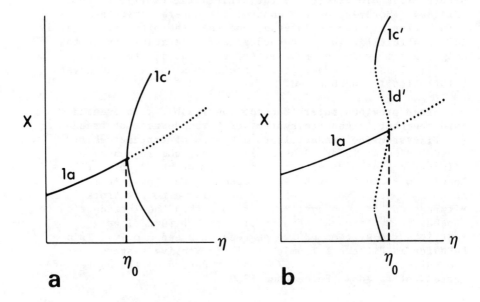

Fig. 4. A direct bifurcation (a) and an inverse bifurcation (b). In the first case the bifurcating family is stable (———) and extends in the direction where the original family is unstable (....). In the second case, the bifurcating family is unstable and extends in the opposite direction. Usually the unstable bifurcating family joins a stable family further away from the original family.

the bifurcations from a family of simple periodic orbits that
we call family 1c' (Fig.5). We have the bifurcations of a
double period family (2d) and of three quadruple period
families (4a, 4b' and 4c).

The collision of bifurcations occurs for $\varepsilon = \varepsilon_{col}$ when
both b_1 and b_2 are equal to -2. We see that the family 4a is
generated from the family 2d for $\varepsilon < \varepsilon_{col}$ and directly from the
family 1c' for $\varepsilon > \varepsilon_{col}$. On the contrary, the family 4b' is
generated from 1c' before the bifurcation and from 2d after
it. Therefore, the interconnections of the various families
are different for $\varepsilon < \varepsilon_{col}$ and $\varepsilon > \varepsilon_{col}$. Furthermore, the
family 4a, which is stable before the collision, is doubly
unstable after it.

The change of the stability character of the various types
of orbits and of their interconnections at collision points is
followed by corresponding changes of the regions covered by the
chaotic nonperiodic orbits.

3) <u>Arnold diffusion</u>. It is well known that, while a set of
closed KAM surfaces around a stable periodic orbit in systems
of two degrees of freedom secures stability for all times, the
same is not true in systems of three or more degrees of
freedom. In such cases, the KAM surfaces do not separate the
phase space and a non-periodic orbit may escape from the
neighborhood of a periodic orbit even if it is linearly
stable. In fact the various chaotic regions in phase space
seem to be connected, and an orbit can wander over large
distances through such regions. This is the so-called "Arnold
diffusion".

However, the time scale for this diffusion may be
extremely large (Lichtenberg and Lieberman 1983). In a
particular case of three coupled oscillators, we found
(Contopoulos et al. 1978) that two stochastic regions do not
communicate for 10^5 periods at least, although these regions
are very close to each other. For an understanding of this
phenomenon, a study of the various types of periodic orbits and
of their stability is quite useful. In particular, the change
of the connections between the various families of orbits, that
we discussed above, may play an important role in increasing
the diffusion rate. This may explain why the diffusion rate is
very slow in certain directions.

The rate at which stochasticity sets in galactic systems
is very important in view of the limited lifetime of
galaxies. In fact the lifetime of a galaxy is roughly equal to

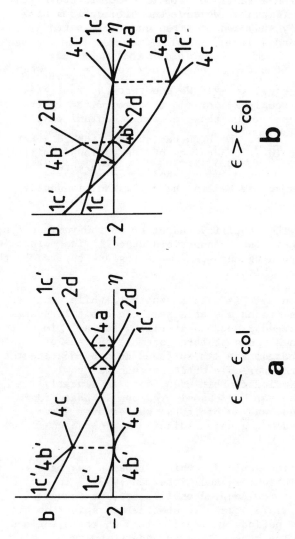

Fig. 5. Characteristics of various families bifurcating from the family of simple periodic orbits 1c. For $\varepsilon > \varepsilon_{col}$ we follow the coordinate x when z=0 as a function of η. (———) stable, (– – –) simply unstable, and (·····) doubly unstable orbits.

the age of the Universe ($1-2 \times 10^{10}$ years), i.e., it is not larger than about 100 rotation periods. Therefore, if the orbit of a star is approximately quasiperiodic, we may ignore stochastic effects, if its time-scale is larger than the age of the Universe. For example, a diffusion that needs more than 10^5 periods to establish itself can be safely ignored.

On the other hand, in a plasma, 10^5 periods is only a small fraction of a second; therefore, such effects may be much more important. The case of planetary motions with periods of a few years, is an intermediate one. Therefore, a more complete study of various systems of three degrees of freedom may give different applications in systems of different scales. In particular, the study of the behavior of various families of periodic orbits should give a better understanding of the mechanisms that lead to chaos in systems of three or more degrees of freedom.

REFERENCES

Benettin, G., Cercignani, C., Galgani, L. Giorgilli, A.,1980, Lett. Nuova Cim. 28, 1.
Contopoulos, G., 1983a, Lett. Nuova Cim. 37, 149.
Contopoulos, G., 1983b, Lett. Nuova Cim. 38, 257.
Contopoulos, G., 1984, preprint.
Contopoulos, G., Galgani, L. and Giorgilli, A., 1978, Phys. Rev. A. 18, 1183.
Contopoulos G., and Zikides, M., 1980, Astron. Astrophys. 90, 108.
Coullet, P. and Tresser, J., 1978, J.Phys. (Paris) 5C, 25.
Feigenbaum, M. J., 1978, J. Stat. Phys. 19, 25.
Feigenbaum, M. J., 1980, Los Alamos Science 1,4.
Greene, J. M., MacKay, R. S., Vivaldi, F. and Feigenbaum, M.J., 1981, Physica 3D, 468.
Heggie, D., 1983, Geles. Mech. 29, 207
Lichtenberg, A.J. and Lieberman, M.A., 1983, Regular and Stochastic Motion, Springer Verlag, New York.
Sellwood, J. A. 1980, Astron. Astrophys. 89, 296.
Sellwood, J.A., 1981, Astron. Astrophys. 99, 362.
Schwarz, M.P., 1981, Astrophys.J. 247, 77.
Thielheim, K.O. and Wollf, H., 1981, Astrophys. J. 245, 39.

NONLINEAR NONRADIAL ADIABATIC STELLAR OSCILLATIONS: NUMERICAL RESULTS FOR MANY-MODE COUPLINGS

W. Däppen

ABSTRACT

The adiabatic nonlinear coupling of many linear radial modes has been extended to nonradial modes. A variational principle for fluid mechanics has been applied to obtain a Hamiltonian formalism. The numerical results presented here show that the nonradial modes can cause a much more irregular motion than the radial modes. First, there are many near-resonant nonradial frequencies, and second, there is a nonlinear velocity-dependent term in the Hamiltonian, which has no radial analogue. In the case of the Sun, the previous results included coupling only for the lowest radial modes, and the coupling coefficients were computed in a polytropic model. Here, radial and nonradial 5-minute modes are coupled, and they have been computed using a 'real' solar model.

Introduction

This paper contains the first numerical results for nonlinear nonradial many-mode interactions. These results follow earlier studies by Perdang and Blacher (1982, 1984a,b) and by Däppen and Perdang (1984a). The principal technique used here, as in the aforementioned studies, is a Hamiltonian particle formalism for the time-dependent amplitudes, which appear in an expansion of the displacement field in terms of linear eigenfunctions. But while in the previous numerical studies only radial modes were coupled, with coupling coefficients computed in a polytropic model, here nonradial modes are included as well, and the coupling coefficients have been computed in a state of the art solar model [Model 1 of Christensen-Dalsgaard (1982); the modes have been produced by Christensen-Dalsgaard's (1981) oscillation program]. Another new feature is the extension of the results to observed high-order five-minute modes. The previous studies only examined the coupling of the few lowest radial modes, which have not been observed yet. As in the previous studies, dissipation is not included. This omission is difficult to justify except perhaps by saying that it is reasonable to begin with the simple adiabatic case when studying nonlinear many-mode interactions. The adiabatic problem is also a first step in a systematic study of dissipative motion [see Perdang (1984) for references]. For the following, the adiabatic hypothesis is taken as an axiom.

To extend the Hamiltonian formalism to the nonradial case, a variational principle for fluid mechanics (Lynden-Bell and Katz, 1981) has been suitably adapted (Däppen and Perdang; 1984a,b). As will be briefly shown below, and can be found in all detail in Däppen and Perdang (1984b), there are several practical and theoretical advantages with the use of a variational principle over a direct use of the equations of motion (Dziembowski, 1982; Buchler and Regev, 1983). Among these advantages I note the existence of conserved quantities: first, energy, which is useful as a controlling device in the numerical computation, and second, mass and entropy, whose conservation laws can be directly built into the formalism. These laws are therefore exactly satisfied and not only up to the chosen order of perturbation theory. A further theoretical advantage of the variational principle used here is that it gives an exact answer to the question of stationary motion (see Perdang, 1984): in the special case of initially absent circulation, there will be no circulation present at later times. In the formalism using the equations of motion, a similar statement is not available.

The results presented here have been computed in a solar model. With the low number of only seven interacting modes, the amplitudes chosen here (about 10^{-3} in relative radius fluctuation, so as to exhibit nonlinear phenomena) are at least three orders of magnitude too large to be relevant for observed solar oscillations. Nevertheless, they follow the trend noted previously (Perdang and Blacher, 1984a; Däppen and Perdang, 1984a), that nonlinear features become visible at much lower amplitudes if the number of interacting modes increases. As far as the Sun is concerned, the present results might only serve to confirm this trend. For some stars, however, such amplitudes are of course realistic.

Method

In the <u>radial</u> case, the Hamiltonian formalism is straightforward (Woltjer, 1935). There is a convenient Lagrangean variable M_r (the mass contained in the sphere of radius r), which enables us to write the integrals of total kinetic and potential energy over a time-independent domain. First, the (Lagrangean) displacement field $\delta r(M_r,t)$ is expanded in the eigenfunctions of the linearized problem, $\delta r_i(M_r)$,

$$\delta r(M_r,t) = \sum_{i=1}^{\infty} q_i(t) \delta r_i(M_r). \tag{1}$$

This expression is then inserted in the integral of the Lagrangean density and one obtains the following particle Lagrangean (Perdang and Blacher, 1982) (μ_i and ω_i denote generalized masses and frequencies of the linear modes)

$$L = \frac{1}{2} \sum_i^{\infty} \mu_i q_i^2 - \frac{1}{2} \sum_i^{\infty} \omega_i^2 q_i^2 - \sum_i^{\infty} V_{ijk} q_i q_j q_k . \tag{2}$$

For the following, it is important to note that the kinetic energy part of (2) is exact. Therefore, all higher order corrections in the Lagrangean (2) only perturb the potential energy. This is due to the simple Lagrangean integration variable. No such choice exists in the nonradial formalism (see below).

To generalize this Hamiltonian formalism, Däppen and Perdang (1984b) modified a variational principle for fluid dynamics (Lynden-Bell and Katz, 1981). Let me just explain the principal ideas. The details can be found in those two papers. The integral of the Lagrangean density is (U denotes thermal energy per unit mass)

$$L = \frac{1}{2} \int_{U(t)} \rho u^2 \, d^3x - \int_{U(t)} U(s(\mathbf{x},t), \rho(\mathbf{x},t)) \, d^3x$$
$$+ \frac{G}{2} \int_{U(t)} \int \frac{\rho(\mathbf{x},t) \rho(\mathbf{x}',t)}{|\mathbf{x} - \mathbf{x}'|} \, d^3x \, d^3x' . \tag{3}$$

This time an Eulerian picture is used, with **u** being the velocity field, **x** referring to an inertial frame, and U(t) denoting the moving boundary of the fluid. If there were a simple expression relating u with ρ and s, and if there were suitable integration variables in which the moving boundary were at rest, (5) would indeed be a suitable Lagrangean. Such a simple expression, however, cannot exist in the general case because **u** has one degree of freedom too many to be expressible by only two scalar fields. An additional scalar field is therefore necessary. Furthermore, the constraints of adiabatic motion (conservation of specific entropy) and of mass conservation would require two Lagrangean multipliers, i.e. two more scalar fields, unless the constraints were built in the variables. Attempts to get rid of these Lagrangean multipliers were made by Katz and Lynden-Bell (1982, 1984), who formulated fluid mechanics as a gauge-invariant field theory. However, these general descriptions are very complicated for practical computations. Fortunately, we do not need the general case for our problem of mode-coupling. Stationary motion (see Perdang, 1984) can be disregarded for a large class of physically meaningful situations. The Lagrangean (3) would become tractable if the motion were restricted to velocity fields that can be expressed by two scalar fields. The Clebsch form of the velocity field (see Perdang, 1984)

$$\mathbf{u} = \alpha \nabla \beta + \nabla \gamma \tag{4}$$

suggests an idea. Lynden-Bell and Katz (1981) have considered the special case in which $\beta = s$, and α and γ are still free. This case is equivalent to

$$\nabla s \cdot \mathrm{curl}\ \mathbf{u} = 0. \tag{5}$$

This constraint still allows some form of circulation, which is expressible by a scalar field (Lynden-Bell and Katz, 1981). For our application of coupling of dynamical modes, we could restrict ourselves to the case of no circulation at all, i.e. to

$$\int_{\mathscr{C}} \mathbf{dl} \cdot \mathbf{u} = 0, \tag{6}$$

for all closed loops $\mathscr{C}$ lying on surfaces of constant specific entropy. By virtue of Stokes' theorem, this integral condition implies (5), and allows a closed expression for **u** as a function of ρ and s (see Däppen and Perdang, 1984b). Fortunately, the restriction (6) not only makes the problem much more treatable, but it also gives an important theoretical answer to the problem of toroidal modes. Toroidal modes were noted first by Perdang (1968). They are the linear analogue of circulation. In a spherically symmetric hydrostatic equilibrium, they are neutral stationary modes belonging to frequency zero. A nonradial expansion analogous to (1) must contain toroidal modes as well, because without them the linear modes could not be a complete set. Therefore, in a formulation where the amplitude equations are derived directly from the equations of motion, leaving out the toroidal modes is incorrect. In the variatonal formalism [restricted by (6)], conservation of circulation is built-in. Stationary motion is absent in all orders of perturbation theory and there are no stationary modes analogous to toroidal modes in this formalism. Thus, not only has one been able to throw out the toroidal modes in a simple way, but one has the proof that this procedure is consistent within the variational formalism.

After these theoretical remarks, let me come back to the practical use of the variational principle. Since (6) implies (5), u can be written in Clebsch form (4) (with the unknown fields θ and ν)

$$\mathbf{u} = \theta \nabla s + \nabla \nu. \qquad (7)$$

Using entropy conservation

$$\dot{s} + \mathbf{u} \cdot \nabla s = 0, \qquad (8)$$

one deduces

$$\theta = - (\dot{s} + \nabla \nu \cdot \nabla s)/|\nabla s|^2. \qquad (9)$$

Mass conservation then serves to solve for the only remaining unknown field ν

$$\mathrm{div}(\rho \nabla \nu - \rho \frac{\nabla \nu \cdot \nabla s}{\nabla s \cdot \nabla s} \nabla s) = - \dot{\rho} + \mathrm{div}(\rho \dot{s} \frac{\nabla s}{|\nabla s|^2}). \qquad (10)$$

This Poisson-type equation has a unique solution (Lynden-Bell and Katz, 1981) which can be obtained by iteration (Däppen and Perdang, 1984b). Since we are interested only in the lowest nonlinear order, one iteration suffices (note that the right hand side of (10) is already first order). It is this nonlinear correction to u that gives a higher order kinetic energy contribution which is absent in (2). With this, one has found an explicit Lagrangean density $L(s,\dot{s},\nabla s;\rho,\dot{\rho},\nabla \rho)$ which has to be inserted in (3).

The last problem is the constraint of mass conservation, because for a given entropy field $s(\mathbf{x},t)$ only those density fields $\rho(\mathbf{x},t)$ that preserve mass in all shells bounded by two surfaces of constant entropy are admitted. One way of solving this problem would be to introduce a Lagrangean multiplier, i.e. an additional scalar field. With a suitable choice of integration variables and field variables, however, this constraint can be incorporated in the variational principle, and by the same token a time-independent domain of integration in (3) is obtained (see Däppen and Perdang, 1984a,b).

The independent integration variables chosen here are (s,θ,ϕ) instead of the usual polar coordinates. With them, the domain of integration is obviously at rest. The dependent field variable replacing density is the stratification field

$$\mu = 4\pi\rho r^2 \left(\frac{\partial r}{\partial s}\right)_{\theta,\phi,t} , \qquad (11)$$

with which conservation of mass is incorporated by admitting only those μ that have no component of zero angular momentum. Since all computations here are performed using an expansion in spherical harmonics, the constraint of mass conservation is satisfied if the $\ell = 0$ term is omitted. No Lagrangean multiplier has to be used then. The other dependent field is $r(s,\theta,\phi,t)$. Note that the entropy field $s(r,\theta,\phi,t)$ cannot be used because s is the conserved independent integration variable. The inverse function, r, carrying the same information must be used instead.

To go over to the particle formalism and the Lagrangean analogue of (2), the nonlinear time-dependent displacement of the basic state fields r and μ are again expanded into the eigenfunctions of the linearized problem, δr_i and $\delta\mu_i$

$$\begin{pmatrix}\delta r \\ \delta\mu\end{pmatrix} = \sum_{i=1}^{\infty} q_i(t) \begin{pmatrix}\delta r_i \\ \delta\mu_i\end{pmatrix} . \qquad (12)$$

This expansion is inserted into the Lagrangean (3), from which one obtains the analogue of (2)

$$L = \frac{1}{2}\sum_1^\infty \mu_i \dot{q}_i^2 - \frac{1}{2}\sum_1^\infty \omega_i^2 q_i^2 + \sum_1^\infty T_{ijk} q_i \dot{q}_j \dot{q}_k$$
$$- \sum_1^\infty V_{ijk} q_i q_j q_k . \qquad (13)$$

Note the presence of the nonlinear part in the kinetic energy which has no analogue in the radial case.

To compute the coupling coefficients T_{ijk} and V_{ijk} is rather lengthy, requiring 14 single and 5 double integrals over products of three factors δr_i, $\delta\mu_i$ or their derivatives. These eigenfunctions δr_i and $\delta\mu_i$ could be computed using the linearized equations of motions resulting from the Lagrangean density (3). A simpler way is to establish the connection with the usual linearized formalism of stellar pulsation theory (see, e.g. Cox, 1980), in which the displacement field is written as

$$\delta R = \text{Re}\left\{\left[\Xi_i(r)\, Y_i(\theta,\phi)\, e_r + \sqrt{\ell(\ell+1)} \cdot H_i(r) \cdot \left(\frac{\partial Y_i}{\partial \theta} e_\theta + \frac{1}{\sin\theta} \frac{\partial Y_i}{\partial \phi} e_\phi\right)\right]\right\} e^{i\omega_i t}. \quad (14)$$

Here, ω_i is the frequency of mode, Ξ_i and H_i denote radial and horizontal components e_r, e_θ, e_ϕ are the unit vectors in polar coordinates, and $Y_i(\theta,\phi)$ is the spherical harmonic of the (ℓ,m) pair of the mode i. As is shown in Däppen and Perdang (1984b), $\delta r_i = \Xi_i$ and $\delta\mu_i = -H_i$, which simplifies the application of the variational principle tremendously, because δr_i and $\delta\mu_i$ can be computed in a standard stellar pulsation model (Christensen-Dalsgaard, 1981).

Results and Discussion

To obtain a first idea how the inclusion of nonradial modes changes the results of nonlinear mode-coupling, experiments involving two sets of each seven modes have been carried out. The first set contains seven consecutive radial five-minute modes, the second set contains three radial modes coupled with two modes of $\ell = 1$ and two modes of $\ell = 2$. The two sets of modes with their frequencies are shown in table 1. As suggested by the asymptotic formula for solar oscillation frequencies [see e.g. Gough (1984)] (α, β and δ being of order unity),

$$\nu = \nu_0\{(n + \ell/2 + \delta) - [\alpha\ell(\ell+1) - \beta]/[n + \ell/2 + \delta]\}, \quad (15)$$

the radial frequencies in table 1 are indeed nearly equally spaced. The distribution of the nonradial frequencies, however, is mainly changed by $\ell/2$ term which produces a twice as dense frequency distribution than in the radial case.

The numerical analysis of the mode-coupling has been carried out with the technique described in Perdang and Blacher (1982). Thus, from a randomly given initial condition, the system of coupled oscillators [governed by the Lagrangean (13)] is followed over several hundred periods. A good numerical integration program must be used and care must be especially taken not to use numerical techniques based on functions $A(t)\sin(\omega t)$ [with some slowly varying $A(t)$], because such a procedure might preclude interesting irregular or chaotic

MODES OF FIGURE 1				MODES OF FIGURE 2			
n	ℓ	ν mHz	P min	n	ℓ	ν mHz	P min
22	0	3.168	5.261	22	0	3.168	5.261
23	0	3.304	5.044	23	0	3.304	5.044
24	0	3.441	4.843	24	0	3.441	4.843
25	0	3.578	4.648	22	1	3.232	5.157
26	0	3.715	4.485	23	1	3.369	4.948
27	0	3.853	4.325	21	2	3.158	5.278
28	0	3.990	4.176	22	2	3.294	5.059

TABLE 1: <u>List of the modes coupled in figures 1 and 2</u>. They are computed with Model 1 of Christensen-Dalsgaard and Gough (1982).

effects. The integration method by Burlisch and Stoer (1964) [an extrapolation method with rational polynomials; see also Wait (1976)] has been found to be very accurate and fast. Since the system is Hamiltonian there is at least one scalar check of the accuracy of the solution. After integrating the system of coupled oscillators, the power spectrum of the surface displacement is computed. While in the radial case no further assumptions have to be made, here detector or filter functions would have to be used to produce a power spectrum that can be compared with observations. Since here only a demonstration of nonlinear effects is made, I have used the simplest possible power spectrum (F is the number of modes taken into account) of

$$\delta R(t) = \sum_{i=1}^{F} q_i(t) \, \Xi_i(R). \qquad (16)$$

More realistic detector functions (Christensen-Dalsgaard and Gough, 1982) will have to be used in future calculations; they will change (16) into a weighted sum with weight factors w_i and will include the horizontal components as well. For the low degrees discussed here, the w_i's are of order unity with any reasonable choice of detector functions.

Figures 1 and 2 show power spectra belonging to different oscillation energies for the pure set of seven radial modes and the mixed set of seven radial and nonradial modes, respectively. In figure 1, from top left to bottom right, spectra with higher oscillation energy are shown. The two dominant nonlinear effects are (i) the difference peaks at about $n\nu_o$ ($n = 1,2,3,...$), and (ii) a dramatic frequency shift when the amplitude comes near its critical value. In figure 2, from top left to the bottom via top right, again spectra with rising oscillation energy are displayed. Due to much reduced commensurability of the frequencies, and the new velocity dependent term in (14), the qualitative aspect of these spectra is entirely different from those of figure 1 and is truly chaotic. Note also here the dramatic period shift when the amplitude approaches its critical value (numerical experiments have confirmed that the critical point in the Hamiltonian function (not just in the potential) is indeed not far away: by rising the amplitude only slightly one has caused the particle to escape).

Concluding, I should like to remark that nonradial modes contribute to irregular behaviour much more than radial modes. The amplitudes necessary to exhibit nonlinear phenomena (about 10^{-3}) are already much lower than in the case of <u>three</u> coupled radial modes (about 0.1, see Däppen and Perdang,

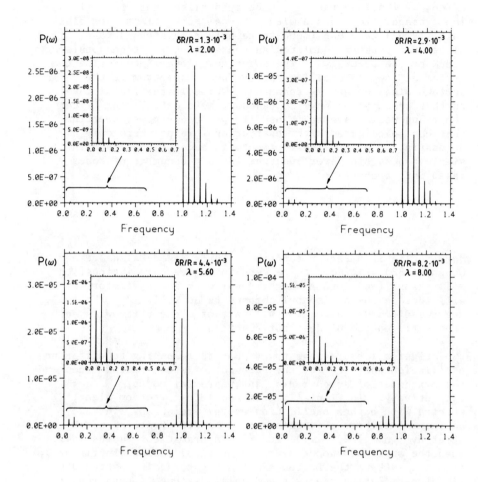

Fig. 1 <u>Seven radial five-minute modes</u>. Power spectrum of the surface displacement for four different oscillation energies (initial conditions: $q_i = 0$, $\dot{q}_i = \lambda a_i$, $i = 1, 2, \ldots 7$; a_i are constants and λ is a scaling parameter). Note the frequency shift for the largest amplitude.

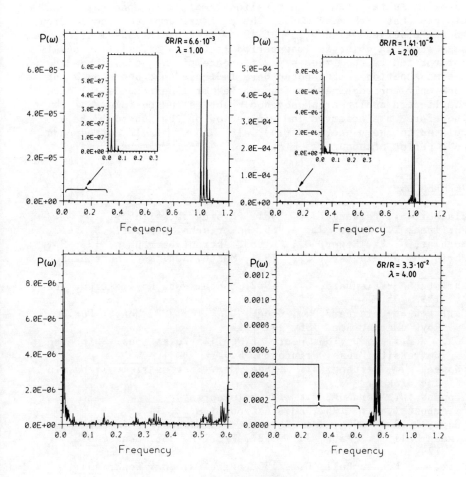

Fig. 2 Three radial and four nonradial five-minute modes. Power spectrum of the [weighted, eq. (16)] surface displacement for three different oscillation energies. Situation and notation as in figure 1.

1984a). As far as nonlinear helioseismology is concerned, one expects that with many modes, the critical amplitude could drop to the observed value of about 10^{-6}. If, as it is sometimes suggested, an observed long period in the solar oscillations is interpreted by the frequency difference of two modes, then the enormous period shifts noted here indicate that one should not look only for combination frequencies of linear modes. A more complicated nonlinear phenomenon might be responsible. Therefore, the observed 160 min period of the Sun could be related to the ν_o-term in (15), although ν_o itself corresponds to a period of about 125 min.

REFERENCES

Blacher, S., Perdang, J.: 1981, Physica 3D, 512-529.
Burlisch, R., Stoer, J.: 1966, Numerische Mathematik 8, 1-13.
Buchler, J. R., Regev, O.,: 1983, Astron. Astrophys. 123, 331.
Christensen-Dalsgaard, J.: 1981, Month. Not. Roy. astron. Soc. 194, 229.
Christensen-Dalsgaard, J.: 1982, Month. Not. Roy. astron. Soc. 199, 735.
Christensen-Dalsgaard, J., Gough, D. O.: 1982, Month. Not. Roy. astron. Soc. 198, 141-171.
Cox, J. P.: 1980, The Theory of Stellar Pulsations, Princeton University Press, Princeton, N.J.
Däppen, W., Perdang, J.: 1984a, Mem. Soc. astr. Italiana (in the press).
Däppen, W., Perdang, J.: 1984b (preprint).
Dziembowski, W.: 1982, Acta Astronomica 32, 147.
Gough, D. O.: 1984, Mem. Soc. astr. Italiana (in the press).
Lynden-Bell, D., Katz, J.: 1981, Proc. R. Soc. Lond. A378, 179-205.
Katz, J., Lynden-Bell, D.: 1982, Proc. R. Soc. Lond. A381, 263-274.
Katz, J., Lynden-Bell, D.: 1984 (preprint).
Perdang, J.: 1968, Astrophysics and Space Science 1, 355-371.
Perdang, J.: 1984, these proceedings.
Perdang, J., Blacher, S.: 1982, Astron. Astrophys. 112, 35-48.
Perdang, J., Blacher, S.: 1984a, Month. Not. Roy. astron. Soc. (in the press).
Perdang, J., Blacher, S.: 1984b, Astron. Astrophys. (in the press).
Wait, R.: 1976, Extrapolation Methods, in 'Modern Numerical Methods for Ordinary Differential Equations', ed. G. Hall and J. M. Watt, Oxford University Press, 105-115.
Woltjer, J.: 1935, Month. Not. Roy. astron. Soc. 95, 260.

CHAOTIC OSCILLATIONS IN A SIMPLE STELLAR MODEL-
A MECHANISM FOR IRREGULAR VARIABILITY

Oded Regev and J. Robert Buchler

ABSTRACT

We propose a mechanism which can cause stellar pulsations of an unexpectedly complex nature. The mechanism is based on the existence of multiple hydrostatic equilibria of a stellar model with a given entropy profile. If two such dynamically stable equilibria are in thermal imbalance, a situation can arise in which the star will be thermally driven from one hydrostatic equilibrium to the other and will oscillate (on a dynamical timescale). In such a case, depending on the ratio of the dynamical to the thermal timescale, the model can exhibit relaxation oscillations, chaotic variability or limit cycles. We demonstrate the mechanism by a simple model of a late-type red giant in which the extended envelope contains large overlapping ionization zones.

Introduction

For a given stellar mass and chemical composition, hydrostatic equilibrium solutions (HES) can be sought if the thermal structure of the star is specified - i.e., $s(m)$, the specific entropy profile is known. The equilibrium of the dynamical equation of stellar structure is obtained by setting the acceleration equal to zero in:

$$\frac{d^2R}{dt^2} = -4\pi R^2 \frac{\partial P}{\partial m} - \frac{Gm}{R^2}. \qquad (1)$$

Here m is the mass variable, $R(m)$ the Lagrangian radius and P the pressure arising from an equation of state $P(\rho,s)$ with ρ (the density) given by $\rho^{-1} = 4\pi R \partial R^2/\partial m$.

It is possible that for a given $s(m)$ several HES exist

(see e.g., Kähler 1979). White dwarf, neutron star and the intermediate (dynamically unstable) branches of zero entropy stars constitute a well known example. A necessary condition for the existence of such multiple HES is that the adiabatic exponent Γ_1 drop below 4/3 over a sufficiently important region of the star (Ledoux 1946), as is indeed the case in the above mentioned example.

Consider a stellar model, which for a given s(m), has two dynamically stable HES (potential minima) and a dynamically unstable one (potential maximum). If the star is in <u>thermal imbalance</u>, when in the stable HES, it will be driven <u>out of the</u> HES on a thermal timescale. Thermal imbalance means that the right hand side of the <u>thermal equation</u> of stellar structure,

$$\frac{ds}{dt} = \frac{1}{T} (q - \frac{\partial L}{\partial m}) , \qquad (2)$$

does not vanish and thus, s changes in time. Here the term in parentheses is the heat input due to local heat sources (q) and energy flux ($\partial L/\partial m$). It is possible then for the star to be thermally driven back and forth between the two dynamically stable HES, while dynamically oscillating when trying to settle down on them. For such a situation to occur, it is, of course, necessary that the two HES are "close" enough - i.e., the thermal energy required to transfer the star between the HES is available. In addition, the thermal evolution has to push the star in the right direction.

Reasonable candidates for the behavior described above are late-type red giants. These stars have very extended envelopes with overlapping H and He ionization zones. It is thus possible that Γ_1 drops below 4/3 over an important enough region and the envelope becomes dynamically unstable (see e.g., Tuchman, Sack and Barkat 1978). It can be expected that the above mentioned mechanism will operate in such situations and give rise to relaxation oscillations (Paczynski and Ziolkowski 1968). If the dynamical timescale, t_{dyn} is of the order of the thermal one, t_{th}, irregular variability may arise.

In the next section we shall analyze a simple one-zone model which approximates qualitatively a late-type red giant and exhibits various types of oscillations. Afterwards we discuss the similarity of our model to well studied simple oscillators which are known to exhibit chaotic behavior and we comment on the astrophysical relevance of our model.

<u>The Model</u>

Our model of mass M consists of one variable zone of constant mass m with no energy sources, bounded externally by a

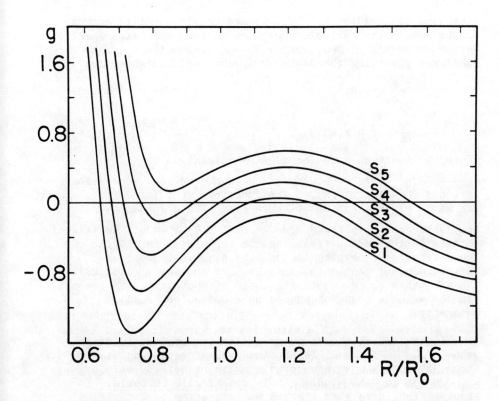

Fig. 1   Oscillator acceleration − g as a function of radius for various values of the entropy. g is in arbitrary units and R in units of R_o (the middle HES for S_3). Three equilibria are apparent for S_2, S_3 and S_4.

time dependent radius R. By s we denote the specific entropy of the zone. The variable zone lies on top of a fixed inert core with mass $M_c = M-m$, radius R_c and luminosity L_c. The equations governing the behavior of the variable zone are:

$$\frac{d^2R}{dt^2} = g(R,s) ,\qquad (3)$$

$$\frac{ds}{dt} = \varepsilon\, h(R,s) ,\qquad (4)$$

where the function g is the total acceleration
$g \equiv - GM/R^2 + 4\pi R^2 P(\rho,s)/m$ and h is the entropy generation rate $h \equiv (L_c-L)/m/T(\rho,s)$. The functions $P(\rho,s)$ and $T(\rho,s)$ are given by an equation of state taking into account ionization and ρ is found from $4\pi \rho(R^3 - R_c^3) = 3m$. Additional details of the model, including the expression for L (the luminosity), the opacity law and the parameters for our reference model can be found in Buchler and Regev (1982,**BR**). The parameter ε, to denote the ratio of the dynamic to thermal timescales, has been introduced as a convenient scaling parameter.

If the variable zone is an ionization region, Γ_1 drops below 4/3. It is possible to find parameters giving rise to three HES at fixed s in a certain range of entropy values $s_{c1} < s < s_{c2}$. The middle HES is <u>dynamically unstable.</u> In Figure 1 the force g is plotted as a function of R for five different entropy values. Clearly, for the middle three values of the entropy three HES exist. Outside the critical entropy range there is only one solution. The variation of HES radii with entropy is shown in Figure 2 (bottom). At the critical entropies one of the stable branches R_+ (or R_-) merges with the unstable branch R_u and disappears. An alternative is to say that the oscillator potential has two troughs in the critical entropy range with the unstable equilibrium corresponding to a central hump of the potential. At the critical entropies, one of the troughs and the hump merge and the other trough survives.

When the HES radii are inserted into the thermal equation (4), one obtains Figure 2 (top). The parameters of the thermal equation (see BR) are chosen such as to put the central dynamically unstable hydrostatic solution into thermal balance (fixed point of system [3] and [4]) and to make it, in addition, <u>secularly unstable.</u> The latter condition guarantees that the sign of the thermal imbalance for the two stable HES is such as to push them toward the point where they disappear. The parameters of the model allow enough freedom to also impose <u>vibrational stability</u> (in thermal imbalance) on the

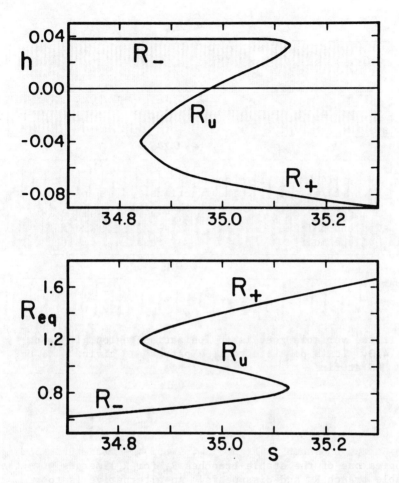

Fig. 2 Bottom – Variation of the HES radii (in units of R_o) with entropy (in units of the gas constant).
Top – Rate of change of the entropy – h (in arbitrary units) along the HES branches.

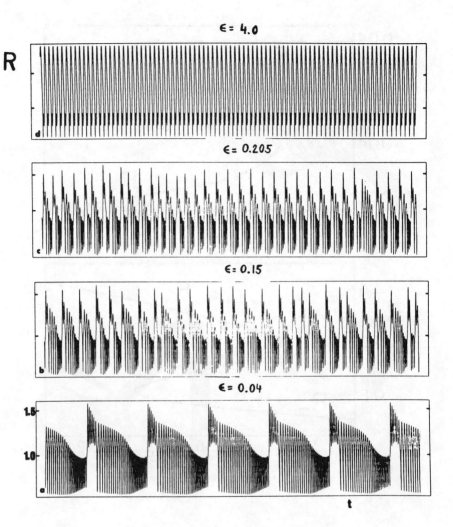

Fig. 3 Radius of the variable zone (in units of R_o) as a function of time for different values of ε. Time scale for one oscillation in the lowest plot is $\sim$ 1 yr.

CHAOTIC OSCILLATIONS IN A SIMPLE STELLAR MODEL 291

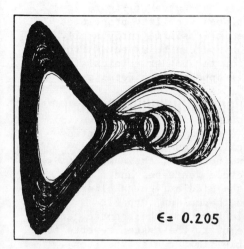
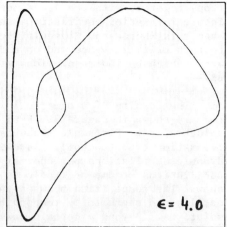
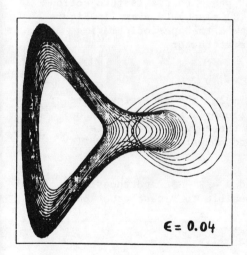
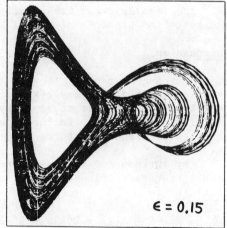

Fig. 4 Projection of phase space trajectories of the solutions of figure 3 onto the $(\dot{R},R)$ plane.

two stable branches. The Z shape of the curve of Figure 2 (top) together with the above mentioned stability properties insure relaxation oscillations when ε is sufficiently small. When ε is increased, irregular behavior can be expected from a certain critical value and on and up to another critical value above which periodic solutions will appear. For details see BR.

The results of a numerical integration of the system (3) - (4) are shown in Figure 3. R(t) is shown for four selected values of ε. For small ε the oscillations are of the relaxation type (bottom). As ε is increased, the number of dynamic oscillations per thermal cycle decreases and oscillations become aperiodic as the middle figures clearly show. The competition between the thermal and dynamic oscillators seems to be responsible here for the chaotic solutions. Beyond a certain value of ε, the system reverts to periodic behavior, now of the stable limit cycle type (top).

The corresponding sequence of phase portraits (projection of the trajectory onto the (dR/dt,R) plane) is shown in Figure 4. The complexity of trajectories in the aperiodic regime suggests the presence of a strange attractor.

Discussion

Although our model is very simple, the analytical form of the functions $g(R,s)$ and $h(R,s)$ is complicated. It is thus worthwhile to attempt to simplify these functions (while retaining their basic features) and compare them with well-studied simple oscillators.

By introducing $x = R - R_o$ and $\lambda = S - S_o$ and choosing R_o and S_o in a specific way it is possible to recast equation (3) into the form

$$\ddot{x} = A(\lambda) + B(\lambda)x + D(\lambda)x^3 + \ldots \qquad (5)$$

in such a way that for $\lambda = 0$ the potential is symmetric around $x = 0$. One can see also that the coefficients must have expansions of the form

$$A(\lambda) = A_1\lambda + O(\lambda^2),$$
$$B(\lambda) = B_o + B_1\lambda + O(\lambda^2),$$
$$D(\lambda) = D_o + D_1\lambda + O(\lambda^2), \qquad (6)$$

where terms of $O(\lambda^2)$ or higher have been dropped. By limiting ourselves to $O(x^3)$ in (5) and lowest order in (6), we obtain exactly the Moore and Spiegel (1966) oscillator potential - MS (Marzec and Spiegel 1980, Baker, Moore and Spiegel 1971), although the latter was obtained in a totally different context than MS's. Our thermal equation (4) is, however, more complicated.

The MS oscillator has been studied in detail by Spiegel and collaborators as a function of its parameters and has been shown to exhibit a large variety of behavior. In particular, the aperiodic oscillations have been associated with the existence of a strange attracting set. The structural similarity of the present model with the MS oscillator makes the existence of a strange attracting set highly probable in our case.

The nonlinear mechanism which is responsible for the oscillation in our model is different from the classical κ and γ mechanisms in stellar pulsation theory (Baker criteria). Indeed, a linear stability analysis of our model would indicate a secular thermal evolution but vibrational stability.

It is clear that our naive one-zone model cannot adequately deal with the shift of the actual ionization front through the stellar envelope. However, we are interested only in the effect of ionization on dynamical stability. In addition, the whole outer region in the low-mass red giants is generally convective, and the entropy and density profiles are rather flat, acting approximately like a single entropy zone. It is our hope, thus, that the one-zone model could well represent the underlying mechanism for the behavior of a large class of late-type red giants. The model is far too crude to allow anything but a vague correlation of models with observation. The construction and study of realistic multizone models is thus necessary in order to ascertain the prediction of the one-zone oscillator. Such a study is now in progress.

REFERENCES

Baker, N.H., Moore, D.H. and Spiegel, E.A., 1966, A.J. __71__, 845.
Baker, N.H., Moore, D.H. and Spiegel, E.A., 1971, Quart. J. Mech. Appl. Math. __24__, 389.
Buchler, J.R. and Regev, O., 1982, Ap. J. __263__, 312 (BR).
Kähler, H., 1979, Astron. Astrophys, __75__, 207.
Ledoux, P., 1946, Ap. J. __104__, 333.
Marzec, C.J. and Spiegel, E.A., 1980, SIAM J. Appl. Math. __38__, 403.
Moore, D.W. and Spiegel, E.A., 1966, Ap. J. __143__, 871 (MS).
Paczynski, B. and Ziolkowski, J., 1968, Acta Astron. 18, 255.
Tuchman, Y., Sack, N. and Barkat, Z.K., 1978, Ap. J. __219__, 183.

X-RAY BURSTERS - THE HOT ROAD TO CHAOS?

Odev Regev and Mario Livio

ABSTRACT

Helium shell burning, at the base of an accreted envelope on the surface of a neutron star, is described in terms of a simple two zone model.

Two separate limit cycles are identified, one of which is responsible for shell flashes in red giants and the second provides the underlying mechanism for X-ray burst sources.

When the impact of the burst luminosity on the accretion rate is introduced, the system is shown to exhibit a complex temporal behaviour. As the basic accretion rate is increased, the system undergoes a sequence of bifurcations, starting from strictly periodic bursts and finally reaching a chaotic behavior.

I. Introduction

The purpose of the present work is to study the properties of Helium burning (the triple α process) in a thin shell, using a simple two zone model which incorporates, however, the important physical processes (see also Barranco, Buchler and Livio 1980, Paczynski 1983a,b).

The circumstances in which such Helium shell burning is important are, in particular, encountered in thermonuclear runaways on the surface of neutron stars. These thermonuclear events are believed to be the underlying mechanism of X-ray burst sources (Lewin and Joss 1983 and references therein). Helium shell burning is also important in red giants (Schwarzschild and Härm 1967).

II. The Two Zone Model

The structure of the model used, is described schematically in Fig. 1. It consists of an inert core of fixed mass M_c, radius R_c and temperature T_c and an inert outer atmosphere of temperature T_{at}. The accreted envelope is divided into two time dependent zones:

(1) The thin burning shell at the base of the accreted matter, of mass Δm_1 and temperature T_1.
(2) A buffer zone which consists of the accreted envelope (excluding the burning region) of mass Δm_2, temperature T_2.

The heat equations for the time dependent zones can be written as

$$T_1 \frac{ds_1}{dt} = \varepsilon_1(\rho_1, T_1) + \frac{1}{\Delta m_1}\left[K_c(T_c^4 - T_1^4) - K_1(T_1^4 - T_2^4)\right] \tag{1}$$

$$T_2 \frac{ds_2}{dt} = \frac{1}{\Delta m_2}\left[K_1(T_1^4 - T_2^4) - K_2(K_2^4 - T_{at}^4)\right] \tag{2}$$

where ε_1 is the nuclear energy generation rate in the triple α reaction and where we have assumed radiative energy transfer between neighboring zones, so that

$$K_i = \frac{(4\pi R_i^2)^2 ac}{3\kappa_i \Delta m_i^*} \tag{3}$$

Δm_i^* being the edge centered masses.

We now make the following assumptions:

(a) Instantaneous hydrostatic adjustment. This is performed following the method of Henyey and Ulrich (1972), neglecting the adjustment of the very thin burning shell and cross terms (see Barranco et al. 1980).
(b) The mass of the burning shell is assumed to be some fixed fraction μ of the envelope mass.

Under these assumptions the equations (1) and (2) assume the nondimensional form (Regev and Livio 1984, Livio and Regev 1984)

$$\frac{dT_1}{dt} = \alpha\left[\frac{\varepsilon_1}{\varepsilon_*} - \gamma_{c_1}\frac{1}{\Delta m_2}(T_1^4 - T_c^4) - \frac{1}{\Delta m_2^2(1+\mu)}(T_1^4 - T_2^4)\right]$$

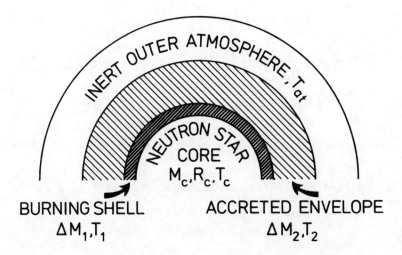

Fig. 1. The structure of the two zone model.

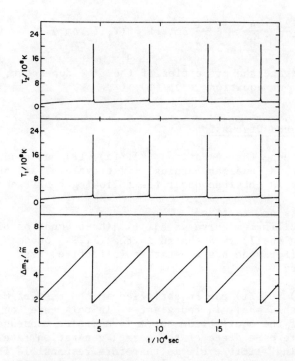

Fig. 2. The zone temperatures (T_1 and T_2) and the accreted envelope mass (Δm_2) as a function of time for the standard model.

$$\equiv F_1(T_1, T_2, \Delta m_2) \qquad (4)$$

$$\frac{dT_2}{dt} = \alpha\chi(1 - \frac{T_2}{T_{h2}})[\frac{1}{\Delta m_2^2(1+\mu)} (T_1^4 - T_2^4) - \gamma_{12} \frac{1}{\Delta m_2^2} (T_2^4 - T_{at}^4)]$$

$$\equiv F_2(T_1, T_2, \Delta m_2) \qquad (5)$$

where the symbols have the same meaning as in Regev and Livio (1984).

We now add an equation of mass conservation

$$\frac{d(\Delta m_2)}{dt} = a - \varepsilon_1 \mu \Delta m_2 Y/Q \qquad (6)$$

where a is the accretion rate, Y is the mass fraction of Helium and Q is the energy liberated per gram of Helium that is burnt. In nondimensional form this equation reads

$$\frac{d(\Delta m_2)}{dt} = \frac{a}{a_*} - \beta \frac{\varepsilon_1}{\varepsilon_*} \Delta m_2 Y \equiv F_3(T_1, T_2, \Delta m_2) \qquad (7)$$

We now investigate the properties of the time dependent, nonlinear, system of equations (4), (5), (7).

III. Results and Discussion

We first note that equations (4), (5), (7) can produce, as a special case, a "nuclear burning" limit cycle. This oscillatory behaviour is obtained under the following conditions:

(i) The mass Δm_2 is <u>fixed</u>.
(ii) The equilibrium <u>curve</u> of eq. (4) (heat equation of burning shell) is S shaped in the (T_1, T_2) phase space.
(iii) The buffer zone has a negative (effective) heat capacity.

Conditions (i) - (iii) can be satisfied in the case of Helium burning in a thin shell in red giants. In this case the S shaped form of the equilibrium curve is a direct consequence of the temperature dependence of the energy generation rate. The "nuclear burning" limit cycle is therefore responsible for shell flashes in red giants (Schwarzschild and Härm 1967, Buchler and Perdang 1979).

Returning now to the general case, represented by the full set of equations, we find the following results:

(a) <u>Constant Accretion Rate</u>

For accretion rates between two critical values $a_{2c} < a < a_{1c}$, we obtain a bursting behavior as shown in Fig. 2. The temperature between bursts rises very slowly, mainly by heat transfer from the core. When the accumulated mass reaches a critical value, the thermonuclear reaction runs away. For accretion rates above a_{1c} or below a_{2c} the system settles on a fixed point and steady burning is obtained. The properties of the bursts and their profile (Fig. 3) are very similar to those obtained in full numerical calculations (Ayasli and Joss 1982).

This behavior is a consequence of an "accretion-burning" limit cycle, which can be demonstrated by analyzing the rate of mass consumption in nuclear burning, as a function of the accreted mass (at the equilibrium temperature). This is shown in Fig. 4. It is easy to show that the two branches marked with S are stable while the one marked with U is unstable. Bursts are obtained for accretion rates between the two values bordering the unstable branch.

(b) <u>Inclusion of impact of burst luminosity on accretion rate</u>

The burst luminosity can affect the accretion rate by being the source of two feedback mechanisms. A positive feedback is obtained from the fact that the burst can generate an evaporative mass loss from the secondary star (Arons 1973, Osaki 1984). We include this effect in the accretion rate, in the simplest possible form, by introducing a factor proportional to $(L_b - L_o)$, where L_b is the burst luminosity and L_o is some threshold required to induce mass loss. A negative feedback is obtained from the fact that the radiation pressure, generated by the burst luminosity, can produce an impulsive velocity change in the accretion disk, thereby reducing the accretion rate. We incorporate this effect, again in a simple way, by introducing a factor proportional to $(1 - L_b/L_m)$ where L_m provides a scaling for the luminosity, that ensures that the system remains in the bursting regime. The accretion rate (see eq. (7)) is therefore written as

$$\frac{a}{a_*} f \left(\frac{L_b - L_o}{L_m}\right)\left(1 - \frac{L_b}{L_m}\right) \qquad (8)$$

where the parameter f now determines the basic (nondimensional) accretion rate. We now study the temporal behavior of the system for increasing values of f. For low values of f, bursts

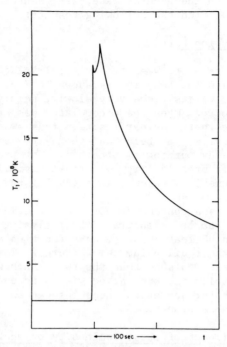

Fig. 3. A detailed burst structure for the standard model. The double peak is clearly seen.

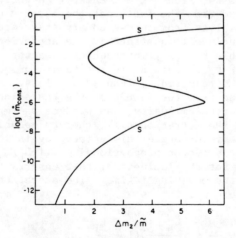

Fig. 4. Logarithm of the mass consumption rate during nuclear burning at the equilibrium temperature (T_1^{eq}) as a function of the accreted envelope mass for the standard model. The stable branches are marked by S and the unstable branch by U.

X-RAY BURSTERS – THE HOT ROAD TO CHAOS?

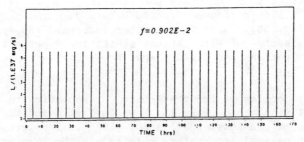

Figure 5(a): The luminosity as a function of time. The control parameter f is determined by the accretion rate (see text). One-period behaviour is apparent.

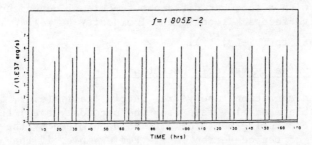

Figure 5(b): The same as 5(a). Two-period behaviour is apparent.

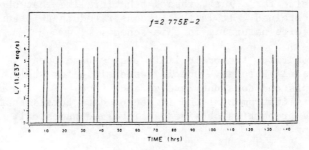

Figure 5(c): The same as 5(a). Four-period behaviour is apparent.

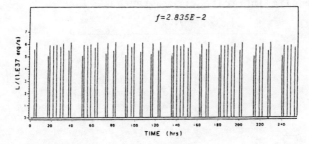

Figure 5(d): The same as 5(a). An irregular (chaotic) bursting behaviour is apparent.

that are strictly periodic are obtained (Fig. 5a). As f is increased, a bifurcation occurs and bursts with two periods are obtained (Fig. 5b). When f is further increased the system undergoes a series of bifurcations exhibiting 4 periods (Fig. 5c), then 8 periods and finally what seems like a chaotic behavior (Fig. 5d).

The projection of the trajectory corresponding to Fig. 5d onto the (T_1, T_2) phase space is shown in Fig. 6 and is very reminiscent of the Rössler attractor (1976). The return map, showing the maximal temperature obtained in the burning shell at consequtive bursts is shown in Fig. 7. Its one dimensional nature represents the causal behavior of the bursts. We would like to point out that the behavior indicating chaos can be obtained, with a somewhat different functional dependence of the positive feedback loop on the luminosity as well (e.g., proportional to $(L_b - L_o)^{1/2}$).

(c) <u>Intrinsically variable accretion rate</u>

If the accretion rate varies intrinsically (e.g., because of a variable mass flow rate from the secondary star) then, of course, the bursting behavior will be determined by the time dependence of the accretion rate. For example, if the accretion rate has some value a_o, $a_{2c} < a_o < a_{1c}$ for $t \leqslant t_1$ and then it jumps to $a_1 > a_{1c}$ for $t_1 < t \leqslant t_2$, returning to the original value for $t > t_2$, then a behavior like that shown in Fig. 8 is obtained. It is thus demonstrated that "standstills" can occur quite naturally in this scheme.

To summarize, two basic limit cycles have been identified in Helium shell burning. One is responsible for shell flashes in red giants and does not involve accretion. For the second, accretion is instrumental and it can produce bursts that have many of the properties obtained in involved numerical computations.

It has been demonstrated that a chaotic behavior can originate from an extremely simple underlying mechanism and does not require intricate circumstances.

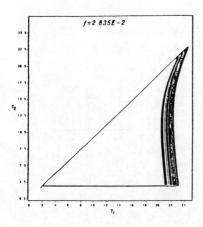
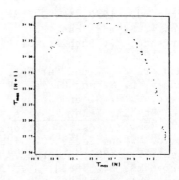

Fig. 6. The projection of the trajectory in phase space of eqs. (4), (5), onto the (T_1, T_2) plane, for the case of Fig. 5(d).

Fig. 7. The maximal temperature in the burning zone at a burst as a function of the same quantity at the previous burst.

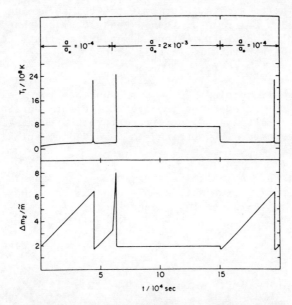

Fig. 8. The burning shell temperature (T_1) and the accreted envelope mass (Δm_2) as a function of time for a model with a variable accretion rate. The accretion rate for each time interval is marked in the upper part of the figure.

REFERENCES

Arons, J., 1973, Ap. J. 184, 539.
Ayasli, S. and Joss, P. C., 1982, Ap. J. 256, 637.
Barranco, M., Buchler, J. R. and Livio, M., 1980, Ap. J. 242, 1226.
Buchler, J. R. and Perdang, J., 1979, Ap. J. 231, 524.
Henyey, L. G. and Ulrich, R. K., 1972, Ap. J. 173, 109.
Lewin, W. H. G. and Joss, P. C., 1983, in Accretion Driven Stellar X-Ray Sources, W. H. G. Lewin and E. P. J. van den Heuvel, eds., Cambridge University Press.
Livio, M. and Regev, O., 1984, submitted to Ap. J. Lett.
Osaki, Y., 1984, preprint.
Paczynski, B., 1983a, Ap. J. 264, 282.
Paczynski, B., 1983b, Ap. J. 267, 315.
Regev, O. and Livio, M., 1984, Astr. Ap. 134, 123.
Rossler, O. E., 1976, Phys. Lett. 57A, 397.
Schwarzschild, M. and Harm, R., 1967, Ap. J. 150, 961.

COMPRESSIBLE MHD TURBULENCE: AN EFFICIENT MECHANISM TO HEAT STELLAR CORONAE

Marco Pettini, Luigi Nocera, and Angelo Vulpiani

ABSTRACT

The nonlinear propagation of a linearly polarized Alfven wave along a uniform magnetic field is studied taking into account the back reaction of sound oscillations that have been excited by the wave itself. The Alfven wave propagates through a uniform, viscous, resistive, magnetized and weakly compressible plasma in typical astrophysical conditions. The back reaction mechanism yields a nonlinear cascade of Alfven waves on small length scales.

The model equation obtained is a wave equation in 1+1 dimension with nonlinear dispersive terms and a damping term. Numerical simulations show that the excitation of Alfven waves at a large scale yields space Fourier spectra for the magnetic and velocity field perturbations which are power law spectra ("turbulent energy cascade").

The discovered results are very promising with respect to the turbulent heating problem where the laminar heating is negligible.

1. Introduction

The possibility of heating the solar corona (and, in general, the coronae around some stars), by dissipation of Alfven waves has been considered for a long time now (see e.g., the review by Kuperus et al. [1]).

These waves turn out to be particularly abundant in the solar corona (see Ref. 2 for the estimated fluxes), though it is very difficult to have them dissipated due to the very small electrical resistivity and viscosity of the solar coronal plasma. Also, the dissipation rate at wavenumber k is proportional to k^2 and efficient dissipation then requires large k; but it turns out that there are 'few' large k waves when they are generated at the base of the solar atmosphere. Altogether then, the dissipated power

$$Q = Q_0 \int E(k) k^2 dk \qquad (1.1)$$

(E(k) being the spectrum of the wave energy) is too low to meet the solar heating requirements ($Q \simeq 10^{-5} - 10^{-3}$ erg cm^{-3} s^{-1}, as estimated in ref. 2) unless a mechanism generating large k waves out of small k waves is operating.

A turbulent cascade might be the answer to this problem and indeed, order of magnitude calculations [2,3] show that a Kolmogorov spectrum for a developed turbulence in (1.1) provides the required heating rate. In general, for an isotropic homogeneous turbulence,

$$Q \simeq 4\pi Q_0 \int_{k_0}^{k_d} k^2 E(k) dk \quad , \qquad (1.2)$$

where k_0 refers to the lowest wavenumber wave propagating in the plasma (usually k_0 equals the inverse of the scale length) and k_d is the upper bound of the inertial range. The latter, of course, depends upon the efficiency of the energy transfer from large k waves to small k waves.

A main theoretical task, then, is to find E(k) out of some nonlinear wave equation describing the propagation of Alfven waves. In this contribution we mainly concentrate on a one dimensional model for spatial turbulence. The most famous one dimensional model out of the Navier-Stokes equation is the Burgers equation, but this one can hardly be thought of as a one dimensional model for fluid turbulence; in fact it is well known that the Burgers equation is analytically solvable. As far as MHD equations are concerned, to the best of our knowledge, a one dimensional dynamical model for MHD turbulence is not available. The development of such a model, the study of its spectral properties and of its chaotic behaviour are reported here.

2. Fundamental Physical Process

We consider a uniform mass of plasma at rest, with infinite viscosity, finite resistivity $1/\sigma$, with constant mean pressure p_0 and constant mean density ρ_0. A uniform magnetic field $\mathbf{B} = B_0 \mathbf{e}_z$ fills the volume occupied by the plasma.

We assume shear oscillations excited along the x direction by perturbations at $z = 0$.

The perturbed magnetic and velocity fields are assumed to be

$$\mathbf{B} = B_0 \mathbf{e}_z + \delta B_x(z,t) \mathbf{e}_x \quad , \tag{2.1}$$

$$\mathbf{v} = 0 + \delta v_x(z,t) \mathbf{e}_x + \delta v_z(z,t) \mathbf{e}_z \quad ,$$

$$\rho = \rho_0 + \delta \rho \quad ,$$

$$p = p_0 + \delta p \quad .$$

An intuitive picture (see Fig. 1) of the physical process originating the nonlinear energy cascade toward small spatial scales is provided by the following naive computations. The magnetic field perturbation $\delta_1 B_x$ yields a current perturbation $\delta_1 J_y$ which is roughly

$$\delta_1 J_y \sim \frac{\partial \delta_1 B_x}{\partial z} \quad , \tag{2.2}$$

then, the coupling between $\delta_1 B_x$ and $\delta_1 J_y$ in the momentum equation produces a velocity perturbation $\delta_1 v_z$ given by

$$\frac{\partial \delta_1 v_z}{\partial t} \sim \delta_1 J_y \cdot \delta_1 B_x \quad ; \tag{2.3}$$

here we are assuming non zero divergence for the velocity field, i.e. compressibility. Now the sound (slow) wave perturbation, $\delta_1 v_z$, can in turn couple with $\delta_1 B_x$ to produce a new magnetic field perturbation, $\delta_2 B_x$, given by the induction equation

$$\frac{\partial \delta_2 B_x}{\partial t} \sim \frac{\partial}{\partial z}(\delta_1 v_z \cdot \delta_1 B_x) \quad . \tag{2.4}$$

Putting $\delta_1 B_x = \sin(k_0 z - \omega_0 t)$ we find

$$\delta_1 J_y \sim \cos(k_0 z - \omega_0 t) \quad ,$$

$$\delta_1 v_z \sim \cos(2 k_0 z - 2 \omega_0 t) \quad , \tag{2.5}$$

$$\delta_2 B_x \sim \cos(3 k_0 z - 3 \omega_0 t) + f(k_0, 2 k_0) \quad ,$$

$$\cdots\cdots\cdots\cdots\cdots\cdots\cdots\cdots\cdots \quad ,$$

so that the following mechanism is acting

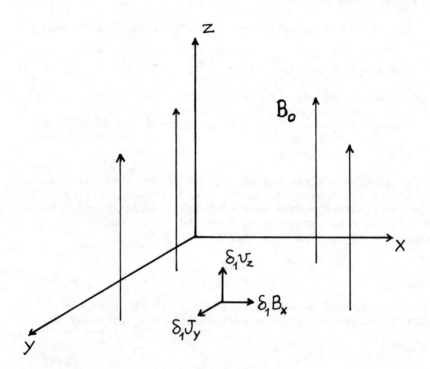

Fig. 1   Schematic geometry adopted.

ALFVEN WAVE → SOUND WAVE → ALFVEN WAVE → SOUND WAVE
                (SLOW)   (SLOW)

k_0 $\qquad\qquad 2k_0 \qquad\qquad 3k_0 \qquad\qquad 4k_0$,

$L_0 = \frac{2\pi}{k_0} \qquad \frac{1}{2}L_0 \qquad\qquad \frac{1}{3}L_0, \qquad\qquad \frac{1}{4}L_0,$

This is a very simple and natural nonlinear energy cascade that can be described in more detail by the complete MHD equations. Unfortunately, no simple argument can be put forth to infer the overall efficiency of the above sketched process, but this is just our main task.

3. The Model Equations and Numerical Results

The following equations are used

$$\frac{\partial \delta\rho}{\partial t} = -\rho_0 \frac{\partial \delta v_z}{\partial z} - \frac{\partial}{\partial z}(\delta\rho\, \delta v_z) \qquad , \qquad (3.1)$$

$$\rho T \frac{d\delta s}{dt} = \frac{4}{3}\eta_0 \left(\frac{\partial \delta v_z}{\partial z}\right)^2 \qquad , \qquad (3.2)$$

$$(\rho_0 + \delta\rho)\left\{\frac{\partial \delta v_x}{\partial t} + \delta v_z \frac{\partial \delta v_x}{\partial z}\right\} = \frac{B_0}{4\pi}\frac{\partial \delta B_x}{\partial z} \qquad , \qquad (3.3)$$

$$(\rho_0 + \delta\rho)\left\{\frac{\partial \delta v_z}{\partial t} + \delta v_z \frac{\partial \delta v_z}{\partial z}\right\} =$$

$$-\frac{\partial}{\partial z}\left(\delta p + \frac{\delta B_x^2}{8\pi}\right) + \eta_0 \frac{\partial^2 \delta v_z}{\partial z^2} \qquad , \qquad (3.4)$$

$$\frac{\partial \delta B_x}{\partial t} = B_0 \frac{\partial \delta v_x}{\partial z} - \frac{\partial}{\partial z}(\delta v_z \delta B_x) + \mu \frac{\partial^2 \delta B_x}{\partial z^2} \qquad . \qquad (3.5)$$

Eq. (3.1) is the usual continuity equation; equation (3.2) is the energy equation, δs is the entropy variation per unit mass and only the viscous heating has been taken into account because an estimate of the ratio between the Joule and the viscous heating rates (Q_J and Q_v respectively) gives

$$\frac{Q_J}{Q_v} \simeq \frac{\left(\frac{c}{4\pi}\nabla\times\mathbf{B}\right)^2}{\sigma\eta_0(\nabla\cdot\delta\mathbf{v})^2} \ll 1 \qquad (3.6)$$

for typical coronal conditions ($T\sim 10^6 K$), equal electron and proton temperatures and $\delta B_x \sim 1G$, $\delta v_z \sim 0.3$ Km s^{-1}. Moreover, by far the most important component of the viscosity tensor [4] is the z-z component η_0. The latter fact also implies that the shear velocity field evolves without dissipation, as it is evident in eq. (3.3). Equation (3.4) is the momentum equation for the evolution of δv_z and eq. (3.5) is the induction equation where $\mu = c^2/4\pi\sigma$.

From observations of line broadening in the solar atmosphere [5] the ratio $\varepsilon = (\delta B_x/B_0)^2$ can be estimated to be in the range $10^{-3} - 10^{-1}$. Using $\varepsilon^{1/2}$ as ordering parameter we get, up to $O(\varepsilon^{3/2})$, $(d\delta s/dt) = 0$ so that $\delta p = v_s^2 \delta \rho$ can be assumed up to the same order. Adimensional quantities can be introduced measuring the velocities in Alfven velocity units V_A, the magnetic field in units of B_0, introducing a typical length scale L and time units as V_A/L. If $\mu = \delta v_z$, $v = \delta v_z$, $w = \delta B_x$ in dimensionless units equations (3.3), (3.4), (3.5) become

$$\frac{\partial v}{\partial t} - \frac{\partial w}{\partial z} + \varepsilon\{\mu \frac{\partial v}{\partial z} + \frac{\delta\rho}{\rho_0} \frac{\partial v}{\partial t}\} = 0, \quad (3.7)$$

$$\Box^2 \mu - Re^{-1}\frac{\partial^3 \mu}{\partial t \partial z^2} + \frac{1}{2}\frac{\partial^2 w^2}{\partial t \partial z} = 0, \quad (3.8)$$

$$\Box^2 w - Rm^{-1}\frac{\partial^3 w}{\partial t \partial z^2} + \varepsilon\{\frac{\partial^2 \mu w}{\partial t \partial z} + \frac{\partial}{\partial z} \mu \frac{\partial v}{\partial z}\} = 0, \quad (3.9)$$

where Re and Rm are respectively the viscous and magnetic Reynolds numbers.

From these equations, with a simple approximation strictly valid only to order ε^0, we can get [6]

$$\Box^2 w = \tilde{\mu}\frac{\partial^3 w}{\partial t \partial z^2} - \frac{1}{6}(\beta-1)^{-1}\varepsilon\{2\frac{\partial^2}{\partial z \partial t} - \frac{\partial^2}{\partial z^2}\}w^3 \quad (3.10)$$

where β is the plasma β. This equation is obtained under the following two assumptions: $\mu = (\beta-1)^{-1} w^2/2$ and $v = -w$ so that the space Fourier spectrum $|\tilde{\mu}(k)|^2$ can be obtained from $|\tilde{w}(k)|^2$. The spectrum $|\tilde{\mu}(k)|^2$ is the relevant one because the heating rate is related to $(\partial_z v_z)^2$ (ref. [3]). The relation between these spectra can be easily worked out writing

$$\tilde{\mu}(k) = \frac{1}{2}(\beta-1)^{-1}\int dq \, \tilde{w}(k-q)\tilde{w}(q) \quad (3.11)$$

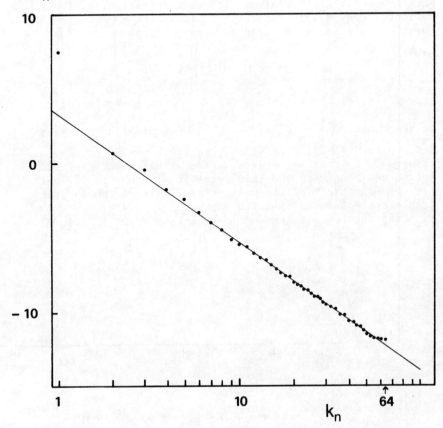

Fig. 2 Power spectrum of $w(z,t)$. Initial conditions used: $\dot{w}_j(0)=0$, $w_j(0)=A \sin(k_0 j \Delta z)$ with $A=30.$, $k_0 = \frac{2\pi}{N\Delta z}$, $j = 1,..,N$, $N = 128$, $\Delta z = 1/128$, $\Delta t = 0.1z$ (for numerical stability reasons), and $(\beta-1)^{-1}\epsilon = 10^{-4}$. Setting the damping coefficient μ equal to zero a real singularity develops in a finite time t_* and the spectrum reported here is observed up to times $t \lesssim t_*$. When $\tilde{\mu} > 0$ (and in this case $\tilde{\mu}=0.01$ has been adopted) the real singularity no longer develops and the spectrum is stable in time.

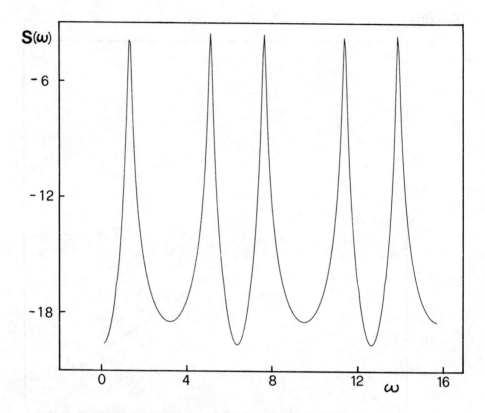

Fig. 3a Power spectrum of $w(z_0,t)$. $S(\omega)$ is the logarithm of $|\tilde{w}(z_0,\omega)|^2$. Initial conditions are the same reported for Fig. 2. Now $A=1.$, $(\mu-1)^{-1}\varepsilon=0.05$ while the other parameters are left unchanged. This spectrum has been obtained with a trance of 1024 values $w(z_0,t)$ sampled out of about $1.3 \cdot 10^5$ time integration steps starting from $t=0$.
Note the nonlinear splitting around the lowest linear frequency $\omega_1=2\pi$.

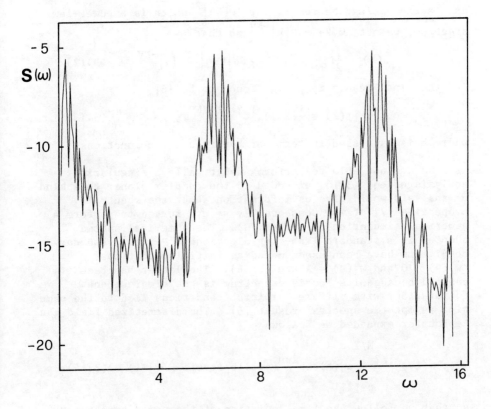

Fig. 3b Same conditions of Fig. 3a. The spectrum has been obtained with a mesh of 1024 points out of about $1.3 \cdot 10^5$ time integration steps after a transient of $1.5 \cdot 10^6$ integration steps. Noisy line broadening as well as a low frequency component are evident marks of a chaotic behavior.

and, anticipating the result for $|\tilde{w}(k)|^2$ which is a power-law spectrum, we put $\tilde{w}(k) = C|k|^{-NV}$ so that

$$\tilde{\mu}(k) = \frac{1}{2}(\beta-1)^{-1}C^2 \int_{-\infty}^{\infty} dq |k-q|^{-NV} |q|^{-NV} \qquad (3.12)$$

and the final result is easily found to be [6]

$$\tilde{\mu}(k) = \frac{1}{2}(\beta-1)^{-1}C^2 b k^{1-2NV}, \qquad (3.13)$$

where b is expressed in terms of hypergeometric functions.

We are not able to perform analytically a singularity analysis of eq. (3.10) extended to the complex plane; this kind of analysis would give us information about the spectral properties [7] of w. Consequently we are forced to perform a numerical simulation of eq. (3.10). The numerical scheme adopted is a standard leap-frog algorithm. Periodic boundary conditions have been assumed and as initial conditions $\dot{w}_j(0) = 0$ and $w_j(0) = A \sin(k_0 j \Delta)$. The latter has been assumed because its Fourier spectrum is the simplest one. Computations with different initial conditions lead to the same time asymptotic spectral result [6]. The discretized field can be Fourier expanded as follows

$$w_j(t) = (2\pi)^{-1/2} \sum_{n=1}^{N/2} \alpha_n(t) \cos\frac{2\pi n}{N}(j-1) + b_n(t) \sin\frac{2\pi n}{N}(j-1) \qquad (3.14)$$

so that we follow the time evolution of the power spectrum

$$W(k_n, t) = |\alpha_n(t)|^2 + |b_n(t)|^2$$

until an asymptotic spectrum is reached.

The results of numerical computations are (see Fig. 2)

$$W(k_n) = |\tilde{w}(k_n)|^2 \sim k_n^{-\alpha}, \quad \alpha \simeq 3.6 \qquad , \qquad (3.15)$$

so that

$$U(k_n) = |\tilde{u}(k_n)|^2 \sim k_n^{-\beta}, \quad \beta \simeq 5.2 \qquad . \qquad (3.16)$$

As our model is one dimensional we also get

$$E(k) \sim k^{-5.2} \qquad .$$

This is still far from the Kolmogorov spectrum $E(k) \sim k^{-5/3}$, which would meet the energetic requirements to heat the solar corona [2] if any such assumption could be reasonable in presence of magnetic fields, compressibility and non-homogeneity of the coronal plasma. Anyway, in three dimensions the described physical process should be more efficient so that a lower bound on the 3D spectrum can be inferred to be

$$E(k_n) = k_n^{d-1} U(k_n) \sim k_n^{-3.2}$$

A much better result ($E(k_n) \sim k_n^{-1.7}$ as a 3D lower bound on the cascade efficiency) can be found by means of a numerical study of the complete set of equations (3.7), (3.8) and (3.9) (ref. [6]).

The solutions of eq. (3.10) are not very chaotic in space even if the power law tail of the spectrum, independent of initial conditions, provides a good model for the "turbulent inertial range" of the energy cascade. On the other hand, a remarkable property of eq. (3.10) is that it displays temporal chaos. This effect is observed in real physical systems, such as fluids, also when only few spatial modes are interacting [8]. In our equation chaos shows up only on very long time scales. We take a long record of values of the field at some space point labelled by $\bar{m}$, i.e., $\{w_{\bar{m}}(t_i)\}$ and then we perform a temporal Fourier transform. After a usually long transient some sort of "phase turbulence" grows up and in Fig. 3b a noisy line broadening together with a conspicous low frequency component is clearly shown. In Fig. 3a the Fourier spectrum performed with a record of data which belong to the transitory is reported. The above reported behavior is a very interesting one for a nonlinear wave equation, but it is not clear whether any relationship can be found with the heating problem. A deeper analysis of this point will be carried on in the future.

REFERENCES

[1] M. Kuperus, J.A. Ionson, D.S. Spicer, "On the theory of coronal heating mechanisms", Ann. Rev. Astron. Astrophys. (1981).

[2] J.V. Hollweg, "Coronal heating by waves", Solar Wind V, edited by M. Neugebauer, NASA Conference Publication 2280, (1984).

[3] P.K. Browning, E. Priest, to appear in Astron. Astrophys. (1984).

[4] S.I. Braginskij, "Transport processes in a plasma", Rev. Plasma Phys. 1, 205 (1965).

[5] H. Gail, E. Sedlmayr, G. Traving, Astron. Astrophys. 44, 421 (1975).

[6] L. Nocera, M. Pettini, A. Vulpiani, submitted to Astron. Astrophys.

[7] See for instance: U. Frisch, "The analytic structure of turbulent flows", 6th Kyoto Summer Institute on Chaos and Statistical Mechanics, September 1983, to appear as Springer Lecture Notes in Physics, and references cited therein.

[8] S. Ciliberto, J.P. Gollub, Phys. Rev. Lett. 52, 922 (1984).

INDEX

Ablowitz M.J. : 161
Accretion disc : 126,299
Accretion rate : 299,301
AC Herculis : 34,35
Acoustic instability : $\overline{X}$
Acoustic wave-guide : 129
AD CMi : 38
Adiabatic oscillations : 4
Aikawa T. : 36,74,139
Aizenman M.L. : 37
Alfvén H. : 223,224,305,306, 310
Allen D. : 34
Am stars : 12
Andronov A.A. : 92
Anharmonicity parameter : 143
Anosov : 245
Appenzeller I. : 5
Ap stars : 12
Argelander F. : 15,30,33
Arneodo A. : 74,98,102,106, 115,118,125
Arnold V.I. : 46,50,107,114, 117,186,219,226,242,245, 246,249,250,259,265,269
Arons J. : 299
Asymptotic frequency : 4
Asymptotics : 4,57,88,137,144, 166,168,187,188,223,232, 265,279,312
Attractor : 100,102,111,117, 124,130,131,147,152,153, 161,168,185,187,188,189, 190,195,197,206,207,209, 211,213,214,216,217,218, 302
Auvergne M. : 76,120
Ayasli S. : 299
AY Sgr : 25
AX Vel : 25

Baade's test : 37
Baglin A. : 76,120
Bailey S.I. : 18
Baker N.H. : 75,76,119,162, 293

Balasz-Detre J. : 25
Balona L.A. : 35
Band : 174,176
Barbanis B. : 56,58
Barkat Z.K. : 286
Baroclinic flow : 172
Barranco M. : 295,296
Bates R.H.T. : 34
Bearzi B. : 252
Beat Cepheids : 25,26,30,36,38, 75,160
Beat phenomenon : 37
Becker F. : 25
Belousov-Zhabotinskii reaction : 171
Belserene E.P. : 22
Beltrami : 226,230,232
Bemporad A. : 23
Benettin G. : 260
Bergé P. : 100
Berry M. : 72,78,88
Beta Canis Majoris stars : 6,37
Beta Cepheids : 6,36,37,139
Betelgeuse : 33
Bifurcation : $\overline{X}$,67,68,92,95,103, 108,109,111,112,117,121, 173,174,175,176,178,189, 259,263,264,265,266,267, 268,269,295,302
Billiard : 51,88,246
Billingsley P. : 209
Birkhoff G.D.: 55,246,249,250
Bjerknes V. : 2
Blacher S. : 57,59,60,62,63,64, 65,78,273,274,275,279
Blazhko S.N. : 24,25
BL Herculis stars : 17,158
Bocchieri P. : 252
Bogdanov : 117,121
Bohr N. : 2
Boltzmann L. : 246,251,254,255
Bonazzola S. : 43,45
Bos R.J. : 39
Boury A. : 4
Boussinesq equations : 119
Bowen R. : 187
BP Her : 15

Brandstadter A. : 190
Breger M. : 28,38,39
Bretherton C.S. : 126,129
Breton M. : 4
Broglia P. : 28
Brownian motion : 166,193,195
Buchler J.R. : $\overline{X}$,74,75,76,77, 120,137-163,274,285-293, 295,298
Bump Cepheids : 15,36,151,162
Burgers equation : 306
Burki G. : 28,160
Burlisch R. : 281
Buzyna G. : 190

Cacciari C. : 16,18
Cahn J.H. : 36
Campbell L. : 61
Campbell W.W. : 19
Canonical transformations : 51,55
Cantor set : 205,207
Carbon star : 28
Carr J. : 106
Carson T.R. : 158
Castor J.I. : 139,142
Cataclysmic variables : 12,89
Cattaneo F. : 121
Celnikier L.M. : 124
Center manifold : 106,119,131, 140,148
Cepheids $\overline{X}$,15,16,17,18,19,23, 24,25,26,30,37,88,137,138, 154,158,160
Ceraski : 28
Cercignani : 254
Chaikin S.E. : 92
Chandrasekhar S. : 2,3,67,68
Chapman-Kolmogorov equation : 178
Chaotic motion, pulsation, signal : 12,52,57,59,60,61, 63,65,72,75,76,77,78,152, 168,171,218,254,259,263, 265,269,279,285,286,301, 302,314
Chemical reaction : 167,168
Cherewick T.A. : 37
Chertoprud V.E. : 121
Childress S. : $\overline{X}$,121,223-244

Chirikov B.V. : 52,252,253
Christensen-Dalsgaard J. : 40, 42,273,279,280,281
Christy R.F. : 36,154,156
Circulation : 69,70,274,276,277
Claverie, A. : 40
Clebsch form, representation : 69,276,277
Coarse graining : 193,194,219
Codimension : 74
Collet P. : 187
Collins G.W. : 3
Compatibility conditions : 145, 148
Compressible plasma : 305
Conservative flow pattern : 66
Constantinescu D.M. : 67
Contopoulos G.:$\overline{IX}$,56,58,245,259-271
Convection : 5,6,7,36,89,119,121, 126,129,141,154
Cooling sequence : 14
Couette-Taylor flow : 172,196, 197
Coullet P.H. : 74,108,113,115, 117,118,119,125,140,146, 148,263
Coutts C.M. : 26
Cowling : 3
Cox A.N. : 151,158,160
Cox J.P. : 6,17,36,45,139,141, 279
Cragg T.A. : 27
Cram L.E. : 129
Criticality : 107
Crutchfield J.P. : 176
Curtiss : 19
Cyclonic events : 242

Däppen W. : $\overline{X}$,69,70,71,72,273-284
Davis M.S. : 30,32
Dedekind : 67
Defouw R.J. : 119
Demarque P. : 18
Detector function : 281
Determinacy : 186
Detre L. : 25
Deubner F.L. : 40,129
Deupree R.G. : 36

INDEX

DI Car : 28
Diffusion : 126,178,224,235,236
Dimension, dimensionality,
 Hausdorff dimension : 172,
 187,188,195,196,197,200,
 207,208,211,218,219
Dirichlet : 67
Dissipative chaos : 7,75,76,
 100,194
Double mode Cepheids : 25
Doubly-diffusive convection :
 $\overline{X}$,121
Drazin P.G. : 118
Dubois M.: 179,181
Dwarf Cepheids : 38,160
Dynamical time-scale : 22,73,
 138,139,285,286,288
Dynamo : $\overline{X}$,121,223,224,225,226,
 230,$\overline{2}$32,241,242,243
Dziembowski W. : 74,139,154,
 160,274

Eckmann J.P. : 187
Eddington A. : $\overline{X}$,2,20,31,46,72
Eddy J.A. : 121,122
Edmonds A.R. : 116
Eggen O.J. : 88
Embedding : 168
Equipartition : 251,252,253,254
Ergodic, ergodicity : 197,239,
 245,246,248,251
Erratic fluctuations, erratic
 motions : 20,129,166
Evans J.W. : 129
Evans N.R. : 28

Fabricius : 137
Faraday : 102
Farmer J.D. : 176,187
Fast modes : 104
Faulkner D.J. : 25,26
Fautrell Y. : 121
Feedback : 299,302
Feigenbaum : 219,263,267
Fermi E. : 245,246,247,251
Fernie J.D.: 23,29,30
Fischer P.L. : 33
Fluctuation, fluctuation time-
 scale : 22,188

Flux ropes, flux sheets, flux
 tubes : 223,232,233,234,
 236,241,242
Fontaine G. : 43,45
Ford J. : 52,252
Fossat E. : 42
Fourier series, components,
 spectra : 20,24,51,54,63,
 104,247,253,305,310,312,315
Fowler A. : 102
Fowler R.H. : 2
Fractal, fractal dimension : 100,
 187,197,200,207,219
Frederickson P. : 207
Friedman B. : 104,107,112
Furenlid I. : 37

G 117-1315A : 43
Gabriel M. : 42
Galactic disc : 126
Galactic model : 259,265
Galerkin approximation : 104
Galgani L. : $\overline{X}$,245-257
Galloway D.J. : 241
Gaposchkin S. : 14,33
GD 226-29 : 44
GD 358 : 44,45
GD 385 : 43,45
Geisel T. : $\overline{IX}$,165-183
Gelly B. : 42
Geometric acoustics : 79
Gibbs measure : 198
Gilman C. : 15
Gilmore R. : 114
Giuli R.T. : 141
g-modes : 6,7,40,42,45,89,161
Goodricke : 15
Gough D.O. : 40,279,280,281
Goupil M.J. : 74,75,140,149
Granulation : 241
Grassberger P. : $\overline{IX}$,172,193-221
Great Sequence : 13,14
Grebogi C. : 186
Grec. G. : 38,40
Greene J.M. : 260
Guckenheimer J. : $\overline{IX}$,106,156,
 172,185-191,243
Gudzenko L.I. : 121
Guerrero G. : 28

Hagen J.G. : 1,18,30,31
Haken H. : 104
Hamiltonian systems : 47-60,219
    245 - 255,260-271,273-284
Hao B.L. : 100
Harding : 18
Hard sphere : 245-246,248
Härm R. : 5,17,295,298
Harvey J.W. : 40
Hayashi H. : 125
HD 161796 : 28,29,30,75
Hedlung : 246
Heggie D. : 263
Heiles C. : $\overline{IX}$,56,58,59,61,245,
    252
Helicity : 226,228,243
Helium shell burning : 295,302
Helmholtz H. von : 2,254
Hénon M. : $\overline{IX}$,56,58,59,61,201,
    206,210,215,216,217,230,
    245,246,249,252
Henyey L.G. : 296
Hertz H. : 2
Hertzsprung : 15,24
Heteroclinic : 225,228,229,243
Hill H.A. : 39
Hirakawa K. : 125
Hodson S.W. : 26,158,160
Hoffmeister C. : 25
Hofmeister E. : 17
Holmes P. : 106,140,185
Homoclinic : 156,246
Hopf bifurcation : 30,121,155,
    156
HR7308 : 15,28
Huberman B.A. : 176
Hudson H.S. : 40,41
Huerre P. : 126
Hunt G.W. : 2
Hurst H.E. : 87
Hypergeometric function : 312
Hysteresis : 30,156,161

Information, information flow :
    193,195,196,200,202,205,
    207,208,211,218,219
Instability strip : 17,161
Integrable, integral of motion :
    50,88,245,247,248,249,250,
    259,260

Intermittency : 121,125,166,179,
    180,181
Intermittent chaos : 165,178,179
Invariant torus : 247
Irregular dynamics : 165,168
Irregular motion, irregular
    variability, irregular
    signal : 52,60,75,78,89,
    137,162,165,166,169,171,
    172,179,181,279,285,286,
    292,301
Irregular period fluctuations :
    26
Ishizuka S. : 125
Isotropic homogeneous turbulence:
    306
Izenman A.J. : 123
Izraelev F.M. : 252,253

Jacchia L. : 61
Jacobi : 67,68,260
Jakobson : 187
Jarzebowski T. : 37
Jeans J.M. : 2,4,92,255
Jones C.A. : 121
Joss P.C. : 124,295,299
Joule heating rate : 309

Kähler H. : 286
Kaluza-Klein theory : 132
KAM : 52,54,55,88,252,269
Kaniel S. : 6
Kantz H. : 199
Kaplan J.L. : 198,207,219
Katz J. : 72,274,275,276,277
Kelvin-Helmholtz circulation :
    69,70
Kepler S.O. : 42,43,44
Khinchin axioms : 208,211
King D.S. : 158
Kirch : 30
Klapp J. : 74,149
Kolmogorov A.N.: 52,193,195,198,
    200,219,245,246,247,248,
    249,250,251
Kolmogorov spectrum : 306,312
Kolmogorov unstable : 52
Korkina E.I. : 226,242
Kovacz G. : 74,139,148
Kovetz A. : 6

INDEX

Kraft R.P. : 17
Krogdahl W.S. : 72
Kuhn T. : 246
Kukarkin B.V. : 14,17,30
Kullback information gain : 208
Kuperus M. : 305
Kwee K.K. : 24,26

Labyrinthine instability : 67
Lacy C.H. : 34
Lagrangian point : 261
Lamb H. : 92
Landau amplitude equation : 73,75,77
Landau-Hopf equation : 111,118
Landau L. : 73,74,119,121
Landolt : 42
Laplace P.S. : 194,218
Lebovitz N.R. : 6,103
Ledoux P. : $\overline{X},\overline{XII},1-9,14,37,48,73,137,160,286$
Legendre function : 104
Leibacher J.W. : 129
Leighton R.B. : 40
Lennard-Jones potential : 252
Lesh J.R. : 37
Lewin W.G.H. : 124,125,295
Lichtenberg A.J. : 269
Lieberman M.A. : 269
Lifshitz E.M. : 73
Limit cycles : 46,72,73,74, 75,100,111,117,121,152, 156,166,285,292,295,298, 299,302
Limit mass : 5
Lindblad B. : 2
Lindstedt-Poincaré method : 144
Linear series : 4
Liouville-Arnold theorem : 50
Li T. : 218
Livio M. : $\overline{X}$,295-304
Lloyd C. : $\overline{37}$
Lloyd Evans T. : 28
Logistic map : 172,216
Loinger : 252
Lomb N.R. : 37
Long-period variables : 30
Lorentz stresses : 224

Lorenz E.N. : $\overline{X}$,7,75,119,123,206, 210,218,$\overline{219}$,245
Ludendorff H. : 15,31
Lyapunov exponent : 176,177,178, 189,193,198,200,211,218, 219,252
Lynden-Bell D. : 3,72,274,275, 276,277

M3 : 26
Mackey-Glass equation : 210,216
Maclaurin : 67
Madore B.E. : 26
Magnenat P. : 61
Magnetic activity : $\overline{X}$
Magnetic buoyancy : $\overline{121}$
Magnetic diffusivity : 223
Magnetic field : 223-243,305-315
Magnetoconvection : 119
Makarenko E.N. : 15
Malkus W.V.R. : 121
Mandelbrot B.B. : 61,100,197
Manneville P. : 179,181
Mantegazza L. : 34,35
Marsden J.E. : 156
Marshall H. : 125
Martinet L. : 61
Martin W.C. : 24
Marzec C.J. : 123,293
Maunder intermission : 121
Mayer-Kress G. : 176
Mayor M. : 28
McCracken M. : 156
McGraw J.T. : 14,42
McLaughlin J.B. : 119
Melnikov : 245
Mengel J.G. : 17
Merrill P.W. : 30
Method of multiple scales : 144
MHD turbulence : 305,306
Michard R. : 129
Michel L. : 19,68
Michelson A.A. : 34
Milne : 3
Mira Ceti : 137
Mira stars : 30,31,33,36,61
Mixing : 197,211,212
Modal selection : 161
Mode-switching : 26,30

Modulation chaos : 125
Mook D.T. : 144
Moon H.T. : 126
Moore D.W. : 75,76,98,119,162, 293
Morel P.J. : 120
Morguleff N. : 38,41
Moser J. : 52,245
Multi-periodic variability : 11,25,38,42
Multiple equilibria : 285
Multiple instabilities : 111
Multiple timescales : 149
Multi-valued period-light curve relation : 17
Murdin P. : 34

Nayfeh A.H. : 144
N-body experiments : 263
Negative heat capacity : 298
Nekhoroshev N.N. : 246,247, 249,250,251,254
Nernst W. : 246,255
Neutron star : 286,295
Newcomb S. : 20
Newell A.C. : 126
NGC 3201 : 16,18
NGC 6649 : 26
Niva G.D. : 26
Nocera L. : $\overline{X}$,305-316
Noise : 165,171-182,186,214, 215,216,250,314
Non-adiabatic non-linear oscillations : 72
Non-adiabatic perturbation formalism : 140
Non-integrable Hamiltonian : 50
Non-linear coupling timescale : 20
Non-linear mode coupling : 22,49,55,273
Non-linear oscillations : 45, 46,47,49,273
Non-radial non-linear oscillations : 65,68,69,160, 273
Non-selfadjoint eigenvalue problem : 4

Normal form of amplitude equations, of differential equations : 74,75,106,107, 117,119,120
Normal form of a Hamiltonian : 51
Nuclear time-scale : 19
Nullity : 107

Observational errors : 26,33,65, 186
O-C plot : 26,27
Off-resonance parameter : 143
Olbers : 31
Oosterhoff P.T. : 24
Orszag S.A. : 119
Osaki Y. : 299
Osborn W. : 26
Ostlie D.A. : 36
Ostriker J.P. : 3
Overstability : 91,109,111,117, 124,126,153

Packard N.H. : 214
Paczynski B. : 4,286,295
Padé : 120
Papaloizou J.C. : 6,77,78
Parisi : 253
Parker E.N. : 121,242
Parkhurst J.A. : 1
Pasta : 245,246,251
Pattern formation : 126
Payne-Gaposchkin C. : 14,33
Pease F.G. : 34
Pekeris C. : 3
Percy J.R. : 28,30,36,37,38
Perdang J. : $\overline{IX},\overline{X}$,1-9,11-89, 120,139,162,273,274,275, 276,278,279,281,298
Period doubling : 172,176,179, 181,260,263
Period-luminosity relation : 138
Period-mass-radius relation : 31
Perturbation : 137,138,139,143, 144,246,247,248,249,250, 259,260,274,307
Pesnell W.D. : 139

INDEX

Petersen J.O. : 36
Pettini M. : $\overline{X}$,305-316
Phase turbulence : 315
Pike C.D. : 37
Plakidis S. : 20,31
Planck M. : 255
Plummer H.C. : 19
p-modes : 40,72,160
Poincaré-Birkhoff-Siegel-
  Kolmogorov-Arnold-
  Moser results on
  Hamiltonian oscillations:
  50
Poincaré H. : $\overline{IX}$,4,50,58,67,
  239,246,$\overline{247}$,248,249
Poincaré recurrence theorem :
  58
Poincaré surface of section :
  60,62,169,170,171
Poisson equation : 277
Poloidal field, mode : 66,69,
  70,160
Pomeau Y. : 100,121,179,181
Power-law spectrum : 312
Press W.H. : 120
Preston G.W. : 25,34
Pringle J.E. : 6
Proctor M.R.E. : 241
Procyon : 42
Pseudo-integrable Hamiltonian :
  52,88
Pulsar : 95,124
Pulsation time : 19

Quantum chaos : 78,79
Quantum zero-point energy :
  255
Quasi-adiabatic approximation :
  139
Quasi-integrable Hamiltonian :
  50,52
Quasi-periodic : 39,51,57,60,
  63,98,269

Radicati L.A. : 68
Random noise : 168
Random processes : 46
Random walk : 179
Rapid burster : 92,124

R Aql : 30,31,32
Rastorgouev A.S. : 17
R Aur : 33
Ray equations : 78-79
Rayleigh : 129,255
Rayleigh-Bénard experiment :
  180,216
R Cam : 33
Recurrence, recurrence time :
  185,189
Recurrent trajectories : 58
Redekopp L.G. : 126
Red giants : 286,293,295,298
Red semiregulars : 33
Red Sequence : 12,14,30,34
Regev O. : $\overline{X}$,74,76,77,139,154,
  162,2$\overline{74}$,285-293,295-304
Regular Sequence : 12,14,15,34
Reid W.H. : 118
Relaxation oscillation : 72,89,
  111,123,285,292
Renyi information : 193,208,209,
  211
Renzini A. : 24,36,46
Repetitive pattern : 189
Resonance, resonant : 36,54,55,
  57,77,146,151,153,158,161,
  247,250,261,273
Response theory : 176
Return map : 171,172,179,180,
  302
Reynolds number : 196,223,240,
  241,310
Rhodes E.J. : 129
R Hya : 36
Richens P.J. : 88
Riemann : 67
r-modes : 6
Robinson E.L. : 14,42
Roche lobe : 89
Rosenfeld L. : 1
Rosenzweig : 67
Rosino L. : 24,34
Rosseland S. : 1,2
Rössler O.E. : 75,168,302
Rotation : 6,43,66,67,68,72,124,
  141,161
Roux J.L. : 171
R R Lyrae : 16,18,19,24,25,26,
  31,34,36,46,75

RR Scorpii : 31,32
R Scuti : 34
R Tri : 33
RU Cam : 27,28,75
Ruelle D. : 102,168,205
Russell H.N. : 4,19
Rutten R.J. : 129
Ruzmaikin A.A. : 121
R Vir : 33
RV Nor : 28
RV Tauri stars : 34,36,65

Sack N. : 286
Sandford R.F. : 28
Sandig H.U. : 33
Sauvenier-Goffin E. : 6
Schaltenbrand R. : 25
Scherrer P.H. : 39
Schmidt : 15
Schmidt E.G. : 36,151
Schneller H. : 33
Schuster H.G. : 100
Schutz B. : 3
Schwarz M.P. : 263
Schwarzschild M. : 5,6,7,17,295, 298
Scotti A. : 252
Secular instability, stability: 4,288
Secular period change : 19,24, 31,36,37,75
Segel L.A. : 126
Self-gravitating figures : 3, 92
Sellwood J.A. : 263
Semi-attractor : 200,216,217
Semi-convection : 5,36
Sensitivity to initial conditions, noise : 98
Severny A.B. : 39
Shaar R.J. : 103
Shannon : 208,209
Shapiro S.L. : 124
Shapley H. : 1,18
Shaw R. : 176,177
Shell flashes : 295
Shil'nikov L.P. : 98
Shobbrook R.R. : 25,26,37,38, 39

Short period RR Lyrae : 38
Siegel C.L. : 55,246
Simon N.R. : 36,139,151,160
Sinai : 187,195,245
Singularity, singular perturbation : 108,311,312
Slave modes : 104
Slow modes : 104,112,113,120
Smale horseshoe : 218
Solar corona : 305
Solar cycle : 92,95,98,120,121, 131
Solar Maximum Mission : 40,41
Solar waves : 125
Solar 5 minute oscillation : 4,11,39-40,63,72,129,273, 274,282,283
Solar 160 min oscillation : 39,72,284
Solitary waves : 126,129
Sound wave : 309
South Pole solar observations : 40,41
Soward A.M. : $\overline{X}$,223-244
Sparrow C. : 119
Speckle interferometry : 31,34
Sperra S.W. : 19
Spherical harmonics : 278,279
Spica : 87
Spiegel E.A. : $\overline{X}$,7,74,75,76, 91-135,140,146,148,162, 293
Standard deviation of local periods : 21,33
Standstill : 302
Stagnation point : 223,224,225, 228,229,230,235,243
Stationary (steady) velocity field, stationary motion : 67,68,69,72,88,223,274,276, 277
Steiner J.M. : 124
Stein J. : 1
Stein R.F. : 129
Stellar coronae : 305
Stellar stability : 1,8
Stellingwerf R.F. : 138
Sterne T.E. : 20,21,26,31,32
Stix M. : 241

Stobie R.S. : 25,38,39
Stochasticity, stochastic motion, stochastic orbit : 12,245,252,253,259,263, 264,267,269
Stoer J. : 281
Stokes : 129,276
Stormer C. : 2
Stover R.J. : 43
Strange attractor : 75,76,100, 102,120,123,131,152,156, 165,166,168,169,171,185, 188,189,194,198,292
Stratification field : 278
Sturm-Liouville problem : 142
SU Dra : 19
S UMa : 31,33
Sun : 11,39,65,129,130, 273,274,284
SU UMa stars : 12
SV And : 33
SV Vul : 15,23
SW And : 19
SW Dra : 19
Sweigart A.V. : 18,24,36,46
Swings P. : 1,2
Szabados L. : 15,23,24,26, 27,28
Szeidl B. : 24,26,

Takens F. : 168,214
Takeuti M. : 36,74,139
Talbot R.J. : 5
Tammann G. : 25
Tassoul J.L. : 88
T Cas : 33
Teukolsky S.A. : 124
T Gem : 31
Thermal instability : 4
Thermal time-scale : 19,22, 73,77,138,139,285,288
Thermohaline convection : 119
Thermonuclear runaway : 295
Thielheim K.O. : 263
Thompson J.M.T. : 2
Thomson J.J. : $\overline{IX}$
Threshold energy : 57,59
Timing noise : 124

T Monocerotis : 23,24
Topological entropy : 200,209
Toroidal field, toroidal mode : 66,69,71,89,276,277
Traub W.A. : 42
Tresser J. : 263
TU Cas : 26
Tuchman Y. : 286
Turbulent energy cascade : 305,306
Turbulent inertial range : 315
Tsesevitch V.P. : 25

U Gem : 12
Ulam S. : 245,246,251
Ulrich R.K. : 129,296
Universality, universal law, universal property : 174, 178,243,260,263
Unno W. : 129
Unpredictability , unpredictability of deterministic behaviour : 95,193,194, 198,218
U Per : 33
U TrA : 25,26

Vandakurov Y. : 4,74,139
Van der Pol oscillator : 73
Van Horn : 120
Variable stars : 1,11
Variational formalism, principle : 3,48,66,71,273 274,275,277,279
Vauclair G. : 43,45
Vibrational instability, stability : 5,6,17,45, 76,78,117,153,288,293
Vidal C. : 100
Virial theorem : 3
Viscous heating rate : 309
Vortex lines : 223
Vorticity : 68
Vulpiani A. : $\overline{X}$,305-316

Wait R. : 281
Walker G.H. : 52
Walker M.F. : 37
Walraven T. : 2,14,73,137,160
Weidemann : 120

Weiss N.O. : 121,242
Welch D.L. : 30
Wersinger J.M. : 152
Wesfried J.E. : 126
Wesselink A.J. : 46,72
White Dwarfs (DA,DB) : 6,14,
    38,42,45,89,120,286
Whitehead J.A. : 126
White R. : 120
Whitney C.A. : 58
Whitney embedding theorem :
    168
Wiener process : 220
Willson L.A. : 31,33
Winget D.E. : 43,44,45
Wizinovitch P. : 38
Wolf A. : 102,123
Wolff H. : 263
Wolf number : 123
Woltjer J. : 36,46,47,48,66,
    71,72,73,74,89,275
Woodard M. : 40,41
Wood P.R. : 33
W Sgr : 19
W Virginis : 16,17,26,28

X Cam : 31
X-ray burster : 92,124,125,
    165,182,295
XZ Cygni : 24

Yorke J.A. : 198,207,218,219
Young A. : 37
Yueh W.R. : 139

Zaitseva G.V. : 27,28
Zaleski S. : 126
Zaslavski : 52
Z Cam : 12
Ziebarth K. : 5
Zikides M. : 263
Ziolkowski J. : 286
ZZ Ceti stars : 42,43,120

α C Mi : 42
α C Vn stars : 12
α Orionis : 33
α Vir : 37,87

δ Cephei : 15,16,23,24,137
δ Ceti : 37
δ Scuti : 35,37,38,61
ζ Gem : 15,19,25,26,27,61
θ Tuc : 39
μ Cephei : 33,87
o Ceti : 31,137
ρ Pup : 38
χ Cyg : 30,31,32
ω Cen : 17,24

3C273 : 92,120
12 Lac : 37
14 Aur : 38,41,61
21 Mon : 39
44 Tau : 38,41,61
47 Tuc : 33
89 Her : 28,29,75